MATLAB®
A Practical Introduction to Programming and Problem Solving

Second Edition

MATLAB®
A Practical Introduction to Programming and Problem Solving

Second Edition

Stormy Attaway
Department of Mechanical Engineering
Boston University

ELSEVIER

AMSTERDAM • BOSTON • HEIDELBERG • LONDON
NEW YORK • OXFORD • PARIS • SAN DIEGO
SAN FRANCISCO • SINGAPORE • SYDNEY • TOKYO
Butterworth-Heinemann is an imprint of Elsevier

Butterworth-Heinemann is an imprint of Elsevier
225 Wyman Street, Waltham, MA 02451, USA
The Boulevard, Langford Lane, Kidlington, Oxford, OX5 1GB, UK

Notices
Knowledge and best practice in this field are constantly changing. As new research and experience
broaden our understanding, changes in research methods, professional practices, or medical treatment
may become necessary.

   Practitioners and researchers must always rely on their own experience and knowledge in evaluating
and using any information, methods, compounds, or experiments described herein. In using such
information or methods they should be mindful of their own safety and the safety of others, including
parties for whom they have a professional responsibility.

   To the fullest extent of the law, neither the Publisher nor the authors, contributors, or editors, assume
any liability for any injury and/or damage to persons or property as a matter of products liability,
negligence or otherwise, or from any use or operation of any methods, products, instructions, or ideas
contained in the material herein.

MATLAB® is a trademark of TheMathWorks, Inc., and is used with permission. TheMathWorks does not
warrant the accuracy of the text or exercises in this book. This book's use or discussion of MATLAB®
software or related products does not constitute endorsement or sponsorship by TheMathWorks of a
particular pedagogical approach or particular use of the MATLAB® software.

MATLAB® and Handle Graphics® are registered trademarks of TheMathWorks, Inc.

Library of Congress Cataloging-in-Publication Data
Attaway, Stormy.
   MATLAB®: a practical introduction to programming and problem solving / Stormy
Attaway. — 2nd ed.
       p.   cm.
Includes index.
ISBN 978-0-12-385081-2
   1. Numerical analysis—Data processing. 2. MATLAB. 3. Computer programming.
I. Title.
QA297.A87 2011
518.028553—dc22   2011015032

British Library Cataloguing-in-Publication Data
A catalogue record for this book is available from the British Library.

For information on all Butterworth–Heinemann publications
visit our Web site at *www.elsevierdirect.com*

Printed in the United States
11   12   13   14   15   10 9 8 7 6 5 4 3 2 1

This book is dedicated to my husband, Ted de Winter.

Contents

Preface

Motivation

The purpose of this book is to teach fundamentals of programming concepts and skills needed for basic problem solving, all using MATLAB® as the vehicle. MATLAB is a powerful software package that has built-in functions to accomplish a diverse range of tasks, from mathematical operations to three-dimensional imaging. Additionally, MATLAB has a complete set of programming constructs that allows users to customize programs to their own specifications.

The many books that introduce MATLAB come in two basic flavors: those that demonstrate the use of the built-in functions in MATLAB, with a chapter or two on some programming concepts; and those that cover only the programming constructs without mentioning many of the built-in functions that make MATLAB efficient to use. Someone who learns just the built-in functions will be well-prepared to use MATLAB, but would not understand basic programming concepts. That person would not be able to then learn a language such as C++ or Java without taking another introductory course or reading another book on the programming concepts. Conversely, anyone who learns only programming concepts first (using any language) would tend to write highly inefficient code using control statements to solve problems, not realizing that in many cases these are not necessary in MATLAB.

This book instead takes a hybrid approach, introducing both the programming and efficient uses. The challenge for students is that it is nearly impossible to predict whether they will in fact need to know programming concepts later or whether a software package such as MATLAB will suffice for their careers. Therefore, the best approach for beginning students is to give them both: the programming concepts and the efficient built-in functions. Since MATLAB is very easy to use, it is a perfect platform for this approach to teaching programming and problem solving.

Since programming concepts are critically important to this book, emphasis is not placed on the time-saving features that evolve with every new MATLAB release. For example, in current versions of MATLAB, statistics on variables are available readily in the Workspace Window. This is not shown in any detail in the book, since whether this feature is available depends on the software version, and because of the desire to explain the concepts in the book.

Modifications in Second Edition

Changes in the second edition of this book include:

- Vectorized code has been made into a separate chapter to emphasize the importance of using MATLAB efficiently.
- There are expanded examples on:
  - Low-level file input functions
  - Plots
  - Graphical user interfaces
  - Vectorized code, including functions **diff**, **meshgrid**, **tic**, and **toc**
- Use of MATLAB version R2011a
- Concepts used in image processing, such as three-dimensional matrices and unsigned integers, are now introduced early, in Chapter 1.
- Modified and new end-of-chapter exercises.
- The introduction to Handle Graphics was moved to Chapter 11, Advanced Plotting Techniques.
- Discussion of symbolic mathematics was moved to Chapter 15, Advanced Mathematics.
- Improved labeling of plots.
- Improved standards for variable names and documentation.
- Added **end** to the end of all functions.

Key Features
Side-by-Side Programming Concepts and Built-in Functions

The most important, and unique, feature of this book is that it teaches programming concepts and the use of the built-in functions in MATLAB side by side. It starts with basic programming concepts such as variables, assignments, input/output, selection, and loop statements. Then throughout the rest of the book, many times a problem will be introduced and then solved using the "programming concept" and also using the "efficient method." This will not be done in every case to the point that it becomes tedious, but just enough to get the ideas across.

Systematic Approach

Another key feature is that the book takes a very systematic, step-by-step approach, building on concepts throughout the book. It is very tempting in a MATLAB text to show built-in functions or features early on with a note that says "we'll do this later." This does not happen in this edition; all functions are covered before they are used in examples. Additionally, basic programming concepts will be explained carefully and systematically. Very basic concepts, such as looping to calculate a sum, counting in a conditional loop, and error-checking, are not found in many texts but will be covered here.

File Input/Output

Many applications in engineering and the sciences involve manipulating large data sets that are stored in external files. Most MATLAB texts at least mention the **save** and **load** functions, and in some cases selected lower-level file input/output functions as well. Since file input and output is so fundamental to so many applications, this book will cover several low-level file input/output functions, as well as reading from and writing to spreadsheet files. Later chapters will also deal with audio and image files. These file input/output concepts are introduced gradually: first **load** and **save** in Chapter 2, then lower-level functions in Chapter 9, and finally sound and images in Chapter 14.

User-Defined Functions

User-defined functions are a very important programming concept, and yet many times the nuances and differences among concepts such as types of functions and function calls versus function headers can be very confusing to beginning programmers. Therefore, these concepts are introduced gradually. First, functions that calculate and return one single value—arguably the easiest type of functions to understand—are demonstrated in Chapter 2. Later, functions that return no values and functions that return multiple values are introduced in Chapter 6. Finally, advanced function features are shown in Chapter 10.

Advanced Programming Concepts

In addition to the basics, some advanced programming concepts, such as string manipulation, data structures (e.g., structures and cell arrays), recursion, anonymous functions, and variable number of arguments to functions, are covered. Sorting, searching, and indexing are also addressed. All of these are again approached systematically; for example, cell arrays are covered before they are used in file input functions and as labels on pie charts.

Problem-Solving Tools

In addition to the programming concepts, some basic mathematics necessary for solving many problems will be introduced. These will include statistical functions, solving sets of linear algebraic equations, and fitting curves to data. The use of complex numbers and some calculus (integration and differentiation) will also be introduced. The basic math will be explained and the built-in functions in MATLAB to perform these tasks will be described.

Plots, Imaging, and Graphical User Interfaces

Simple two-dimensional plots are introduced very early in the book in Chapter 2 so that plot examples can be used throughout. Chapter 11 then shows more plot types, and demonstrates customizing plots and how the graphics properties are handled in MATLAB. This chapter makes use of strings and cell arrays to customize labels. Also, there is an introduction to image processing and the basics necessary to understand programming graphical user interfaces (GUIs) in Chapter 14.

Vectorized Code

Efficient uses of the capabilities of the built-in operators and functions in MATLAB are demonstrated throughout the book. However, to emphasize the importance of using MATLAB efficiently, vectorized code is treated in a separate chapter. Techniques, such as preallocating vectors and using logical vectors, are featured, as well as methods of determining how efficient the code is.

Layout of Text

The book consists of two parts. The first part covers programming constructs and demonstrates the programming method versus efficient use of built-in functions to solve problems. The second part covers tools that are used for basic problem solving, including plotting, image processing, and mathematical techniques to solve systems of linear algebraic equations, fit curves to data, and perform basic statistical analyses. The first six chapters cover the very basics in MATLAB and in programming, and are all prerequisites for the rest of the book. After that, many chapters in the problem-solving section can be introduced when desired, to produce a customized flow of topics in the book. This is true to an extent, although the order of the chapters has been chosen carefully to ensure that the coverage is systematic.

The individual chapters are described here, as well as which topics are required for each chapter. Part I, Introduction to Programming Using MATLAB, includes the following chapters.

Chapter 1: Introduction to MATLAB covers expressions, operators, characters, variables, and assignment statements. Scalars, vectors, and matrices are all introduced as are many built-in functions that manipulate them.
Chapter 2: Introduction to MATLAB Programming introduces the idea of algorithms and scripts. This includes simple input and output, and commenting. Scripts are then used to create and customize simple plots, and to do file input and output. Finally, the concept of a user-defined function is introduced with only the type of function that calculates and returns a single value.

Chapter 3: Selection Statements introduces relational expressions and their use in **if** statements, with **else** and **elseif** clauses. The **switch** statement is also demonstrated, as is the concept of choosing from a menu. Also, functions that return **logical true** or **false** are introduced.

Chapter 4: Loop Statements introduces the concepts of counted (**for**) and conditional loops (**while**). Many common uses, such as summing and counting, are covered. Nested loops are also introduced. Some more sophisticated uses of loops, such as error-checking and combining loops and selection statements, are also covered.

Chapter 5: Vectorized Code introduces the idea of "vectorizing" code, which essentially means rewriting code that uses loops to more efficiently make use of built-in functions, and the fact that operations can be done on vectors and matrices in MATLAB. Functions that are useful in vectorizing code are emphasized in this chapter. Functions that time the speed of code are also introduced.

Knowledge of the concepts presented in the first five chapters is assumed throughout the rest of the book.

Chapter 6: MATLAB Programs covers more on scripts and user-defined functions. User-defined functions that return more than one value and also that do not return anything are introduced. The concept of a program in MATLAB, which consists of a script that calls user-defined functions, is demonstrated with examples. A longer menu-driven program is shown as a reference, but could be omitted. Subfunctions and scope of variables are also introduced, as are some debugging techniques.

This program concept is used throughout the rest of the book.

Chapter 7: String Manipulation covers many built-in string manipulation functions as well as converting between string and number types. Several examples include using custom strings in plot labels and input prompts.

Chapter 8: Data Structures: Cell Arrays and Structures introduces two main data structures: cell arrays and structures. Once structures are covered, more complicated data structures, such as nested structures and vectors of structures, are also introduced. Cell arrays are used in several applications in later chapters, such as file input in Chapter 9, variable number of function arguments in Chapter 10, and plot labels in Chapter 11, and are therefore considered important and are covered first. The rest of the chapter on structures can be omitted.

Chapter 9: Advanced File Input and Output covers lower-level file input/output statements that require opening and closing the file. Functions that can read the entire file at once as well as those that require reading one line at a time are introduced, and examples that demonstrate the differences in their use are shown. Additionally, reading from and writing to spreadsheet files and also *.mat* files that store MATLAB variables are introduced. Cell arrays and string functions are used extensively in this chapter.

Chapter 10: Advanced Functions covers more advanced features of and types of functions such as anonymous functions, nested functions, and recursive functions. Function handles and their use both with anonymous functions and function functions are introduced. The concept of having a variable number of input and/or output arguments to a function is introduced; this is implemented using cell arrays. String functions are also used in several examples in this chapter. The section on recursive functions is at the end and may be omitted.

Part II, Advanced Topics for Problem Solving with MATLAB, contains the following chapters.

Chapter 11: Advanced Plotting Techniques continues with more on the plot functions introduced in Chapter 2. Various two-dimensional plot types, such as pie charts and histograms, are introduced, as is customizing plots using cell arrays and string functions. Three-dimensional plot functions as well as selected functions that create the coordinates for specified objects are demonstrated. The notion of Handle Graphics® is covered, and selected graphics properties, such as line width and color, are introduced. Applications that involve reading data from files and then plotting use both cell arrays and string functions.

Chapter 12: Matrix Representation of Linear Algebraic Equations introduces a basic method that can be used in MATLAB to solve systems of equations using a matrix representation. First, matrix and vector operations and matrix definitions are described. This section can be covered at any point after Chapter 5. Then, matrix solutions using the Gauss-Jordan and Gauss-Jordan elimination methods are described. This section includes mathematical techniques and also the MATLAB functions that implement them.

Chapter 13: Basic Statistics, Sets, Sorting, and Indexing starts with some of the built-in statistical and set operations in MATLAB. Since some of these require a sorted data set, methods of sorting are described. Finally, the concepts of indexing into a vector and searching a vector are introduced. Sorting a vector of structures and indexing into a vector of structures are described, but these sections can be omitted. A recursive binary search function is in the end and may be omitted.

Chapter 14: Sights and Sounds briefly discusses sound files and introduces image processing. An introduction to programming graphical user interfaces is also given, including the creation of a button group. Nested functions are used in the GUI examples. A **patch** function example uses a structure.

Chapter 15: Advanced Mathematics covers three basic topics: curve fitting, complex numbers, and integration and differentiation in calculus. Finally, some of the Symbolic Math Toolbox functions are shown, including those that solve equations. This method returns a structure as a result.

Pedagogical Features

There are several pedagogical tools that are used throughout this book that are intended to make it easier to learn the material. A list of **Key Terms** covered in each chapter, in sequence, is on the first page.

First, the book takes a conversational tone with sections called **Quick Question!** These are designed to stimulate thought about the material that has just been covered. A question is posed, and then the answer is given. It will be most beneficial to the reader to try to think about the question before reading the answer! In any case, these sections should not be skipped over as the answers often contain very useful information.

Practice problems are given throughout the chapters. These are very simple problems that serve as drills of the material just covered.

When certain problems are introduced, they are solved both using **The Programming Concept** and **The Efficient Method**. This facilitates understanding the built-in functions and operators in MATLAB as well as the underlying programming concepts. The Efficient Method boxes highlight methods that will save time for the programmer, and in many cases are faster to execute in MATLAB, as well.

Additionally, to aid the reader:

- Identifier names (variables and user-defined functions) are shown in *italics* (as are filenames and file extensions).
- MATLAB function names are shown in **bold**.
- Reserved words are shown in **bold and underline**.
- Key important terms are shown in ***bold and italic***.

The end-of-chapter summary contains, where applicable, several sections:

- **Common Pitfalls**: A list of common mistakes that are made, and how to avoid them.
- **Programming Style Guidelines**: To encourage the creation of "good" programs that others can actually understand, the programming chapters have guidelines that make programs easier to read and understand, and therefore easier to work with and to modify.
- **MATLAB Reserved Words**: A list of the reserved key words in MATLAB. Throughout the text, these are shown in bold, underlined type.

- **MATLAB Functions and Commands**: A boxed list of the MATLAB built-in functions and commands covered in the chapter, in the order covered. Throughout the text, these are shown in bold type.
- **MATLAB Operators**: A boxed list of the MATLAB operators covered in the chapter in the order covered.
- **Exercises**: A comprehensive set of exercises, ranging from the rote to more engaging applications.

Additional Book Resources

A companion web site is available with downloadable *.m* files for all examples in the text, at *www.elsevierdirect.com/9780123850812*. Other book-related resources will also be posted on the web site from time to time.

Additional teaching resources are available for faculty using this book as a text for their course(s). Please visit *www.textbooks.elsevier.com* to register for access to:

- Instructor solutions manual for end-of-chapter problems
- Electronic figures from the text for creation of lecture slides
- Downloadable M-files for all examples in the text

Acknowledgments

I am indebted to many, many family members, colleagues, mentors, and to numerous students.

Throughout the last 24 years of coordinating and teaching the basic computation courses for the College of Engineering at Boston University, I have been blessed with many fabulous students as well as graduate teaching fellows and undergraduate teaching assistants. There have been hundreds of teaching assistants over the years, too many to name individually, but I thank them all for their support.

In particular, the following teaching assistants were very helpful in reviewing drafts of the original manuscript and suggesting examples: Edy Tan, Megan Smith, Brandon Phillips, Carly Sherwood, Ashmita Randhawa, Mike Green, Kevin Ryan, and Brian Hsu. For this Second Edition, Brian Hsu and Paul Vermilion suggested several revisions. Brian Hsu, Jake Herrmann, and Ben Duong contributed exercises. Kevin Ryan created the script to produce the cover illustrations.

A number of colleagues have been very encouraging through the years. I would especially like to thank my former and current department chairmen, Tom Bifano and Ron Roy, for their support and motivation, and Tom for his GUI example suggestions. I am also indebted to my mentors at Boston University, Bill Henneman of the Computer Science Department, and Merrill Ebner of the Department of Manufacturing Engineering, as well as to Bob Cannon from the University of South Carolina.

I would like to thank all the reviewers of the proposal and drafts of this book. Their comments have been extremely helpful and I hope I have incorporated their suggestions to their satisfaction. In addition to several anonymous reviewers, the reviewers for this edition include:

- Peter Bernard, University of Maryland
- Sanjukta Bhowmick, Pennsylvania State University
- Chris Brown, University of Rochester
- Steven Brown, University of Delaware

- Anthony Muscat, University of Arizona
- Charles Riedesel, University of Nebraska, Lincoln
- Jeff Ringenberg, The University of Michigan
- Richard Ulrich, University of Arkansas

Also, I thank those at Elsevier who helped to make this book possible, including Joseph Hayton, Publisher; Fiona Geraghty, Editorial Project Manager; Marilyn Rash, Project Manager; Eric DeCicco, Cover Designer/Illustrator; and Tim Pitts, a Publisher at Elsevier in the United Kingdom.

Finally, thanks go to all members of my family, especially my parents Roy Attaway and Jane Conklin, both of whom encouraged me at an early age to read and to write. Thanks also to my husband Ted de Winter for his encouragement and good-natured taking care of the weekend chores while I worked on this project!

The photo of Ted fishing in the image-processing section was taken by Wes Karger.

Introduction to Programming Using MATLAB

CHAPTER 1

Introduction to MATLAB

KEY TERMS

prompt
programs
script files
variables
assignment statement
assignment operator
user
initializing
incrementing
decrementing
identifier names
reserved words
key words
mnemonic
default
unary
operand
binary
scientific notation
exponential notation
precedence
associativity

nesting
call a function
arguments
returning values
constants
types
classes
double precision
floating point
unsigned
characters
strings
type casting
saturation arithmetic
random numbers
seed
pseudorandom
character encoding
character set
vectors
matrices
row vector

column vector
scalar
elements
array
array operations
iterate
step value
concatenating
index
subscript
index vector
transposing
subscripted indexing
unwinding a matrix
linear indexing
vector of variables
empty vector
deleting elements
three-dimensional
    matrices

CONTENTS

MATLAB® is a very powerful software package that has many built-in tools for solving problems and developing graphical illustrations. The simplest method for using the MATLAB product is interactively; an expression is entered by the user and MATLAB immediately responds with a result. It is also possible to write scripts and programs in MATLAB, which are essentially groups of commands that are executed sequentially.

This chapter will focus on the basics, including many operators and built-in functions that can be used in interactive expressions. Means of storing values, including vectors and matrices, will also be introduced.

1.1 GETTING INTO MATLAB

MATLAB is a mathematical and graphical software package with numerical, graphical, and programming capabilities. It has built-in functions to perform many operations, and there are toolboxes that can be added to augment these functions (e.g., for signal processing). There are versions available for different hardware platforms, in both professional and student editions.

When the MATLAB software is started, a window opens in which the main part is the Command Window (see Figure 1.1). In the Command Window, you should see:

>>

The >> is called the *prompt*. In the Student Edition, the prompt instead is:

EDU>>

In the Command Window, MATLAB can be used interactively. At the prompt, any MATLAB command or expression can be entered, and MATLAB will immediately respond with the result.

It is also possible to write *programs* in MATLAB that are contained in *script files* or M-files. Programs will be introduced in Chapter 2.

The following commands can serve as an introduction to MATLAB and allow you to get help:

- **info** will display contact information for the product.
- **demo** has demos of some of the features of MATLAB.
- **help** will explain any command; **help help** will explain how help works.
- **helpbrowser** opens a Help Window.
- **lookfor** searches through the help for a specific word or phrase. (*Note:* This can take a long time.)

To get out of MATLAB, either type **quit** at the prompt, or choose File, then Exit MATLAB from the menu.

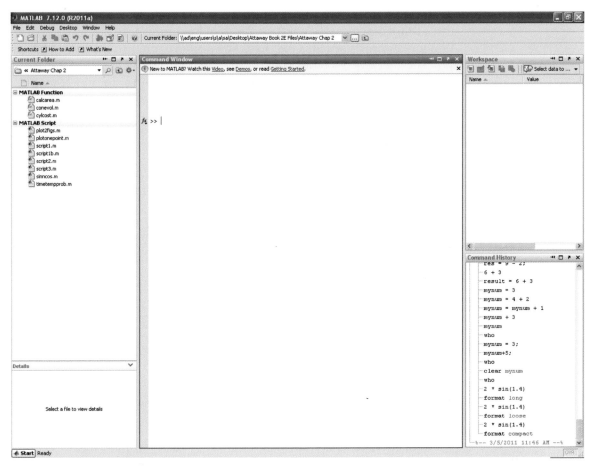

FIGURE 1.1 MATLAB Command Window

1.2 THE MATLAB DESKTOP ENVIRONMENT

In addition to the Command Window, there are several other windows that can be opened and may be opened by default. What is described here is the default layout for these windows in Version R2011a, although there are other possible configurations. Different versions of MATLAB may show other configurations by default, and the layout can always be customized. Therefore, the main features will be briefly described here.

Directly above the Command Window, there is a pull-down menu for the Current Folder. The folder that is set as the Current Folder is where files will be saved.

To the right of the Command Window is the Workspace Window on top and the Command History Window on the bottom. The Command History Window shows commands that have been entered, not just in the current session (in the current Command Window), but previously as well. (The Workspace Window will be described in the next section.) To the left of the Command Window is the Current Folder Window. This shows the files that are stored in the Current Folder. These can be grouped by type, and sorted by name. If a file is selected, information about that file is shown on the bottom.

This default configuration can be altered by clicking on Desktop, or using the icons at the top right corner of each window. These include an "x" that will close a particular window, and a curled arrow that in its initial state pointing to the upper right allows one to undock that window. Once undocked, clicking on the curled arrow pointing to the lower right will dock the window again.

1.3 VARIABLES AND ASSIGNMENT STATEMENTS

To store a value in a MATLAB session, or in a program, a *variable* is used. The Workspace Window shows variables that have been created. One easy way to create a variable is to use an *assignment statement*. The format of an assignment statement is

```
variablename = expression
```

The variable is always on the left, followed by the = symbol, which is the *assignment operator* (unlike in mathematics, the single equal sign does *not* mean equality), followed by an expression. The expression is evaluated and then that value is stored in the variable. For example, this is the way it would appear in the Command Window:

```
>> mynum = 6
mynum =
     6
>>
```

Here, the *user* (the person working in MATLAB) typed "mynum = 6" at the prompt, and MATLAB stored the integer 6 in the variable called *mynum*, and then displayed the result followed by the prompt again. Since the equal sign is the assignment operator, and does not mean equality, the statement should be read as "mynum gets the value of 6" (*not* "mynum equals 6").

Note that the variable name must always be on the left, and the expression on the right. An error will occur if these are reversed.

```
>> 6 = mynum
??? 6 = mynum
    |
Error: The expression to the left of the equals sign is not a valid
target for an assignment.

>>
```

Putting a semicolon at the end of a statement suppresses the output. For example,

```
>> res = 9 - 2;
>>
```

This would assign the result of the expression on the right side the value 7 to the variable *res*; it just doesn't show that result. Instead, another prompt appears immediately. However, at this point in the Workspace Window the variables *mynum* and *res* and their values can be seen.

The spaces in a statement or expression do not affect the result, but make it easier to read. The following statement, which has no spaces, would accomplish exactly the same thing as the previous statement:

```
>> res=9-2;
```

Note
In the remainder of the text, the prompt that appears after the result will not be shown.

MATLAB uses a default variable named *ans* if an expression is typed at the prompt and it is not assigned to a variable. For example, the result of the expression $6 + 3$ is stored in the variable *ans*.

```
>> 6 + 3
ans =
    9
```

This default variable is reused any time just an expression is typed at the prompt.

A shortcut for retyping commands is to hit the up arrow ↑, which will go back to the previously typed command(s). For example, if you decided to assign the result of the expression $6 + 3$ to the variable "result" instead of using the default *ans*, you could hit the up arrow and then the left arrow to modify the command rather than retyping the entire statement.

```
>> result = 6 + 3
result =
    9
```

This is very useful, especially if a long expression is entered with an error, and it is desired to go back to correct it.

To change a variable, another assignment statement can be used, which assigns the value of a different expression to it. Consider, for example, the following sequence of statements:

```
>> mynum = 3
mynum =
         3

>> mynum = 4 + 2
mynum =
         6

>> mynum = mynum + 1
mynum =
         7
```

In the first assignment statement, the value 3 is assigned to the variable *mynum*. In the next assignment statement, *mynum* is changed to have the value of the expression 4 + 2, or 6. In the third assignment statement, *mynum* is changed again, to the result of the expression *mynum* + 1. Since at that time *mynum* had the value 6, the value of the expression was 6 + 1, or 7.

At that point, if the expression *mynum* + 3 is entered, the default variable *ans* is used since the result of this expression is not assigned to a variable. Thus, the value of *ans* becomes 10 but *mynum* is unchanged (it is still 7). Note that just typing the name of a variable will display its value.

```
>> mynum + 3
ans =
        10

>> mynum
mynum =
         7
```

1.3.1 Initializing, incrementing, and decrementing
Frequently, values of variables change. Putting the first or initial value in a variable is called *initializing* the variable.

Adding to a variable is called *incrementing*. For example, the statement

```
mynum = mynum + 1
```

increments the variable *mynum* by 1.

QUICK QUESTION!

How can 1 be subtracted from the value of a variable called *num*?

Answer:

```
num = num - 1;
```

This is called *decrementing* the variable.

1.3.2 Variable names

Variable names are an example of *identifier names*. We will see other examples of identifier names, such as file names, in future chapters. The rules for identifier names are:

- The name must begin with a letter of the alphabet. After that, the name can contain letters, digits, and the underscore character (e.g., *value_1*), but it cannot have a space.
- There is a limit to the length of the name; the built-in function **namelengthmax** tells what this maximum length is.
- MATLAB is case-sensitive, which means there is a difference between upper- and lowercase letters. So, variables called *mynum*, *MYNUM*, and *Mynum* are all different (although this would be confusing and should not be done).
- Although underscore characters are valid in a name, their use can cause problems with some programs that interact with MATLAB, so some programmers use mixed case instead (e.g., *partWeights* instead of *part_weights*)
- There are certain words called *reserved words*, or *key words*, that cannot be used as variable names.
- Names of built-in functions can be but should not be used as variable names.

Additionally, variable names should always be *mnemonic*, which means that they should make some sense. For example, if the variable is storing the radius of a circle, a name such as *radius* would make sense; *x* probably wouldn't.

The Workspace Window shows the variables that have been created in the current Command Window and their values.

The following commands relate to variables:

- **who** shows variables that have been defined in this Command Window (this just shows the names of the variables)
- **whos** shows variables that have been defined in this Command Window (this shows more information on the variables, similar to what is in the Workspace Window)
- **clear** clears out all variables so they no longer exist
- **clear** *variablename* clears out a particular variable
- **clear** *variablename1 variablename2* ... clears out a list of variables (note: separate the names with spaces)

If nothing appears when **who** or **whos** is entered, that means there aren't any variables! For example, in the beginning of a MATLAB session, variables could be created and then selectively cleared (remember that the semicolon suppresses output).

```
>> who
>> mynum = 3;
>> mynum + 5;

>> who
Your variables are:
ans   mynum

>> clear mynum
>> who
Your variables are:
ans
```

1.4 EXPRESSIONS

Expressions can be created using values, variables that have already been created, operators, built-in functions, and parentheses. For numbers, these can include operators such as multiplication, and functions such as trigonometric functions. An example of such an expression is:

```
>> 2 * sin(1.4)
ans =
    1.9709
```

1.4.1 The format function and ellipsis

The *default* in MATLAB is to display numbers that have decimal points with four decimal places, as shown in the previous example. (The default means if you do not specify otherwise, this is what you get.) The **format** command can be used to specify the output format of expressions.

There are many options, including making the format **short** (the default) or **long**. For example, changing the format to **long** will result in 15 decimal places. This will remain in effect until the format is changed back to **short**, as demonstrated in the following.

```
>> format long
>> 2 * sin(1.4)
ans =
   1.970899459976920

>> format short
>> 2 * sin(1.4)
ans =
    1.9709
```

The **format** command can also be used to control the spacing between the MATLAB command or expression and the result; it can be either **loose** (the default) or **compact**.

```
>> format loose
>> 5 * 33
ans =

   165

>> format compact
>> 5 * 33
ans =
   165
>>
```

Especially long expressions can be continued on the next line by typing three (or more) periods, which is the continuation operator, or the **ellipsis**. To do this, type part of the expression followed by an ellipsis, then hit the Enter key and continue typing the expression on the next line.

```
>> 3 + 55 − 62 + 4 − 5 ...
+ 22 − 1

ans =
   16
```

1.4.2 Operators

There are in general two kinds of operators: *unary* operators, which operate on a single value, or *operand*, and *binary* operators, which operate on two values or operands. The symbol "-", for example, is both the unary operator for negation and the binary operator for subtraction.

Here are some of the common operators that can be used with numerical expressions:

+	addition
−	negation, subtraction
*	multiplication
/	division (divided by, e.g., 10/5 is 2)
\	division (divided into, e.g., 5\10 is 2)
^	exponentiation (e.g., 5^2 is 25)

In addition to displaying numbers with decimal points, numbers can also be shown using *scientific or exponential notation*. This uses e for the exponent of 10 raised to a power. For example, $2 * 10^4$ could be written two ways:

```
>> 2 * 10^4
ans =
     20000

>> 2e4
ans =
     20000
```

Operator precedence rules

Some operators have *precedence* over others. For example, in the expression 4 + 5 * 3, the multiplication takes precedence over the addition, so first 5 is multiplied by 3, then 4 is added to the result. Using parentheses can change the precedence in an expression:

```
>> 4 + 5 * 3
ans =
     19

>> (4 + 5) * 3
ans =
     27
```

Within a given precedence level, the expressions are evaluated from left to right (this is called *associativity*).

Nested parentheses are parentheses inside of others; the expression in the *inner parentheses* is evaluated first. For example, in the expression 5− (6 * (4 + 2)), first the addition is performed, then the multiplication, and finally the subtraction, to result in −31. Parentheses can also be used simply to make an expression clearer. For example, in the expression ((4 + (3 * 5)) −1), the parentheses are not necessary, but are used to show the order in which the parts of the expression will be evaluated.

For the operators that have been covered thus far, the following is the precedence (from the highest to the lowest):

( )	parentheses
^	exponentiation
−	negation
*, /, \	all multiplication and division
+, −	addition and subtraction

PRACTICE 1.1

Think about what the results would be for the following expressions, and then type them in to verify your answers:

```
4 ^ 2 − 1

4 ^ (2 − 1)

2\ 3

4 * 2 − 9/3

5 − − 3
```

1.4.3 Built-in functions and help

There are many built-in functions in MATLAB. The **help** command can be used to identify MATLAB functions, and also how to use them. For example, typing **help** at the prompt in the Command Window will show a list of **help topics** that are groups of related functions. This is a very long list; the most elementary help topics appear at the beginning.

For example, one of these is listed as **matlab\elfun**; it includes the elementary math functions. Another of the first help topics is **matlab\ops**, which shows the operators that can be used in expressions.

To see a list of the functions contained within a particular help topic, type **help** followed by the name of the topic. For example,

```
>> help elfun
```

will show a list of the elementary math functions. It is a very long list, and it is broken into trigonometric (for which the default is radians, but there are equivalent functions that instead use degrees), exponential, complex, and rounding and remainder functions.

To find out what a particular function does and how to call it, type **help** and then the name of the function. For example,

```
>> help sin
```

will give a description of the **sin** function.

To *call* a function, the name of the function is given followed by the *argument(s)* that are passed to the function in parentheses. Most functions then *return* value(s). For example, to find the absolute value of –4, the following expression would be entered:

```
>> abs(−4)
```

which is a *call* to the function **abs**. The number in the parentheses, the −4, is the *argument*. The value 4 would then be *returned* as a result.

QUICK QUESTION!

What would happen if you use the name of a function, for example, *sin*, as a variable name?

Answer: This is allowed in MATLAB, but then *sin* could not be used as the built-in function until the variable is cleared. For example, examine the following sequence:

```
>> sin(3.1)
ans =
     0.0416
>> sin = 45
sin =
    45
```

Continued

QUICK QUESTION!—CONT'D

```
>> sin(3.1)                          >> who
??? Subscript indices must either be real   Your variables are:
positive integers or logicals.       ans
>> who                               >> sin(3.1)
Your variables are:                  ans =
ans sin                                 0.0416
>> clear sin
```

In addition to the trigonometric functions, the **elfun** help topic also has some rounding and remainder functions that are very useful. Some of these include **fix**, **floor**, **ceil**, **round**, **rem**, and **sign**.

The **rem** function returns the remainder from a division; for example, 5 goes into 13 twice with a remainder of 3, so the result of this expression is 3:

```
>> rem(13,5)
ans =
    3
```

QUICK QUESTION!

What would happen if you reversed the order by mistake, and typed the following:

```
rem(5,13)
```

Answer: The **rem** function is an example of a function that has two arguments passed to it. In some cases, the order in which

the arguments are passed does not matter, but for the **rem** function the order does matter. The **rem** function divides the second argument into the first. In this case, the second argument, 13, goes into 5 zero times with a remainder of 5, so 5 would be returned as a result.

Another function in the **elfun** help topic is **sign**, which returns 1 if the argument is positive, 0 if it is 0, and –1 if it is negative. For example,

```
>> sign(-5)
ans =
   -1

>> sign(3)
ans =
    1
```

PRACTICE 1.2

Use the **help** function to determine what the rounding functions **fix**, **floor**, **ceil**, and **round** do. Experiment with them by passing different values to the functions, including some negative, some positive, and some with fractions less than 0.5 and some greater. *It is very important when testing functions that you thoroughly test by trying different kinds of arguments!*

1.4.4 Constants

Variables are used to store values that might change, or for which the values are not known ahead of time. Most languages also have the capacity to store *constants*, which are values that are known ahead of time, and cannot possibly change. An example of a constant value would be **pi**, or π, which is $3.14159\ldots$. In MATLAB, there are functions that return some of these constant values, some of which include:

pi	$3.14159\ldots$
i	$\sqrt{-1}$
j	$\sqrt{-1}$
inf	infinity ∞
NaN	stands for "not a number," such as the result of $0/0$

QUICK QUESTION!

There is no built-in constant for e (2.718), so how can that value be obtained in MATLAB?

Answer: Use the exponential function **exp**; e or e^1 is equivalent to exp(1).

```
>> exp(1)
ans =
    2.7183
```

1.4.5 Types

Every expression, or variable, has a *type* associated with it. MATLAB supports many types, which are called *classes*. A class is essentially a combination of a type and the operations that can be performed on values of that type.

For example, there are types to store different kinds of numbers. For float or real numbers, or in other words numbers with a decimal place (e.g., 5.3), there are two basic types: **single** and **double**. The name of the type **double** is short for *double precision*; it stores larger numbers than the **single** type. MATLAB uses a *floating point* representation for these numbers.

There are many integer types, such as **int8**, **int16**, **int32**, and **int64**. The numbers in the names represent the number of bits used to store values of that type.

For example, the type **int8** uses eight bits altogether to store the integer and its sign. Since one bit is used for the sign, this means that seven bits are used to store actual numbers (0s or 1s). There are also *unsigned* integer types **uint8**, **uint16**, **uint32**, and **uint64**. For these types, the sign is not stored, meaning that the integers can only be positive (or 0).

For example, the type **uint8** stores 2^8 or 256 integers, ranging from 0 to 255. The range of values that can be stored in **int8**, on the other hand, is from −128 to +127. The range can be found for any type by passing the name of the type as a string (which means in single quotes) to the functions **intmin** and **intmax**. For example,

```
>> intmin('int8')
ans =
  -128
>> intmax('int8')
ans =
  127
```

The larger the number in the type name, the larger the number that can be stored in it. We will for the most part use the type **int32** when an integer type is required.

The type **char** is used to store either single *characters* (e.g., 'x') or *strings*, which are sequences of characters (e.g., 'cat'). Both characters and strings are enclosed in single quotes.

The type **logical** is used to store **true/false** values. This will be explained in more detail in Chapter 3.

Variables that have been created in the Command Window can be seen in the Workspace Window. In that window, for every variable, the variable name, value, and class (which is essentially its type) can be seen. Other attributes of variables can also be seen in the Workspace Window. Which attributes are visible by default depends on the version of MATLAB. However, when the Workspace Window is chosen, clicking on View allows the user to choose which attributes will be displayed.

By default, numbers are stored as the type **double** in MATLAB. There are, however, many functions that convert values from one type to another. The names of these functions are the same as the names of the types shown in this section. These names can be used as functions to convert a value to that type. This is called *casting* the value to a different type, or *type casting*. For example, to convert a value from the type **double**, which is the default, to the type **int32**, the function **int32** would be used. Entering the assignment statement

```
>> val = 6 + 3
```

would result in the number 9 being stored in the variable *val*, with the default type of **double**, which can be seen in the Workspace Window. Subsequently, the assignment statement

```
>> val = int32(val);
```

would change the type of the variable to **int32**, but would not change its value. Here is another example using two different variables.

```
>> num = 6 + 3;
>> numi = int32(num);
>> whos
  Name      Size     Bytes    Class     Attributes
  num       1 x 1        8    double
  numi      1 x 1        4    int32
```

One reason for using an integer type for a variable is to save space in memory.

QUICK QUESTION!

What would happen if you go beyond the range for a particular type? For example, the largest integer that can be stored in **int8** is 127, so what would happen if we type cast a larger integer to the type **int8**?

```
>> int8(200)
```

Answer: The value would be the largest in the range, in this case 127. If we instead use a negative number that is smaller than the lowest value in the range, its value would be –128. This is an example of what is called *saturation arithmetic.*

```
>> int8(200)
ans =
  127
>> int8(-130)
ans =
  -128
```

PRACTICE 1.3

- Calculate the range of integers that can be stored in the types **uint16** and **int16**. Use **intmin** and **intmax** to verify your results.
- Enter an assignment statement and view the variable type in the Workspace Window. Then, change its type and view it again.

1.4.6 Random numbers

When a program is being written to work with data, and the data are not yet available, it is often useful to test the program first by initializing the data variables to *random numbers*. There are several built-in functions in MATLAB that generate random numbers, some of which will be illustrated in this section.

Random number generators or functions are not truly random. Basically, the way it works is that the process starts with one number, which is called a *seed*. Frequently, the initial seed is either a predetermined value or it is obtained from the built-in clock in the computer. Then, based on this seed, a process determines the next "random number." Using that number as the seed the next time, another random number is generated, and so forth. These are actually called *pseudorandom*—they are not truly random because there is a process that determines the next value each time.

The function **rand** can be used to generate uniformly distributed random real numbers; calling it generates one random real number in the range from 0 to 1. There are no arguments passed to the **rand** function in its simplest form. Here are two examples of calling the **rand** function:

```
>> rand
ans =
    0.9501

>> rand
ans =
    0.2311
```

The seed for the **rand** function will always be the same each time MATLAB is started, unless the initial seed is changed. In later versions of MATLAB, this is done with the **rng** function:

```
>> rng('shuffle')
```

Note: This is only done once in any given MATLAB session to set the seed; the **rand** function can then be used as shown before any number of times to generate random numbers. For earlier versions of MATLAB, the following can be used instead:

```
>> rand ('state',sum (100*clock))
```

This uses the current date and time that are returned from the built-in **clock** function to set the seed.

Since **rand** returns a real number in the range from 0 to 1, multiplying the result by an integer N would return a random real number in the range from 0 to N. For example, multiplying by 10 returns a real number in the range from 0 to 10, so the expression

```
rand*10
```

would return a result in the range from 0 to 10.

To generate a random real number in the range from *low* to *high*, first create the variables *low* and *high*. Then, use the expression `rand* (high-low)+ low`. For example, the sequence

```
>> low = 3;
>> high = 5;
>> rand * (high-low)+ low
```

would generate a random real number in the range from 3 to 5.

The function **randn** is used to generate normally distributed random real numbers.

Generating random integers

Since the **rand** function returns a real number, this can be rounded to produce a random integer. For example,

```
>> round(rand * 10)
```

would generate one random integer in the range from 0 to 10 (`rand * 10` would generate a random real number in the range from 0 to 10; rounding that will return an integer). Or, one can generate a random integer in a range:

```
>> low = 2;
>> high = 6;
>> round(rand * (high-low)+low)
```

This would generate a random integer in the range from 2 to 6.

PRACTICE 1.4

Generate a random

- real number in the range from 0 to 1
- real number in the range from 0 to 50
- real number in the range from 10 to 20
- integer in the range from 1 to 10

Note

In some versions of MATLAB, there is another built-in function that specifically generates random integers: **randint** in some cases and **randi** in others.

1.5 CHARACTERS AND ENCODING

A character in MATLAB is represented using single quotes (e.g., 'a' or 'x'). The quotes are necessary to denote a character; without them, a letter would be interpreted as a variable name. Characters are put in an order using what is called a *character encoding*. In the character encoding, all characters in the computer's *character set* are placed in a sequence and given equivalent integer values. The character set includes all letters of the alphabet, digits, and punctuation marks; basically, all of the keys on a keyboard are characters. Special characters, such as the Enter key, are also included. So, 'x', '!', and '3' are all characters. With quotes, '3' is a character, not a number.

The most common character encoding is the American Standard Code for Information Interchange, or ASCII. Standard ASCII has 128 characters, which have equivalent integer values from 0 to 127. The first 32 (integer values 0 through 31) are nonprinting characters. The letters of the alphabet are in order, which means 'a' comes before 'b', then 'c', and so forth.

The numeric functions can be used to convert a character to its equivalent numerical value (e.g., **double** will convert to a double value, and **int32** will convert to an integer value using 32 bits). For example, to convert the character 'a' to its numerical equivalent, the following statement could be used:

```
>> numequiv = double ('a')
numequiv =
     97
```

This stores the **double** value 97 in the variable *numequiv*, which shows that the character 'a' is the 98th character in the character encoding (since the equivalent numbers begin at 0). It doesn't matter which number type is used to convert 'a'; for example,

```
>> numequiv = int32 ('a')
```

would also store the integer value 97 in the variable *numequiv*. The only difference between these will be the type of the resulting variable (**double** in the first case, **int32** in the second).

The function **char** does the reverse; it converts from any number to the equivalent character:

```
>> char (97)
ans =
a
```

Note

Quotes are not shown when a character is displayed.

Since the letters of the alphabet are in order, the character 'b' has the equivalent value of 98, 'c' is 99, and so on. Math can be done on characters. For example, to get the next character in the character encoding, 1 can be added either to the integer or the character:

```
>> numequiv = double ('a');

>> char (numequiv + 1)
ans =
b

>> 'a' + 2
ans =
    99
```

Note the difference in the formatting (the indentation) when a number is displayed versus a character:

```
>> var = 3
var =
    3
>> var = '3'
var =
3
```

MATLAB also handles strings, which are sequences of characters in single quotes. For example, using the **double** function on a string will show the equivalent numerical value of all characters in the string:

```
>> double('abcd')
ans =
    97    98    99    100
```

To shift the characters of a string "up" in the character encoding, an integer value can be added to a string. For example, the following expression will then shift by 1:

```
>> char('abcd' + 1)
ans =
bcde
```

PRACTICE 1.5

- Find the numerical equivalent of the character 't'.
- Find the character equivalent of 112.

1.6 VECTORS AND MATRICES

Vectors and *matrices* are used to store sets of values, all of which are the same type. A vector can be either a *row vector* or a *column vector*. A matrix can be visualized as a table of values. The dimensions of a matrix are $r \times c$, where r is the number of rows and c is the number of columns. This is pronounced "r by c." If a vector has n elements, a row vector would have the dimensions $1 \times n$, and a column vector would have the dimensions $n \times 1$.

A *scalar* (one value) has the dimensions 1×1. Therefore, vectors and scalars are actually just special cases of matrices. Here are some diagrams showing, from left to right, a scalar, a column vector, a row vector, and a matrix.

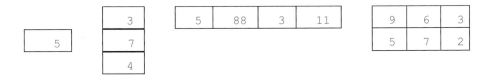

The scalar is 1×1, the column vector is 3×1 (three rows by one column), the row vector is 1×4 (one row by four columns), and the matrix is 2×3 (two rows by three columns). All of the values in these matrices are stored in what are called *elements*.

MATLAB is written to work with matrices; the name MATLAB is short for "matrix laboratory." Since MATLAB is written to work with matrices, it is very easy to create vector and matrix variables, and there are many operations and functions that can be used on vectors and matrices.

A vector in MATLAB is equivalent to what is called a one-dimensional *array* in other languages. A matrix is equivalent to a two-dimensional array. Usually, even in MATLAB, some operations that can be performed on either vectors or matrices are referred to as *array operations*. The term "array" is also frequently used to mean generically either a vector or a matrix.

1.6.1 Creating row vectors

There are several ways to create row vector variables. The most direct way is to put the values that you want in the vector in square brackets, separated by either spaces or commas. For example, both of these assignment statements create the same vector *v*:

```
>> v = [1    2    3    4]
v =
    1   2   3   4

>> v = [1,2,3,4]
v =
    1   2   3   4
```

Both of these create a row vector variable that has four elements; each value is stored in a separate element in the vector.

The colon operator and linspace function

If, as in the preceding examples, the values in the vector are regularly spaced, the *colon operator* can be used to *iterate* through these values. For example, 1:5 results in all of the integers from 1 to 5:

```
>> vec = 1:5
vec =
    1   2   3   4   5
```

Note that in this case, the brackets [ ] are not necessary to define the vector.

With the colon operator, a *step value* can also be specified with another colon, in the form (first:step:last). For example, to create a vector with all integers from 1 to 9 in steps of 2:

```
>> nv = 1:2:9
nv =
    1   3   5   7   9
```

QUICK QUESTION!

What happens if adding the step value would go beyond the range specified by the last, such as

```
1:2:6
```

Answer: This would create a vector containing 1, 3, and 5. Adding 2 to the 5 would go beyond 6, so the vector stops at 5; the result would be

```
    1       3       5
```

How can you use the colon operator to generate the following vector?

| 9 | 7 | 5 | 3 | 1 |

Answer:

```
9:-2:1
```

The step operator can be a negative number, so the resulting sequence is in descending order.

Similarly, the **linspace** function creates a linearly spaced vector; **linspace(x,y,n)** creates a vector with n values in the inclusive range from x to y. For example, the following creates a vector with five values linearly spaced between 3 and 15, including the 3 and 15:

```
>> ls = linspace(3,15,5)
ls =
     3     6     9     12     15
```

Vector variables can also be created using existing variables. For example, a new vector is created here consisting first of all values from *nv* followed by all values from *ls*:

```
>> newvec = [nv ls]
newvec =
     1   3   5   7   9   3   6   9   12   15
```

Putting two vectors together like this to create a new one is called *concatenating* the vectors.

Referring to and modifying elements

The elements in a vector are numbered sequentially; each element number is called the *index*, or *subscript*. In MATLAB, the indices start at 1. Normally, diagrams of vectors and matrices show the indices. For example, for the variable *newvec* created earlier the indices 1 to 10 of the elements are shown above the vector:

```
                    newvec

        1   2   3   4   5   6   7   8   9   10
      +---+---+---+---+---+---+---+---+----+----+
      | 1 | 3 | 5 | 7 | 9 | 3 | 6 | 9 | 12 | 15 |
      +---+---+---+---+---+---+---+---+----+----+
```

A particular element in a vector is accessed using the name of the vector variable and the index or subscript in parentheses. For example, the fifth element in the vector *newvec* is a 9:

```
>> newvec(5)
ans =
     9
```

The expression *newvec(5)* would be pronounced "newvec sub 5," where sub is short for subscript. A subset of a vector, which would be a vector itself, can also be obtained using the colon operator. For example, the following statement would get the fourth through sixth elements of the vector *newvec*, and store the result in a vector variable *b*:

```
>> b = newvec(4:6)
b =
     7    9    3
```

Any vector can be used for the indices into another vector, not just one created using the colon operator. For example, the following would get the first, fifth, and tenth elements of the vector *newvec*:

```
>> newvec([1   5   10])
ans =
     1        9        15
```

The vector [1 5 10] is called an **index vector**; it specifies the indices in the original vector that are being referenced.

The value stored in a vector element can be changed by specifying the index or subscript. For example, to change the second element from the preceding vector *b* to now store the value 11 instead of 9:

```
>> b(2) = 11
b =
     7    11    3
```

By referring to an index that does not yet exist, a vector can also be extended. For example, the following creates a vector that has three elements. By then assigning a value to the fourth element, the vector is extended to have four elements.

```
>> rv = [3   55   11]
rv =
     3        55        11

>> rv(4) = 2
rv =
     3        55        11       2
```

If there is a gap between the end of the vector and the specified element, 0s are filled in. For example, the following extends the variable *rv* again:

```
>> rv(6) = 13
rv =
     3        55        11       2       0       13
```

As we will see later, this is actually not very efficient because it can take extra time.

PRACTICE 1.6

Think about what would be produced by the following sequence of statements and expressions, and then type them in to verify your answers:

```
pv = 2:2:8
pv(4) = 33
pv(6) = 11
prac = pv(3:5)
linspace(4,12,3)
```

1.6.2 Creating column vectors

One way to create a column vector is to explicitly put the values in square brackets, separated by semicolons (rather than commas or spaces):

```
>> c = [1; 2; 3; 4]
c =
    1
    2
    3
    4
```

There is no direct way to use the colon operator to get a column vector. However, any row vector created using any method can be *transposed* to result in a column vector. In general, the transpose of a matrix is a new matrix in which the rows and columns are interchanged. For vectors, transposing a row vector results in a column vector, and transposing a column vector results in a row vector. A built-in operator, the apostrophe, in MATLAB will transpose.

```
>> r = 1:3;
>> c = r'
c =
    1
    2
    3
```

1.6.3 Creating matrix variables

Creating a matrix variable is simply a generalization of creating row and column vector variables. That is, the values within a row are separated by either spaces or commas, and the different rows are separated by semicolons. For example, the matrix variable *mat* is created by explicitly typing values:

```
>> mat = [4  3  1;  2  5  6]
mat =
    4    3    1
    2    5    6
```

There must always be the same number of values in each row. If you attempt to create a matrix in which there are different numbers of values in the rows, the result will be an error message, such as in the following:

```
>> mat = [3 5 7; 1 2]
??? Error using ==> vertcat
CAT arguments dimensions are not consistent.
```

Iterators can also be used for the values in the rows using the colon operator. For example:

```
>> mat = [2:4; 3:5]
mat =
    2    3    4
    3    4    5
```

Different rows in the matrix can also be specified by hitting the Enter key after each row instead of typing a semicolon when entering the matrix values, as in:

```
>> newmat = [2 6 88
33 5 2]

newmat =
    2     6    88
   33     5     2
```

Matrices of random numbers can be created using the **rand** function. If a single value n is passed to **rand**, an $n \times n$ matrix will be created, or passing two arguments will specify the number of rows and columns:

```
>> rand(2)
ans =

    0.2311    0.4860
    0.6068    0.8913

>> rand(1,3)
ans =
    0.7621    0.4565    0.0185
```

Matrices of random integers can be generated using **round**, as previously demonstrated:

```
>> round(rand(2,2)*10)
ans =
    1    9
    4    8
```

MATLAB also has several functions that create special matrices. For example, the **zeros** function creates a matrix of all zeros, and the **ones** function creates a matrix of all ones. Like **rand**, either one argument can be passed (which will be both the number of rows and columns), or two arguments (first the number of rows and then the number of columns).

```
>> zeros(3)
ans =
        0     0     0
        0     0     0
        0     0     0

>> ones(2,4)
ans =

     1     1     1     1
     1     1     1     1
```

Examples of other special matrix functions appear in Chapter 12.

Referring to and modifying matrix elements

To refer to matrix elements, the row and then the column subscripts are given in parentheses (always the row first and then the column). For example, this creates a matrix variable *mat* and then refers to the value in the second row, third column of *mat*:

```
>> mat = [2:4; 3:5]
mat =
     2     3     4
     3     4     5

>> mat(2,3)
ans =
     5
```

This is called *subscripted indexing*; it uses the row and column subscripts. It is also possible to refer to a subset of a matrix. For example, this refers to the first and second rows, second and third columns:

```
>> mat(1:2,2:3)
ans =
     3     4
     4     5
```

Using a colon for the row subscript means all rows, regardless of how many, and using a colon for the column subscript means all columns. For example, this refers to all columns within the first row, or in other words the entire first row:

```
>> mat(1,:)
ans =
     2     3     4
```

This refers to the entire second column:

```
>> mat(:, 2)
ans =
    3
    4
```

If a single index is used with a matrix, MATLAB *unwinds* the matrix column by column. For example, for the matrix *intmat* created here, the first two elements are from the first column, and the last two are from the second column:

```
>> intmat = [100 77; 28 14]
intmat =
    100    77
     28    14

>> intmat(1)
ans =
    100

>> intmat(2)
ans =
     28

>> intmat(3)
ans =
     77

>> intmat(4)
ans =
     14
```

This is called *linear indexing*. It is usually much better style when working with matrices to use subscripted indexing.

An individual element in a matrix can be modified by assigning a new value to it.

```
>> mat = [2:4; 3:5];
>> mat(1,2) = 11
mat =
    2    11     4
    3     4     5
```

An entire row or column could also be changed. For example, the following replaces the entire second row with values from a vector.

Note

Since the entire row is being modified, a vector with the correct length must be assigned.

```
>> mat(2,:) = 5:7
mat =
    2    11     4
    5     6     7
```

To extend a matrix, an individual element could not be added since that would mean there is no longer the same number of values in every row. However, an entire row or column could be added. For example, the following would add a fourth column to the matrix.

```
>> mat(:,4) = [9 2]'
mat =
     2    11     4     9
     5     6     7     2
```

Just as we saw with vectors, if there is a gap between the current matrix and the row or column being added, MATLAB will fill in with zeros.

```
>> mat(4,:) = 2:2:8
mat =
     2    11     4     9
     5     6     7     2
     0     0     0     0
     2     4     6     8
```

1.6.4 Dimensions

The **length** and **size** functions in MATLAB are used to find dimensions of vectors and matrices. The **length** function returns the number of elements in a vector. The **size** function returns the number of rows and columns in a vector or matrix. For example, the following vector *vec* has four elements so its length is 4. It is a row vector, so the size is *1 × 4*.

```
>> vec = -2:1
vec =
    -2   -1    0    1

>> length(vec)
ans =
     4

>> size(vec)
ans =
     1    4
```

To create the following matrix variable *mat*, iterators are used on the two rows and then the matrix is transposed so that it has three rows and two columns, or in other words the size is *3 × 2*.

```
>> mat = [1:3; 5:7]'
mat =
     1    5
     2    6
     3    7
```

The **size** function returns the number of rows and then the number of columns, so to capture these values in separate variables we put a vector of two variables on the left of the assignment. The variable *r* stores the first value returned, which is the number of rows, and *c* stores the number of columns.

Note

This example demonstrates a very important and unique concept in MATLAB: the ability to have a vector of variables on the left side of an assignment.

```
>> [r c] = size(mat)
r =
    3
c =
    2
```

If called as just an expression, the **size** function will return both values in a vector:

```
>> size(mat)
ans =
    3    2
```

For a matrix, the **length** function will return either the number of rows or the number of columns, whichever is largest (in this case the number of rows, 3).

```
>> length(mat)
ans =
    3
```

QUICK QUESTION!

How could you create a matrix of zeros with the same size as another matrix?

Answer: For a matrix variable *mat*, the following expression would accomplish this:

```
zeros(size(mat))
```

The **size** function returns the size of the matrix, which is then passed to the **zeros** function, which then returns a matrix of zeros with the same size as *mat*. It is not necessary in this case to store the values returned from the **size** function in variables.

MATLAB also has a function **numel** that returns the total number of elements in any array (vector or matrix):

```
>> vec = 9:-2:1
vec =
    9    7    5    3    1

>> numel(vec)
ans =
    5

>> mat = [3:2:7; 9 33 11]
```

```
mat =
     3      5      7
     9     33     11

>> numel(mat)
ans =
     6
```

For vectors, this is equivalent to the length of the vector. For matrices, it is the product of the number of rows and columns.

MATLAB also has a built-in expression **end** that can be used to refer to the last element in a vector; for example, *v(end)* is equivalent to *v(length(v))*. For matrices, it can refer to the last row or column. So, for example, using **end** for the row index would refer to the last row.

In this case, the element referred to is in the first column of the last row:

```
>> mat = [1:3; 4:6]'
mat =
     1      4
     2      5
     3      6

>> mat(end,1)
ans =
     3
```

Using **end** for the column index would refer to the last column (e.g., the last column of the second row):

```
>> mat(2,end)
ans =
     5
```

This can only be used as an index.

Changing dimensions

In addition to the transpose operator, MATLAB has several built-in functions that change the dimensions or configuration of matrices, including **reshape**, **fliplr**, **flipud**, and **rot90**.

The **reshape** function changes the dimensions of a matrix. The following matrix variable *mat* is 3 × 4; in other words it has 12 elements.

```
>> mat = round(rand(3,4)*100)
    14     61      2     94
    21     28     75     47
    20     20     45     42
```

Important

In programming applications, it is better to not assume that the dimensions of a vector or matrix are known. Instead, to be general, use either the **length** or **numel** function to determine the number of elements in a vector, and use **size** (and store the result in two variables) for a matrix.

These 12 values could instead be arranged as a *2 × 6* matrix, *6 × 2*, *4 × 3*, *1 × 12*, or *12 × 1*. The **reshape** function iterates through the matrix columnwise. For example, when reshaping *mat* into a *2 × 6* matrix, the values from the first column in the original matrix (14, 21, and 20) are used first, then the values from the second column (61, 28, 20), and so forth.

```
>> reshape(mat,2,6)
ans =
    14     20     28      2     45     47
    21     61     20     75     94     42
```

Note that in these examples *mat* is unchanged; instead, the results are stored in the default variable *ans* each time.

The **fliplr** function "flips" the matrix from left to right (in other words, the left-most column, the first column, becomes the last column and so forth), and the **flipud** function flips up to down.

```
>> mat
mat =
    14     61      2     94
    21     28     75     47
    20     20     45     42

>> fliplr(mat)
ans =
    94      2     61     14
    47     75     28     21
    42     45     20     20

>> mat
mat =
    14     61      2     94
    21     28     75     47
    20     20     45     42

>> flipud(mat)
ans =
    20     20     45     42
    21     28     75     47
    14     61      2     94
```

The **rot90** function rotates the matrix counterclockwise 90 degrees, so for example, the value in the top right corner becomes instead the top left corner and the last column becomes the first row.

```
>> mat
mat =
      14      61       2      94
      21      28      75      47
      20      20      45      42

>> rot90(mat)
ans =
      94      47      42
       2      75      45
      61      28      20
      14      21      20
```

QUICK QUESTION!

Is there a **rot180** function? Is there a **rot-90** function (to rotate clockwise)?

Answer: Not exactly, but a second argument can be passed to the **rot90** function, which is an integer n; the function will rotate $90 * n$ degrees. The integer can be positive or negative. For example, if 2 is passed, the function will rotate the matrix 180 degrees (so, it would be the same as rotating the result of **rot90** another 90 degrees).

```
>> mat
mat =
      14      61       2      94
      21      28      75      47
      20      20      45      42

>> rot90(mat,2)
ans =
      42      45      20      20
      47      75      28      21
      94       2      61      14
```

If a negative number is passed for n, the rotation would be in the opposite direction, that is, clockwise.

```
>> mat
mat =
      14      61       2      94
      21      28      75      47
      20      20      45      42

>> rot90(mat,-1)
ans =
      20      21      14
      20      28      61
      45      75       2
      42      47      94
```

The function **repmat** can be used to create a matrix; **repmat(mat,m,n)** creates a larger matrix that consists of an $m \times n$ matrix of copies of *mat*. For example, here is a 2×2 random matrix:

```
>> intmat = round(rand(2)*100)
intmat =
      50      34
      96      59
```

Replicating this matrix six times as a *3 × 2* matrix would produce copies of *intmat* in this form:

intmat	intmat
intmat	intmat
intmat	intmat

```
>> repmat(intmat,3,2)
ans =
    50    34    50    34
    96    59    96    59
    50    34    50    34
    96    59    96    59
    50    34    50    34
    96    59    96    59
```

1.6.5 Using functions with vectors and matrices

Since MATLAB is written to work with vectors and matrices, an entire vector or matrix can be passed as an argument to a function. MATLAB will evaluate the function on every element, and return as a result a vector or matrix with the same dimensions as the original. For example, we could pass the following vector *vec* to the **abs** function to get the absolute value of every element.

```
>> vec = -3:4
vec =
    -3    -2    -1     0     1     2     3     4

>> abs(vec)
ans =
     3     2     1     0     1     2     3     4
```

The original vector *vec* has eight elements, and since the **abs** function is evaluated for every element, the resulting vector also has eight elements.

This also would be the case for matrices; the result will be the same size as the input:

```
>> mat = round(rand(2,3)*10-5)
mat =
    -3    -2     2
     3     0     4

>> abs(mat)
ans =
     3     2     2
     3     0     4
```

We will see much more on operations and functions of arrays (vectors and matrices) in Chapters 4, 5, and 12.

1.6.6 Empty vectors

An *empty vector* (i.e., a vector that stores no values) can be created using empty square brackets:

```
>> evec = []
evec =
     []

>> length(evec)
ans =
   0
```

Values can then be added to an empty vector by concatenating, or adding, values to the existing vector. The following statement takes what is currently in *evec*, which is nothing, and adds a 4 to it.

```
>> evec = [evec 4]
evec =
   4
```

Note

There is a difference between having an empty vector variable and not having the variable at all.

The following statement takes what is currently in *evec*, which is 4, and adds an 11 to it.

```
>> evec = [evec 11]
evec =
   4    11
```

This can be continued as many times as desired, to build a vector up from nothing. Sometimes this is necessary, although generally it is not a good idea if it can be avoided because it can be quite time consuming.

Empty vectors can also be used to **delete elements from vectors**. For example, to remove the third element from a vector, the empty vector is assigned to it:

```
>> vec = 1:5
vec =
   1    2    3    4    5

>> vec(3) = []
vec =
   1    2    4    5
```

The elements in this vector are now numbered 1 through 4.

Subsets of a vector could also be removed. For example,

```
>> vec = 1:8
vec =
   1    2    3    4    5    6    7    8
```

```
>> vec(2:4) = []
vec =
     1    5    6    7    8
```

Individual elements cannot be removed from matrices, since matrices always have to have the same number of elements in every row.

```
>> mat = [7 9 8; 4 6 5]
mat =
     7    9    8
     4    6    5

>> mat(1,2) = [];
??? Indexed empty matrix assignment is not allowed.
```

However, entire rows or columns could be removed from a matrix. For example, to remove the second column:

```
>> mat(:,2) = []
mat =
     7    8
     4    5
```

Also, if linear indexing is used with a matrix to delete an element, the matrix will be reshaped into a vector.

```
>> mat = [7 9 8; 4 6 5]
mat =
     7    9    8
     4    6    5

>> mat(3) = []
mat =
     7    4    6    8    5
```

PRACTICE 1.7

Think about what would be produced by the following sequence of statements and expressions, and then type them in to verify your answers.

```
m = [ 1:4; 3 11 7 2]

m(2,3)

m(:,3)

m(4)

size(m)
```

```
numel(m)

reshape(m,1,numel(m))

vec = m(1,:)

vec(2) = 5

vec(3) = []

vec(5) = 8

vec = [ vec 11]
```

1.6.7 Three-dimensional matrices

The matrices that have been shown so far have been two dimensional; these matrices have rows and columns. Matrices in MATLAB are not limited to two dimensions, however. In fact, in Chapter 14 we will see image applications in which *three-dimensional matrices* are used. For a three-dimensional matrix, imagine a two-dimensional matrix as being flat on a page, and then the third dimension consists of more pages on top of that one (so, they are stacked on top of each other).

Here is an example of creating a three-dimensional matrix. First, a two-dimensional matrix *mat* is created. It is modified by flipping it up and down and left to right. This new matrix *newm* is made into the third dimension of the original by extending it. Note that we end up with a matrix that has two layers, each of which is 3×5. The resulting three-dimensional matrix has dimensions $3 \times 5 \times 2$.

```
>> mat = reshape(1:15,3,5)
mat =
     1     4     7    10    13
     2     5     8    11    14
     3     6     9    12    15

>> newm = fliplr(flipud(mat))
newm =
    15    12     9     6     3
    14    11     8     5     2
    13    10     7     4     1

>> mat(:,:,2) = newm
mat(:,:,1) =
     1     4     7    10    13
     2     5     8    11    14
     3     6     9    12    15
```

```
mat(:,:,2) =
    15    12     9     6     3
    14    11     8     5     2
    13    10     7     4     1

>> size(mat)
ans =
     3     5     2
```

Three-dimensional matrices can also be created using the **zeros, ones**, and **rand** functions by specifying three dimensions to begin with (e.g., **zeros(3,5,2)**).

Unless specified otherwise, in the remainder of this book "matrices" will be assumed to be two dimensional.

SUMMARY

Common Pitfalls

It is common when learning to program to make simple spelling mistakes and to confuse the necessary punctuation. Examples are given here of very common errors. Some of these include:

- Putting a space in a variable name
- Confusing the format of an assignment statement as

    ```
    expression = variablename
    ```
    rather than
    ```
    variablename = expression
    ```
    The variable name must always be on the left.
- Using a built-in function name as a variable name, and then trying to use the function
- Confusing the two division operators / and \
- Forgetting the operator precedence rules
- Confusing the order of arguments passed to functions—for example, to find the remainder of dividing 3 into 10 using **rem(3,10)** instead of **rem(10,3)**
- Not using different types of arguments when testing functions
- Attempting to create a matrix that does not have the same number of values in each row
- Forgetting to use parentheses to pass an argument to a function (e.g., "fix 2.3" instead of "**fix(2.3)**"). MATLAB returns the ASCII equivalent for each character when this mistake is made. (What happens is that it is interpreted as the function of a string, "**fix('2.3')**".)

Programming Style Guidelines

Following these guidelines will make your code much easier to read and understand, and therefore easier to work with and modify.

- Use mnemonic variable names (names that make sense, such as *radius* instead of *xyz*).
- Although variables named *result* and *RESULT* are different, using the same word(s) for different variables would be confusing.

- Do not use names of built-in functions as variable names.
- If different sets of random numbers are desired, set the seed for the **rand** function.
- If possible, try not to extend vectors or matrices, as it is not very efficient.
- Do not use just a single index when referring to elements in a matrix; instead, use both the row and column indices (use subscripted indexing rather than linear indexing).
- In general, never assume that the dimensions of any array (vector or matrix) are known. Instead, use the function **length** or **numel** to determine the number of elements in a vector, and the function **size** for a matrix:

```
len = length(vec);

[ r c] = size(mat);
```

MATLAB Functions and Commands

info	i	rand
demo	j	rng
help	inf	clock
lookfor	NaN	randint
quit	exp	randi
namelengthmax	single	randn
who	double	linspace
whos	int8	zeros
clear	int16	ones
format	int32	length
sin	int64	size
abs	uint8	numel
fix	uint16	end
floor	uint32	reshape
ceil	uint64	fliplr
round	intmin	flipud
rem	intmax	rot90
sign	char	repmat
pi	logical	

MATLAB Operators

assignment =	multiplication *	negation −
ellipsis ...	exponentiation ^	colon :
continuation ...	divided by /	transpose '
addition +	divided into \	
subtraction −	parentheses ( )	

Exercises

1. Create a variable to store the atomic weight of silicon (28.085).
2. Create a variable *myage* and store your age in it. Subtract one from the value of the variable. Add 2 to the value of the variable.
3. Use the built-in function **namelengthmax** to find out the maximum number of characters that you can have in an identifier name under your version of MATLAB.
4. Explore the **format** command in more detail. Use **help format** to find options. Experiment with **format bank** to display dollar values.
5. Find a **format** option that would result in the following output format:

```
>> 5/16 + 2/7
ans =
   67/112
```

6. Think about what the results would be for the following expressions, and then type them in to verify your answers.

```
25 / 4 * 4
3 + 4 ^ 2
4 \ 12 + 4
3 ^ 2
(5 - 2) * 3
```

7. The combined resistance R_T of three resistors R_1, R_2, and R_3 in parallel is given by

$$R_T = \frac{1}{\dfrac{1}{R_1} + \dfrac{1}{R_2} + \dfrac{1}{R_3}}$$

Create variables for the three resistors and store values in each, and then calculate the combined resistance.

As the world becomes more "flat," it is increasingly important for engineers and scientists to be able to work with colleagues in other parts of the world. Correct conversion of data from one system of units to another (e.g., from the metric system to the American system or vice versa) is critically important.

8. Create a variable *pounds* to store weight in pounds. Convert this to kilograms and assign the result to a variable *kilos*. The conversion factor is 1 kilogram = 2.2 pounds.
9. Create a variable *ftemp* to store a temperature in degrees Fahrenheit (F). Convert this to degrees Celsius (C) and store the result in a variable *ctemp*. The conversion factor is C = (F − 32) * 5/9.
10. Find another quantity to convert from one system of units to another.
11. The function **sin** calculates and returns the sine of an angle in radians. Use **help elfun** to find the name of the function that returns the sine of an angle in degrees. Verify that calling this function and passing 90 degrees to it results in 1.
12. A vector can be represented by its rectangular coordinates *x* and *y* or by its polar coordinates *r* and θ. The relationship between them is given by the equations:

```
x = r * cos (θ)
y = r * sin (θ)
```

Assign values for the polar coordinates to variables *r* and *theta*. Then, using these values, assign the corresponding rectangular coordinates to variables *x* and *y*.

13. Wind often makes the air feel even colder than it is. The wind chill factor (WCF) measures how cold it feels with a given air temperature T (in degrees Fahrenheit) and wind speed (V, in miles per hour). One formula for the WCF is:

$$WCF = 35.7 + 0.6\ T - 35.7(V^{0.16}) + 0.43\ T(V^{0.16})$$

- Create variables for the temperature T and wind speed V, and then using this formula calculate the WCF.

14. Use **help elfun** or experiment to answer the following questions:
   ▪ Is fix(3.5) the same as floor(3.5)?
   ▪ Is fix(3.4) the same as fix(−3.4)?
   ▪ Is fix(3.2) the same as floor(3.2)?
   ▪ Is fix(−3.2) the same as floor(−3.2)?
   ▪ Is fix(−3.2) the same as ceil(−3.2)?

15. Find MATLAB expressions for the following:

$$\sqrt{19}$$

$$3^{1.2}$$

```
tan (π)
```

16. Use **intmin** and **intmax** to determine the range of values that can be stored in the types **uint32** and **uint64**.

17. Are there equivalents to **intmin** and **intmax** for real number types? Use **help** to find out.

18. Store a number with a decimal place in a **double** variable (the default). Convert the variable to the type **int32** and store the result in a new variable.

19. Generate a random
   ▪ real number in the range from 0 to 1
   ▪ real number in the range from 0 to 20
   ▪ real number in the range from 20 to 50
   ▪ integer in the range from 0 to 10
   ▪ integer in the range from 0 to 11
   ▪ integer in the range from 50 to 100

20. Open a new Command Window, and type **rand** to get a random real number. Make a note of the number. Then, exit MATLAB and repeat this, again making a note of the random number; it should be the same as before. Finally, exit MATLAB and again open a new Command Window. This time, change the seed before generating a random number; it should be different.

21. In the ASCII character encoding, the letters of the alphabet are in order; for example, 'a' comes before 'b' and also 'A' comes before 'B'. However, which comes first—lowercase or uppercase letters?

22. Shift the string 'xyz' up in the character encoding by two characters.

23. Using the colon operator, create the following row vectors:

    3   4   5   6

    1.0000 1.5000 2.0000 2.5000 3.0000

    5   4   3   2

24. Using the **linspace** function, create the following row vectors:

    4   6   8

    −3   −6   −9 −12 −15

    9   7   5

25. Create the following row vectors twice, using **linspace** and using the colon operator:

    1   2   3   4   5   6   7   8   9   10

    2   7   12

26. Create a variable *myend* that stores a random integer in the range from 8 to 12. Using the colon operator, create a vector that iterates from 1 to *myend* in steps of 3.

27. Using the colon operator and the transpose operator, create a column vector that has the values −1 to 1 in steps of 0.2.

28. Write an expression that refers to only the odd-numbered elements in a vector, regardless of the vector length. Test your expression on vectors that have both an odd and even number of elements.

29. Create a vector variable *vec*; it can have any length. Then, write assignment statements that would store the first half of the vector in one variable and the second half in another. Make sure that your assignment statements are general, and work whether *vec* has an even or odd number of elements. (*Hint:* Use a rounding function such as **fix**.)

30. Generate a *2 × 3* matrix of random
    - real numbers, each in the range from 0 to 1
    - real numbers, each in the range from 0 to 10
    - integers, each in the range from 5 to 20

31. Create a variable *rows* that is a random integer in the range from 1 to 5. Create a variable *cols* that is a random integer in the range from 1 to 5. Create a matrix of all zeros with the dimensions given by the values of *rows* and *cols*.

32. Find an *efficient* way to generate the following matrix:

```
mat =
      7    8    9   10
     12   10    8    6
```

Then, give expressions that will, for the matrix *mat*,
- refer to the element in the first row, third column
- refer to the entire second row
- refer to the first two columns

33. Create a *2 × 3* matrix variable *mymat*. Pass this matrix variable to each of the following functions and make sure you understand the result: **fliplr**, **flipud**, and **rot90**. In how many different ways can you **reshape** it?

34. Create a *4 × 2* matrix of all zeros and store it in a variable. Then, replace the second row in the matrix with a vector consisting of a 3 and a 6.

35. Create a vector *x* that consists of 20 equally spaced points in the range from −π to +π. Create a *y* vector that is **sin(x)**.

36. Create a *3 × 5* matrix of random integers, each in the range from −5 to 5. Get the **sign** of every element.

37. Create a *4 × 6* matrix of random integers, each in the range from −5 to 5; store it in a variable. Create another matrix that stores for each element the absolute value of the corresponding element in the original matrix.

38. Create a *3 × 5* matrix of random real numbers. Delete the third row.

39. Create a vector variable *vec*. Find as many expressions as you can that would refer to the last element in the vector, without assuming that you know how many elements it has (i.e., make your expressions general).

40. Create a matrix variable *mat*. Find as many expressions as you can that would refer to the last element in the matrix, without assuming that you know how many elements, rows, or columns it has (i.e., make your expressions general).

41. Create a three-dimensional matrix and get its **size**.

42. The built-in function **clock** returns a vector that contains six elements: the first three are the current date (year, month, day) and the last three represent the current time in hours, minutes, and seconds. The seconds is a real number, but all others are integers. Store the result from **clock** in a variable called *myc*. Then, store the first three elements from this variable in a variable *today* and the last three elements in a variable *now*. Use the **fix** function on the vector variable *now* to get just the integer part of the current time.

Introduction to MATLAB Programming

KEY TERMS

computer program
scripts
algorithm
top-down design
external file
default input device
prompting
default output device
execute/run
high-level languages
machine language
executable
compiler
source code
object code
interpreter

script files
documentation
comments
input/output (I/O)
user
empty string
error message
formatting
format string
place holder
conversion characters
newline character
field width
leading blanks
trailing zeros
plot symbols

markers
line types
toggle
modes
appending
user-defined functions
control
function header
output arguments
input arguments
function body
function definition
program
local variables

We have now used the MATLAB® product interactively in the Command Window. That is sufficient when all one needs is a simple calculation. However, in many cases, quite a few steps are required before the final result can be obtained. In those cases, it is more convenient to group statements together in what is called a *computer program*.

In this chapter, we will introduce the simplest MATLAB programs, which are called *scripts*. Examples of scripts that customize simple plots will illustrate the concept. Input will be introduced, both from files and from the user.

MATLAB®: A Practical Introduction to Programming and Problem Solving

45

Output to files and to the screen will also be introduced. Finally, user-defined functions that calculate and return values will be described. These topics serve as an introduction to programming, which will be expanded on in Chapter 6.

2.1 ALGORITHMS

Before writing any computer program, it is useful to first outline the steps that will be necessary. An *algorithm* is the sequence of steps needed to solve a problem. In a modular approach to programming, the problem solution is broken down into separate steps, and then each step is further refined until the resulting steps are small enough to be manageable tasks. This is called the *top-down design* approach.

As a simple example, consider the problem of calculating the area of a circle. First, it is necessary to determine what information is needed to solve the problem, which in this case is the radius of the circle. Next, given the radius of the circle, the area of the circle would be calculated. Finally, once the area has been calculated, it has to be displayed in some way. The basic algorithm then is three steps:

- Get the input: the radius
- Calculate the result, the area
- Display the output

Even with an algorithm this simple, it is possible to further refine each of the steps. When a program is written to implement this algorithm, the steps would be:

- Where does the input come from? Two possible choices would be from an *external file* on a disk, or from the user (the person running the program) who enters the number by typing it from the keyboard. For every system, one of these will be the *default input device* (which means, if not specified otherwise, this is where the input comes from!). If the user is supposed to enter the radius, the user has to be told to type in the radius (and, in what units). Telling the user what to enter is called *prompting*. So, the input step actually becomes two steps: prompt the user to enter a radius, and then read it into the program.
- To calculate the area, the formula is needed. In this case, the area of the circle is π multiplied by the radius squared. So, that means the value of the constant for π is needed in the program.
- Where does the output go? Two possibilities are (1) to an external file or (2) to the screen. Depending on the system, one of these will be the *default output device*. When displaying the output from the program, it should always be as informative as possible. In other words, instead of just printing the area (just the number), it should be printed in a nice sentence format. Also, to make the output even more clear, the input should be printed. For example, the output might be the sentence: "For a circle with a radius of 1 inch, the area is 3.1416 inches squared."

For most programs, the basic algorithm consists of the three steps outlined in the preceding:

1. Get the input(s)
2. Calculate the result(s)
3. Display the result(s)

As can be seen here, even the simplest problem solutions can then be further refined.

2.2 MATLAB SCRIPTS

Once a problem has been analyzed, and the algorithm for its solution has been written and refined, the problem is then written in a particular programming language. A *computer program* is a sequence of instructions, in a given language, that accomplishes a task. To *execute*, or *run*, a program is to have the computer actually follow these instructions sequentially.

High-level languages have English-like commands and functions, such as "print this" or "if $x < 5$, do something." The computer, however, can only interpret commands written in its **machine language**. Programs that are written in high-level languages must therefore be translated into machine language before the computer can actually execute the sequence of instructions in the program. A program that does this translation from a high-level language to an *executable* file is called a *compiler*. The original program is called the *source code*, and the resulting executable program is called the *object code*.

By contrast, an *interpreter* goes through the code line-by-line, executing each command as it goes. MATLAB uses what are called either *script* files, or M-files (the reason for this is that the extension on the file name is .*m*). These script files are interpreted, rather than compiled. Therefore, the correct terminology is that these are scripts, not programs. However, the terms are somewhat loosely used by many people, and the documentation in MATLAB itself refers to scripts as programs. In this book, we will reserve the use of the word "program" to mean a set of scripts and functions, as described briefly in Section 2.7 and then in more detail in Chapter 6.

A script is a sequence of MATLAB instructions that is stored in an M-file and saved. The contents of a script can be displayed in the Command Window using the **type** command. The script can be executed, or run, by simply entering the name of the file (without the .*m* extension).

Before creating a script, make sure the Current Folder (called "Current Directory" in earlier versions) is set to the folder in which you want to save your files.

To create a script, click on File, then New, then Script. (In earlier versions of MATLAB, click on File, then New, then M-file.) A new window will appear called the Editor. To create a new script, simply type the sequence of statements (note that line numbers will appear on the left).

When finished, save the file using File and then Save. Make sure that the extension of *.m* is on the file name (this should be the default). The rules for file names are the same as for variables (they must start with a letter, after that there can be letters, digits, or the underscore). For example, we will now create a script called *script1.m* that calculates the area of a circle. It assigns a value for the radius, and then calculates the area based on that radius.

In this text, scripts will be displayed in a box with the name of the M-file on top.

```
script1.m
radius = 5
area = pi * (radius^2)
```

In the Command Window, the contents of the script can be displayed and executed. The **type** command shows the contents of the file named *script1.m*; note that the *.m* is not included.

```
>> type script1
radius = 5
area = pi * (radius^2)
```

There are two ways to view a script once it has been written: either open the Editor Window to view it, or use the **type** command as shown here to display it in the Command Window.

To actually run or execute the script, the name of the file is entered at the prompt (again, without the *.m*). When executed, the results of the two assignment statements are displayed, since the output was not suppressed for either statement.

```
>> script1
radius =
    5
area =
    78.5398
```

Once the script has been executed, you may find that you want to make changes to it (especially if there are errors!). To edit an existing file, there are several methods to open it. The easiest are:

■ Click on File, then Open, and then click on the name of the file.
■ Within the Current Folder Window, double-click on the name of the file in the list of files.

2.2.1 Documentation

It is very important that all scripts be *documented* well, so that people can understand what the script does and how it accomplishes its task. One way of documenting a script is to put *comments* in it. In MATLAB, a comment is anything from a % to the end of that particular line. Comments are completely ignored when the script is executed. To put in a comment, simply type the % symbol at the beginning of a line, or select the comment lines and then click on Text and then Comment and the Editor will put in the % symbols for the comments.

For example, the previous script to calculate the area of a circle could be modified to have comments:

```
script1b.m
% This script calculates the area of a circle

% First the radius is assigned
radius = 5
% The area is calculated based on the radius
area = pi * (radius^2)
```

The first comment at the beginning of the script describes what the script does. Then, throughout the script, comments describe different parts of the script (not usually a comment for every line, however!). Comments don't affect what a script does, so the output from this script would be the same as for the previous version.

The **help** command in MATLAB works with scripts as well as with built-in functions. The first block of comments (defined as contiguous lines at the beginning) will be displayed. For example, for *script1b*:

```
>> help script1b
   This script calculates the area of a circle
```

The reason that a blank line was inserted in the script between the first two comments is that otherwise both would have been interpreted as one contiguous comment, and both lines would have been displayed with **help**. The very first comment line is called the "H1 line"; it is what the function **lookfor** searches through.

PRACTICE 2.1

Write a script to calculate the area of a rectangle. Be sure to comment the script.

2.3 INPUT AND OUTPUT

The previous script would be much more useful if it were more general; for example, if the value of the radius could be read from an external source rather than being assigned in the script. Also, it would be better to have the script print the output in a nice, informative way. Statements that accomplish these tasks are called *input/output* statements, or *I/O* for short. Although, for simplicity, examples of input and output statements will be shown here from the Command Window, these statements will make the most sense in scripts.

2.3.1 Input Function

Input statements read in values from the default or standard input device. In most systems, the default input device is the keyboard, so the input statement reads in values that have been entered by the *user*, or the person who is running the script. To let the user know what he or she is supposed to enter, the script must first prompt the user for the specified values.

The simplest input function in MATLAB is called **input**. The **input** function is used in an assignment statement. To call it, a string is passed that is the prompt that will appear on the screen, and whatever the user types will be stored in the variable named on the left of the assignment statement. For ease of reading the prompt, it is useful to put a colon and then a space after the prompt. For example,

```
>> rad = input('Enter the radius: ')
Enter the radius: 5
rad =
     5
```

If character or string input is desired, 's' must be added as a second argument to the **input** function:

```
>> letter = input('Enter a char: ','s')
Enter a char: g
letter =
g
```

If the user enters only spaces or tabs before hitting the Enter key, they are ignored and an *empty string* is stored in the variable:

```
>> mychar = input('Enter a character: ', 's')
Enter a character:
mychar =
     ' '
```

Note
Although normally the quotes are not shown around a character or string, in this case they are shown to demonstrate that there is nothing inside the string.

However, if blank spaces are entered before other characters, they are included in the string. In the next example, the user hit the space bar four times before entering "go." The **length** function returns the number of characters in the string.

```
>> mystr = input('Enter a string: ', 's')
Enter a string:      go
mystr =
     go

>> length(mystr)
ans =
     6
```

QUICK QUESTION!

What would be the result if the user enters blank spaces after other characters? For example, the user here entered "xyz " (four blank spaces):

```
>> mychar = input('Enter chars: ', 's')
Enter chars: xyz
mychar =
xyz
```

Answer: The space characters would be stored in the string variable. It is difficult to see in the previous example, but is clear from the length of the string.

```
>> length(mychar)
ans =
   7
```

The length can be seen in the Command Window by using the mouse to highlight the value of the variable; the xyz and four spaces will be highlighted.

It is also possible for the user to type quotation marks around the string rather than including the second argument 's' in the call to the **input** function.

```
>> name = input('Enter your name: ')
Enter your name: 'Stormy'
name =
Stormy
```

However, this assumes that the user would know to do this so it is better to signify that character input is desired in the **input** function itself. Also, if the 's' is specified and the user enters quotation marks, these would become part of the string.

```
>> name = input('Enter your name: ','s')
Enter your name: 'Stormy'
name =
'Stormy'
>> length(name)
ans =
     8
```

Note what happens if string input has not been specified, but the user enters a letter rather than a number.

```
>> num = input('Enter a number: ')
Enter a number: t
??? Error using ==> input
Undefined function or variable 't'.

Enter a number: 3
num =
    3
```

MATLAB gave an *error message* and repeated the prompt. However, if *t* is the name of a variable, MATLAB will take its value as the input.

```
>> t = 11;
>> num = input ('Enter a number: ')
Enter a number: t
num =
    11
```

Separate **input** statements are necessary if more than one input is desired. For example,

```
>> x = input ('Enter the x coordinate: ');
>> y = input ('Enter the y coordinate: ');
```

Normally, the results from **input** statements are suppressed with a semicolon at the end of the assignment statements.

PRACTICE 2.2

Create a script that would prompt the user for a temperature, and then 'F' or 'C', and store both inputs in variables. For example, when executed it would look like this (assuming the user enters 85 and then F):

```
Enter the temperature: 85
Is that F or C? : F
```

2.3.2 Output statements: disp and fprintf

Output statements display strings and/or the results of expressions, and can allow for *formatting*, or customizing how they are displayed. The simplest output function in MATLAB is **disp**, which is used to display the result of an expression or a string without assigning any value to the default variable *ans*. However, **disp** does not allow formatting. For example,

```
>> disp ('Hello')
Hello

>> disp (4^3)
    64
```

Formatted output can be printed to the screen using the **fprintf** function. For example,

```
>> fprintf ('The value is %d, for sure!\n', 4^3)
The value is 64, for sure!
>>
```

To the **fprintf** function, first a string (called the *format string*) is passed that contains any text to be printed as well as formatting information for the expressions to be printed. In this example, the %d is an example of format information.

The %d is sometimes called a *place holder*, because it specifies where the value of the expression that is after the string is to be printed. The character in the place holder is called the *conversion character*, and it specifies the type of value that is being printed. There are others, but what follows is a list of the simple place holders:

```
%d  integer (it actually stands for decimal integer)
%f  float (real number)
%c  single character
%s  string
```

The character '\n' at the end of the string is a special character called the *newline character*; what happens when it is printed is that the output that follows moves down to the next line.

QUICK QUESTION!

What do you think would happen if the newline character is omitted from the end of an **fprintf** statement?
Answer: Without it, the next prompt would end up on the same line as the output. It is still a prompt, and so an expression can be entered, but it looks messy as shown here.

```
>> fprintf('The value is %d, surely!',...
4^3)
The value is 64, surely!>> 5 + 3
ans =
    8
```

Note that with the **disp** function, however, the prompt will always appear on the next line:

```
>> disp('Hi')
Hi
>>
```

Also, note that an ellipsis can be used after a string but not in the middle.

QUICK QUESTION!

How can you get a blank line in the output?
Answer: Have two newline characters in a row.

```
>> fprintf('The value is %d\n\nOK!\n',4^3)
The value is 64

OK!
```

This also points out that the newline character can be anywhere in the string; when it is printed, the output moves down to the next line.

Note that the newline character can also be used in the prompt in the **input** statement. For example,

```
>> x = input('Enter the \nx coordinate: ');
Enter the
x coordinate: 4
```

However, that is the only formatting character allowed in the prompt in **input**.

To print two values, there would be two place holders in the format string, and two expressions after the format string. The expressions fill in for the place holders in sequence.

```
>> fprintf('The int is %d and the char is %c\n', ...
   33 - 2, 'x')
The int is 31 and the char is x
```

Note
If the field width is wider than necessary, **leading** blanks are printed, and if more decimal places are specified than necessary, **trailing** zeros are printed.

A *field width* can also be included in the place holder in **fprintf**, which specifies how many characters total are to be used in printing. For example, %5d would indicate a field width of 5 for printing an integer and %10s would indicate a field width of 10 for a string. For floats, the number of decimal places can also be specified; for example, %6.2f means a field width of 6 (including the decimal point and the two decimal places) with two decimal places. For floats, just the number of decimal places can also be specified; for example, %.3f indicates three decimal places, regardless of the field width.

```
>> fprintf('The int is %3d and the float is %6.2f\n',5,4.9)
The int is     5 and the float is   4.90
```

QUICK QUESTION!

What do you think would happen if you tried to print 1234.5678 in a field width of 3 with two decimal places (using the following)?

```
>> fprintf('%3.2f\n', 1234.5678)
```

Answer: It would print the entire 1234, but round the decimals to two places, that is,

```
1234.57
```

If the field width is not large enough to print the number, the field width will be increased. Basically, to cut the number off would give a misleading result, but rounding the decimal places does not change the number by much.

QUICK QUESTION!

What would happen if you use the %d conversion character but you're trying to print a real number?
Answer: MATLAB will show the result using exponential notation.

```
>> fprintf('%d\n',1234567.89)
1.234568e+006
```

Note: If you want exponential notation, this is not the correct way to get it; instead, there are conversion characters that can be used. Use the **help** browser to see this option as well as many others!

There are many other options for the format string. For example, the value being printed can be left-justified within the field width using a minus sign. The following example shows the difference between printing the integer 3 using %5d and using %-5d. The x's below are used to show the spacing.

```
>> fprintf('The integer is xx%5dxx and xx%-5dxx\n',3,3)
The integer is xx    3xx and xx3    xx
```

Also, strings can be truncated by specifying "decimal places":

```
>> fprintf('The string is %s or %.2s\n', 'street', 'street')
The string is street or st
```

There are several special characters that can be printed in the format string in addition to the newline character. To print a slash, two slashes in a row are used, and also to print a single quote two single quotes in a row are used. Additionally, \t is the tab character.

```
>> fprintf('Try this out: tab\t quote '' slash \\ \n')
Try this out: tab    quote ' slash \
```

Printing vectors and matrices

For a vector, if a conversion character and the newline character are in the format string, it will print in a column regardless of whether the vector itself is a row vector or a column vector.

```
>> vec = 2:5;
>> fprintf('%d\n', vec)
2
3
4
5
```

Without the newline character, it would print in a row but the next prompt would appear on the same line:

```
>> fprintf('%d', vec)
2345>>
```

However, in a script, a separate newline character could be printed to avoid this problem. It is also much better to separate the numbers with spaces.

```
printvec.m
```

```
% This demonstrates printing a vector

vec = 2:5;
fprintf('%d ',vec)
fprintf('\n')
```

```
>> printvec
2 3 4 5
>>
```

If the number of elements in the vector is known, that many conversion characters can be specified and then the newline:

```
>> fprintf('%d %d %d %d\n', vec)
2 3 4 5
```

This is not very general, however, and is therefore not preferable.

For matrices, MATLAB unwinds the matrix column by column. For example, consider the following random 2×3 matrix:

```
>> mat = [5 9 8; 4  1 10]
mat =
      5    9    8
      4    1   10
```

Specifying one conversion character and then the newline character will print the elements from the matrix in one column. The first values printed are from the first column, then the second column, and so on.

```
>> fprintf('%d\n', mat)
5
4
9
1
8
10
```

If three of the %d conversion characters are specified, the **fprintf** will print three numbers across on each line of output, but again the matrix is unwound column by column. It again prints first the two numbers from the first column (across on the first row of output), then the first value from the second column, and so on.

```
>> fprintf('%d %d %d\n', mat)
5 4 9
1 8 10
```

If the transpose of the matrix is printed, however, using the three %d conversion characters, the matrix is printed as it appears when created.

```
>> fprintf('%d %d %d\n', mat') % Note the transpose
5 9 8
4 1 10
```

For vectors and matrices, even though formatting cannot be specified, the **disp** function may be easier to use in general than **fprintf** because it displays the result in a straightforward manner. For example,

```
>> mat = [15 11 14; 7 10 13]
mat =

>> disp(mat)
     15    11    14
      7    10    13
```

```
>> vec = 2:5
vec =
     2     3     4     5

>> disp(vec)
     2     3     4     5
```

2.4 SCRIPTS WITH INPUT AND OUTPUT

Putting all of this together now, we can implement the algorithm from the beginning of this chapter. The following script calculates and prints the area of a circle. It first prompts the user for a radius, reads in the radius, and then calculates and prints the area of the circle based on this radius.

script2.m
```
% This script calculates the area of a circle
% It prompts the user for the radius

% Prompt the user for the radius and calculate
% the area based on that radius
fprintf('Note: the units will be inches.\n')
radius = input('Please enter the radius: ');
area = pi * (radius^2);

% Print all variables in a sentence format
fprintf('For a circle with a radius of %.2f inches,\n',...
    radius)
fprintf('the area is %.2f inches squared\n',area)
```

Executing the script produces the following output:

```
>> script2
Note: The units will be inches.
Please enter the radius: 3.9
For a circle with a radius of 3.90 inches,
the area is 47.78 inches squared
```

Note that the output from the first two assignment statements (including the **input**) is suppressed by putting semicolons at the end. That is usually done in scripts, so that the exact format of what is displayed by the program is controlled by the **fprintf** functions.

PRACTICE 2.3

Write a script to prompt the user separately for a character and a number, and print the character in a field width of 4 and the number left justified in a field width of 5 with two decimal places. Test this by entering numbers with varying widths.

2.5 SCRIPTS TO PRODUCE AND CUSTOMIZE SIMPLE PLOTS

MATLAB has many graphing capabilities. Customizing plots is often desired and this is easiest to accomplish by creating a script rather than typing one command at a time in the Command Window. For that reason, simple plots and how to customize them will be introduced in this chapter on MATLAB programming.

Typing **help** will display the help topics that contain graph functions, including **graph2d** and **graph3d**. Typing **help graph2d** would display some of the two-dimensional graph functions, as well as functions to manipulate the axes and to put labels and titles on the graphs.

2.5.1 The plot function

For now, we'll start with a very simple graph of one point using the **plot** function.

The following script, *plotonepoint*, plots one point. To do this, first values are given for the x and y coordinates of the point in separate variables. The point is plotted using a red star (*). The plot is then customized by specifying the minimum and maximum values on first the x and then y axes. Labels are then put on the x-axis, the y-axis, and the graph itself using the function **xlabel**, **ylabel**, and **title**.

All of this can be done from the Command Window, but it is much easier to use a script. The following shows the contents of the script *plotonepoint* that accomplishes this. The x coordinate represents the time of day (e.g., 11am) and the y coordinate represents the temperature in degrees Fahrenheit at that time.

plotonepoint.m

```
% This is a really simple plot of just one point!

% Create coordinate variables and plot a red '*'
x = 11;
y = 48;
plot(x,y,'r*')

% Change the axes and label them
axis([ 9 12 35 55])
xlabel('Time')
ylabel('Temperature')

% Put a title on the plot
title('Time and Temp')
```

In the call to the **axis** function, one vector is passed. The first two values are the minimum and maximum for the x axis, and the last two are the minimum and maximum for the y axis. Executing this script brings up a Figure Window with the plot (see Figure 2.1).

To be more general, the script could prompt the user for the time and temperature, rather than just assigning values. Then, the axis function could be used based on whatever the values of *x* and *y* are, as in the following example:

```
axis([ x−2 x+2 y−10 y+10] )
```

In addition, although they are the x and y coordinates of a point, variables named *time* and *temp* might be more mnemonic than x and y.

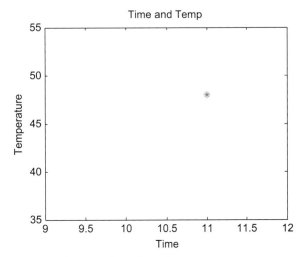

FIGURE 2.1 Plot of one data point

PRACTICE 2.4

Modify the script *plotonepoint* to prompt the user for the time and temperature, and set the axes based on these values.

To plot more than one point, *x* and *y* vectors are created to store the values of the (x,y) points. For example, to plot the points

```
(1,1)
(2,5)
(3,3)
(4,9)
(5,11)
(6,8)
```

first an *x* vector is created that has the x values (since they range from 1 to 6 in steps of 1, the colon operator can be used), and then a *y* vector is created with the y values. The following will create (in the Command Window) *x* and *y* vectors and then plot them (see Figure 2.2).

```
>> x = 1:6;
>> y = [1 5 3 9 11 8];
>> plot(x,y)
```

Note that the points are plotted with straight lines drawn in between. Also, the axes are set up according to the data; for example, the x values range from 1 to 6 and the y values from 1 to 11, so that is how the axes are set up.

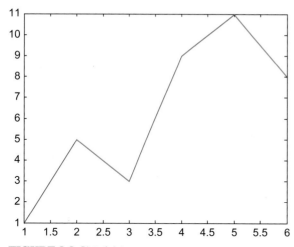

FIGURE 2.2 Plot of data points from vectors

Also, note that in this case the x values are the indices of the y vector (the y vector has six values in it, so the indices iterate from 1 to 6). When this is the case, it is not necessary to create the x vector. For example,

```
>> plot(y)
```

will plot exactly the same figure without using an x vector.

Customizing a plot: color, line types, marker types

Plots can be done in the Command Window, as shown here, if they are really simple. However, many times it is desired to customize the plot with labels, titles, and so on, so it makes more sense to do this in a script. Using the **help** function for **plot** will show the many options such as the line types and colors.

In the previous script *plotonepoint*, the string 'r*' specified a red star for the point type. The possible colors are:

```
b   blue
c   cyan
g   green
k   black
m   magenta
r   red
y   yellow
```

The *plot symbols*, or *markers*, that can be used are:

```
o   circle
d   diamond
h   hexagram
p   pentagram
+   plus
.   point
s   square
*   star
v   down triangle
<   left triangle
>   right triangle
^   up triangle
x   x-mark
```

Line types can also be specified by the following:

```
--   dashed
-.   dash dot
:    dotted
-    solid
```

If no line type is specified, a solid line is drawn between the points, as seen in the last example.

2.5.2 Simple related plot functions

Other functions that are useful in customizing plots include **clf**, **figure**, **hold**, **legend**, and **grid**. Brief descriptions of these functions are given here; use **help** to find out more about them:

> `clf` clears the Figure Window by removing everything from it.

> `figure` creates a new, empty Figure Window when called without any arguments. Calling it as `figure(n)` where *n* is an integer is a way of creating and maintaining multiple Figure Windows, and of referring to each individually.

> `hold` is a toggle that freezes the current graph in the Figure Window, so that new plots will be superimposed on the current one. Just `hold` by itself is a *toggle*, so calling this function once turns the hold on, and then the next time turns it off. Alternatively, the commands `hold on` and `hold off` can be used.

> `legend` displays strings passed to it in a legend box in the Figure Window, in order of the plots in the Figure Window.

> `grid` displays grid lines on a graph. Called by itself, it is a toggle that turns the grid lines on and off. Alternatively, the commands `grid on` and `grid off` can be used.

There are many other plot types. We will see more in Chapter 11, but another simple plot type is a **bar** chart.

For example, the following script creates two separate Figure Windows. First, it clears the Figure Window. Then, it creates an *x* vector and two different *y* vectors (*y1* and *y2*). In the first Figure Window, it plots the *y1* values using a bar chart. In the second Figure Window, it plots the *y1* values as black lines, puts **hold on** so that the next graph will be superimposed, and plots the *y2* values as black circles. It also puts a legend on this graph and uses a grid. Labels and titles are omitted in this case since it is generic data.

plot2figs.m

```
% This creates 2 different plots, in 2 different
% Figure Windows, to demonstrate some plot features

clf
x = 1:5; % Not necessary
y1 = [ 2 11 6 9 3] ;
y2 = [ 4 5 8 6 2] ;
% Put a bar chart in Figure 1
figure(1)
bar(x,y1)
% Put plots using different y values on one plot
% with a legend
figure(2)
plot(x,y1,'k')
hold on
plot(x,y2,'ko')
grid on
legend('y1','y2')
```

Running this script will produce two separate Figure Windows. If there are no other active Figure Windows, the first, which is the bar chart, will be in the one titled "Figure 1" in MATLAB. The second will be in "Figure 2." See Figure 2.3 for both plots.

Note that the first and last points are on the axes, which makes them difficult to see. That is why the **axis** function is frequently used, as it creates space around the points so that they are all visible.

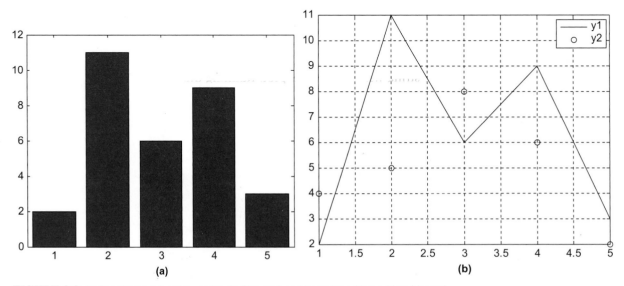

FIGURE 2.3 (a) Bar chart produced by script. (b) Plot produced by script, with a grid and legend.

PRACTICE 2.5

Modify the script using the **axis** function so that all points are easily seen.

The ability to pass a vector to a function and have the function evaluate every element of the vector can be very useful in creating plots. For example, the following script graphically displays the difference between the **sin** and **cos** functions:

sinncos.m

```
% This script plots sin(x) and cos(x) in the same Figure Window
% for values of x ranging from 0 to 2*pi

clf
x = 0: 2*pi/40: 2*pi;
y = sin(x);
plot(x,y,'ro')
hold on
y = cos(x);
plot(x,y,'b+')
legend('sin', 'cos')
xlabel('x')
ylabel('sin(x) or cos(x)')
title('sin and cos on one graph')
```

The script creates an *x* vector; iterating through all of the values from 0 to $2*\pi$ in steps of $2*\pi/40$ gives enough points to get a good graph. It then finds the sine of each *x* value, and plots these points using red circles. The command **hold on** freezes this in the Figure Window so the next plot will be superimposed.

Next, it finds the cosine of each *x* value and plots these points using blue plus symbols (+). The **legend** function creates a legend; the first string is paired with the first plot, and the second string with the second plot. Running this script produces the plot seen in Figure 2.4.

PRACTICE 2.6

Write a script that plots **exp(x)** and **log(x)** for values of *x* ranging from 0.5 to 2.5.

2.6 INTRODUCTION TO FILE INPUT/OUTPUT (LOAD AND SAVE)

In many cases, input to a script will come from a data file that has been created by another source. Also, it is useful to be able to store output in an external file that can be manipulated and/or printed later. In this section,

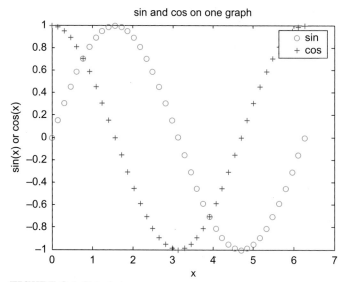

FIGURE 2.4 Plot of **sin** and **cos** in one Figure Window with a legend

we will demonstrate which are the simplest methods used to read from an external data file and also to write to an external data file.

There are basically three different operations, or **modes**, on files. Files can be:

- read from
- written to
- *appended* to

Writing to a file means writing to a file, from the beginning. Appending to a file is also writing, but starting at the end of the file rather than the beginning. In other words, appending to a file means adding to what was already there.

There are many different file types, which use different filename extensions. For now, we will keep it simple and just work with *.dat* or *.txt* files when working with data, or text, files. There are several methods for reading from files and writing to files; we will for now use the **load** function to read and the **save** function to write to files. More file types and functions for manipulating them will be discussed in Chapter 9.

Note

If the file already exists, the **save** function will overwrite the file. Remember that **save** always begins writing from the beginning of a file.

2.6.1 Writing data to a file

The **save** function can be used to write data from a matrix to a data file, or to append to a data file. The format is:

```
save filename matrixvariablename -ascii
```

The "-ascii" qualifier is used when creating a text or a data file. Use the following to create a matrix and then save the values of the matrix variable to a data file called *testfile.dat*:

```
>> mymat = rand(2,3)
mymat =
     0.4565    0.8214    0.6154
     0.0185    0.4447    0.7919

>> save testfile.dat mymat -ascii
```

This creates a file called *testfile.dat* that stores the numbers:

```
     0.4565    0.8214    0.6154
     0.0185    0.4447    0.7919
```

The **type** command can be used to display the contents of the file; note that scientific notation is used:

```
>> type testfile.dat

   4.5646767e-001    8.2140716e-001    6.1543235e-001
   1.8503643e-002    4.4470336e-001    7.9193704e-001
```

2.6.2 Appending data to a data file

Once a text file exists, data can be appended to it. The format is the same as the preceding, with the addition of the qualifier "-append." For example, the following creates a new random matrix and appends it to the file that was just created:

```
>> mat2 = rand(3,3)
mymat =
      0.9218      0.4057      0.4103
      0.7382      0.9355      0.8936
      0.1763      0.9169      0.0579
>> save testfile.dat mat2 -ascii -append
```

This results in the file *testfile.dat* containing the following:

```
   0.4565      0.8214      0.6154
   0.0185      0.4447      0.7919
   0.9218      0.4057      0.4103
   0.7382      0.9355      0.8936
   0.1763      0.9169      0.0579
```

Note
Although technically any size matrix could be appended to this data file, to be able to read it back into a matrix later there would have to be the same number of values on every row (or, in other words, the same number of columns).

PRACTICE 2.7

Prompt the user for the number of rows and columns of a matrix, create a matrix with that many rows and columns of random numbers, and write it to a file.

2.6.3 Reading from a file

Once a file has been created (as in the preceding), it can be read into a matrix variable. If the file is a data file, the **load** function will read from the file *filename.ext* (e.g., the extension might be *.dat*) and create a matrix with the same name as the file. For example, if the data file *testfile.dat* had been created as shown in the previous section, this would read from it, and store the result in a matrix variable called *testfile*:

```
>> clear
>> load testfile.dat
>> who
Your variables are:
```

```
testfile
>> testfile
testfile =
      0.4565    0.8214    0.6154
      0.0185    0.4447    0.7919
      0.9218    0.4057    0.4103
      0.7382    0.9355    0.8936
      0.1763    0.9169    0.0579
```

The **load** command works only if there are the same number of values in each line, so that the data can be stored in a matrix, and the **save** command only writes from a matrix to a file. If this is not the case, lower-level file I/O functions must be used; these will be discussed in Chapter 9.

Example: Load from a file and plot the data

As an example, a file called *timetemp.dat* stores two lines of data. The first line is the times of day, and the second line is the recorded temperature at each of those times. The first value of 0 for the time represents midnight. For example, the contents of the file might be:

```
  0      3      6      9     12     15     18     21
55.5   52.4   52.6   55.7   75.6   77.7   70.3   66.6
```

The following script loads the data from the file into a matrix called *timetemp*. It then separates the matrix into vectors for the time and temperature, and then plots the data using black star (*) symbols.

```
timetempprob.m
% This reads time and temperature data for an afternoon
% from a file and plots the data

load timetemp.dat

% The times are in the first row, temps in the second row
time = timetemp(1,:);
temp = timetemp(2,:);

% Plot the data and label the plot
plot(time,temp,'k*')
xlabel('Time')
ylabel('Temperature')
title('Temperatures one afternoon')
```

Running the script produces the plot seen in Figure 2.5.

To create the data file, the Editor in MATLAB can be used; it is not necessary to create a matrix and **save** it to a file. Instead, just enter the numbers in a new file, and Save As *timetemp.dat*.

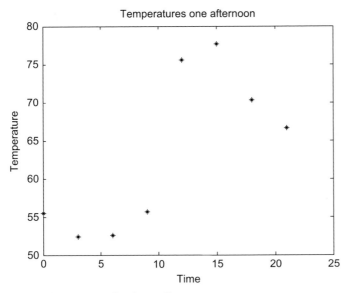

FIGURE 2.5 Plot of temperature data from a file

Note: It is difficult to see the point at time 0 since it falls on the *y*-axis. The **axis** function could be used to change the axes from the defaults shown here.

PRACTICE 2.8

The sales (in billions) for two separate divisions of the XYZ Corporation for each of the four quarters of 2010 are stored in a file called *salesfigs.dat*:

```
1.2 1.4 1.8 1.3
2.2 2.5 1.7 2.9
```

- First, create this file (just type the numbers in the Editor, and Save As *salesfigs.dat*).
- Load the data from the file into a matrix.
- Then, write a script that will
  - separate this matrix into two vectors.
  - create the plot seen in Figure 2.6 (which uses circles and stars as the plot symbols).

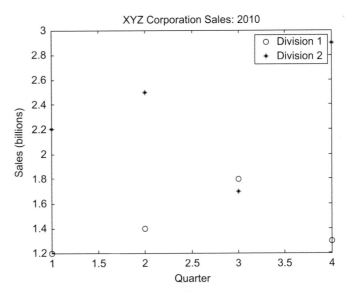

FIGURE 2.6 Plot of sales data from file

QUICK QUESTION!

Sometimes files are not in the format that is desired. For example, a file *expresults.dat* has been created that has some experimental results, but the order of the values is reversed in the file:

```
4   53.4
3   44.3
2   50.0
1   55.5
```

How could we create a new file that reverses the order?

Answer: We can **load** from this file into a matrix, use the **flipud** function to "flip" the matrix up to down, and then **save** this matrix to a new file:

```
>> load expresults.dat
```

```
>> expresults
expresults =
    4.0000   53.4000
    3.0000   44.3000
    2.0000   50.0000
    1.0000   55.5000

>> correctorder = flipud(expresults)
correctorder =
    1.0000   55.5000
    2.0000   50.0000
    3.0000   44.3000
    4.0000   53.4000
>> save neworder.dat correctorder - ascii
```

2.7 USER-DEFINED FUNCTIONS THAT RETURN A SINGLE VALUE

We have already seen the use of many functions in MATLAB. We have used many built-in functions such as **sin**, **fix**, **abs**, and **double**. In this section, *user-defined functions* will be introduced. These are functions that the programmer defines, and then uses, in either the Command Window or in a script.

There are several different types of functions. For now, we will concentrate on the kind of function that calculates and returns a single result, much like built-in functions such as **sin** and **abs**. Other types of functions will be introduced in Chapter 6.

First, let us review some of what we already know about functions, including the use of built-in functions. Although, by now, the use of these functions is straightforward, explanations will be given in some detail here to compare and contrast to the use of user-defined functions.

The **length** function is an example of a built-in function that calculates a single value; it returns the length of a vector. As an example,

```
length(vec)
```

is an expression that represents the number of elements in the vector *vec*. This expression could be used in the Command Window or in a script. Typically, the value returned from this expression might be assigned to a variable:

```
>> vec = 1:3:10;
>> lv = length(vec)
lv =
    4
```

Alternatively, the length of the vector could be printed:

```
>> fprintf('The length of the vector is %d\n', length(vec))
The length of the vector is 4
```

The *call* to the **length** function consists of the name of the function, followed by the *argument* in parentheses. The function receives as input the argument, and returns a result. What happens when the call to the function is encountered is that *control* is passed to the function itself (i.e., the function begins executing). The argument(s) are also passed to the function.

The function executes its statements and does whatever it does (the actual contents of the built-in functions are not generally known or seen by the programmer) to determine the number of elements in the vector. Since the function is calculating a single value, this result is then *returned*, and it becomes the value of the expression. Control is also passed back to the expression that called it in the first place, which then continues (e.g., in the first example the value would then be assigned to the variable *lv*, and in the second example the value was printed).

2.7.1 Function definitions

There are different ways to organize scripts and functions, but for now every function that we write will be stored in a separate M-file, which is why they are commonly called "M-file functions." Although to type in functions in the Editor it is possible to choose File, followed by New, and then Function in MATLAB, it will be easier for now to type in the function by choosing File, then New, and then Script (in older versions there is no choice; it is File, then New, then M-file for both scripts and functions).

A function in MATLAB that returns a single result consists of the following:

- The function *header* (the first line), comprised of:
  - The reserved word **function**
  - Since the function *returns* a result, the name of the *output argument* followed by the assignment operator =
  - The name of the function (IMPORTANT: This should be the same as the name of the M-file in which this function is stored to avoid confusion.)
  - The *input arguments* in parentheses, which correspond to the arguments that are passed to the function in the function call
- A comment that describes what the function does (this is printed when **help** is used)
- The *body* of the function, which includes all statements and eventually must put a value in the output argument
- **end** at the end of the function. (*Note:* This is not necessary in many cases.)

The general form of a *function definition* for a function that calculates and returns one value looks like this:

```
functionname.m
```
```
function outputargument = functionname(input arguments)
% Comment describing the function

Statements here; these must include putting a value in
the output argument

end % of the function
```

For example, the following is a function called *calcarea* that calculates and returns the area of a circle; it is stored in a file called *calcarea.m*.

```
calcarea.m
```
```
function area = calcarea(rad)
% calcarea calculates the area of a circle
% Format of call: calcarea(radius)
% Returns the area

area = pi * rad * rad;
end
```

A radius of a circle is passed to the function to the input argument *rad*; the function calculates the area of this circle and stores it in the output argument *area*.

In the function header, we have the reserved word **function**, then the output argument *area* followed by the assignment operator =, then the name of the function (the same as the name of the M-file), and then the input argument *rad*, which is the radius. Since there is an output argument in the function header, somewhere in the body of the function we must put a value in this output argument. This is how a value is returned from the function. In this case, the function is simple and all we have to do is assign to the output argument *area* the value of the built-in constant **pi** multiplied by the square of the input argument *rad*.

The function can be displayed in the Command Window using the **type** command.

```
>> type calcarea

        function area = calcarea(rad)
        % calcarea calculates the area of a circle
        % Format of call: calcarea(radius)
        % Returns the area

        area = pi * rad * rad;
        end
```

Note
Many of the functions in MATLAB are implemented as M-file functions; these can also be displayed using **type**.

2.7.2 Calling a function

The following is an example of a call to this function in which the value returned is stored in the default variable *ans*:

```
>> calcarea(4)
ans =
   50.2655
```

Technically, calling the function is done with the name of the file in which the function resides. To avoid confusion, it is easiest to give the function the same name as the file name, so that is how it will be presented in this book. In this example, the function name is *calcarea* and the name of the file is *calcarea.m*. The result returned from this function can also be stored in a variable in an assignment statement; the name could be the same as the name of the output argument in the function itself, but that is not necessary. So, for example, either of these assignments would be fine:

```
>> area = calcarea(5)
area =
   78.5398

>> myarea = calcarea(6)
myarea =
   113.0973
```

The value returned from the *calcarea* function could also be printed using either **disp** or **fprintf**:

```
>> disp(calcarea(4))
   50.2655
>> fprintf('The area is %.1f\n', calcarea(4))
The area is 50.3
```

> **Note**
> The printing is not done in the function itself; rather, the function returns the area and then an output statement can print or display it.

Using **help** with the function displays the contiguous block of comments under the function header. It is useful to put the format of the call to the function in this block comment:

```
>> help calcarea
   calcarea calculates the area of a circle
   Format of call: calcarea(radius)
   Returns the area
```

Many organizations have standards regarding what information should be included in the block comment in a function. These can include:

- Name of the function
- Description of what the function does
- Format of the function call
- Description of input arguments
- Description of output argument

- Description of variables used in function
- Programmer name and date written
- Information on revisions

Although this is excellent programming style, for the most part in this book these elements will be omitted simply to save space. Also, documentation in MATLAB suggests that the name of the function should be in all uppercase letters in the beginning of the block comment. However, this can be somewhat misleading in that MATLAB is case-sensitive and typically lowercase letters are used for the actual function name.

2.7.3 Calling a user-defined function from a script

Now, we will modify our script that prompts the user for the radius and calculates the area of a circle to call our function *calcarea* to calculate the area of the circle rather than doing this in the script.

script3.m

```
% This script calculates the area of a circle
% It prompts the user for the radius
radius = input('Please enter the radius: ');
% It then calls our function to calculate the
% area and then prints the result
area = calcarea(radius);
fprintf('For a circle with a radius of %.2f,',radius)
fprintf(' the area is %.2f\n',area)
```

Running this will produce the following:

```
>> script3
Please enter the radius: 5
For a circle with a radius of 5.00, the area is 78.54
```

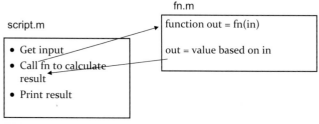

script.m

- Get input
- Call fn to calculate result
- Print result

fn.m

function out = fn(in)

out = value based on in

FIGURE 2.7 General form of a simple program

Simple programs

In this book, a script that calls function(s) is what we will call a MATLAB *program*. In the previous example, the program consisted of the script *script3* and the function it calls, *calcarea*. The general form of a simple program, consisting of a script that calls a function to calculate and return a value, looks like Figure 2.7. It is also possible for a function to call another (whether built-in or user-defined).

2.7.4 Passing multiple arguments

In many cases it is necessary to pass more than one argument to a function. For example, the volume of a cone is given by

$$V = \frac{1}{3}\pi r^2 h$$

where r is the radius of the circular base and h is the height of the cone. Therefore, a function that calculates the volume of a cone needs both the radius and the height:

conevol.m

```
function outarg = conevol(radius, height)
% conevol calculates the volume of a cone
% Format of call: conevol(radius, height)
% Returns the volume

outarg = (pi/3) * radius ^ 2 * height;
end
```

Since the function has two input arguments in the function header, two values must be passed to the function when it is called. The order makes a difference. The first value that is passed to the function is stored in the first input argument (in this case, *radius*), and the second argument in the function call is passed to the second input argument in the function header.

This is very important: The arguments in the function call must correspond one to one with the input arguments in the function header. Here is an example of calling this function. The result returned from the function is simply stored in the default variable *ans*.

```
>> conevol(4,6.1)
ans =
  102.2065
```

In the next example, the result is instead printed with a format of two decimal places.

```
>> fprintf('The cone volume is %.2f\n',conevol(3, 5.5))
The cone volume is 51.84
```

PRACTICE 2.9

Write a script that will prompt the user for the radius and height, call the function *conevol* to calculate the cone volume, and print the result in a nice sentence format. So, the program will consist of a script and the *conevol* function that it calls.

PRACTICE 2.10

For a project, we need some material to form a rectangle. Write a function *calcrectarea* that will receive the length and width of a rectangle in inches as input arguments, and will return the area of the rectangle. For example, the function could be called as shown, in which the result is stored in a variable and then the amount of material required is printed, rounded up to the nearest square inch.

```
>> ra = calcrectarea(3.1, 4.4)
ra =
   13.6400

>> fprintf('need %d sq in\n', ...
      ceil(ra))
need 14 sq in
```

2.7.5 Functions with local variables

The functions discussed thus far have been very simple. However, in many cases the calculations in a function are more complicated, and may require the use of extra variables within the function; these are called *local variables*.

For example, a closed cylinder is being constructed of a material that costs a certain dollar amount per square foot. We will write a function that will calculate and return the cost of the material, rounded up to the nearest square foot, for a cylinder with a given radius and a given height. The total surface area for the closed cylinder is

$$SA = 2\pi r h + 2\pi r^2$$

For a cylinder with a radius of 32 inches, height of 73 inches, and cost per square foot of the material of $4.50, the calculation would be given by the following algorithm:

- Calculate the surface area SA = 2 * π * 32 * 73 + 2 * π * 32 * 32 inches squared.
- Convert the SA from square inches to square feet = SA/144.
- Calculate the total cost = SA in square feet * cost per square foot.

The function includes local variables to store the intermediate results.

```
cylcost.m
function outcost = cylcost(radius, height, cost)
% cylcost calculates the cost of constructing a closed
%   cylinder
% Format of call: cylcost(radius, height, cost)
% Returns the total cost

% The radius and height are in inches
% The cost is per square foot

% Calculate surface area in square inches
surf_area = 2 * pi * radius * height + 2 * pi * radius ^ 2;
```

```
% Convert surface area in square feet and round up
surf_areasf = ceil(surf_area/144);

% Calculate cost
outcost = surf_areasf * cost;
end
```

The following shows examples of calling the function:

```
>> cylcost(32,73,4.50)
ans =
   661.5000

>> fprintf('The cost would be $%.2f\n', cylcost(32,73,4.50))
The cost would be $661.50
```

SUMMARY

Common Pitfalls

- Spelling a variable name different ways in different places in a script or function.
- Forgetting to add the second 's' argument to the **input** function when character input is desired.
- Not using the correct conversion character when printing.
- Confusing **fprintf** and **disp**. Remember that only **fprintf** can format.

Programming Style Guidelines

- Especially for longer scripts and functions, start by writing an algorithm.
- Use comments to document scripts and functions, as follows:
  - Block of contiguous comments at the top to describe a script
  - Block of contiguous comments under the function header for functions
  - Comments throughout any M-file (script or function) to describe each section
- Make sure that the "H1" comment line contains useful information.
- Use your organization's standard style guidelines for block comments.
- Use mnemonic identifier names (names that make sense, e.g., *radius* instead of *xyz*) for variable names and for file names.
- Make all output easy to read and informative.
- Put a newline character at the end of every string printed by **fprintf** so that the next output or the prompt appears on the line below.
- Put informative labels on the x and y axes and a title on all plots.
- Keep functions short—typically no longer than one page in length.
- Suppress the output from all assignment statements in a function.
- Functions that return a value do not normally print the value; it should simply be returned by the function.

MATLAB Reserved Words	
function	end

MATLAB Functions and Commands		
type	ylabel	legend
input	title	grid
disp	axis	bar
fprintf	clf	load
plot	figure	save
xlabel	hold	

MATLAB Operator
comment %

Exercises

1. Write a simple script that will calculate the volume of a hollow sphere,

$$\frac{4\pi}{3}(r_o^3 - r_i^3)$$

where r_i is the inner radius and r_o is the outer radius. Assign a value to a variable for the inner radius, and also assign a value to another variable for the outer radius. Then, using these variables, assign the volume to a third variable. Include comments in the script.

2. The atomic weight is the weight of a mole of atoms of a chemical element. For example, the atomic weight of oxygen is 15.9994 and the atomic weight of hydrogen is 1.0079. Write a script that will calculate the molecular weight of hydrogen peroxide, which consists of two atoms of hydrogen and two atoms of oxygen. Include comments in the script. Use **help** to view the comment in your script.

3. Write an **input** statement that will prompt the user for the name of a chemical element as a string. Then, find the length of the string.

4. Write an **input** statement that will prompt the user for a real number, and store it in a variable. Then, use the **fprintf** function to print the value of this variable using two decimal places.

5. The **input** function can be used to enter a vector, such as:

```
>> vec = input ('Enter a vector: ')
Enter a vector: 4:7
vec =
      4    5    6    7
```

Experiment with this, and determine how the user can enter a matrix.

6. Experiment, in the Command Window, with using the **fprintf** function for real numbers. Make a note of what happens for each. Use **fprintf** to print the real number 12345.6789
   - without specifying any field width
   - in a field width of 10 with four decimal places
   - in a field width of 10 with two decimal places
   - in a field width of 6 with four decimal places
   - in a field width of 2 with four decimal places

7. Experiment in the Command Window with using the **fprintf** function for integers. Make a note of what happens for each. Use **fprintf** to print the integer 12345
   - without specifying any field width
   - in a field width of 5
   - in a field width of 8
   - in a field width of 3

8. Create the following variables:

   ```
   x = 12.34;
   y = 4.56;
   ```

   Then, fill in the **fprintf** statements using these variables that will accomplish the following:

   ```
   >> fprintf(
   x is    12.340
   >> fprintf(
   x is 12
   >> fprintf(
   y is 4.6
   >> fprintf(
   y is 4.6    !
   ```

9. Write a script to prompt the user for the length and width of a rectangle, and print its area with two decimal places. Put comments in the script.

10. Write a script called *echoname* that will prompt the user for his or her name, and then echo print the name in a sentence in the following format (use %s to print it):

    ```
    >> echoname
    What is your name? Susan
    Wow, your name is Susan!
    ```

11. Write a script called *echostring* that will prompt the user for a string, and will echo print the string in quotes:

    ```
    >> echostring
    Enter your string: hi there
    Your string was: 'hi there'
    ```

12. In the metric system, fluid flow is measured in cubic meters per second (m^3/s). A cubic foot per second (ft^3/s) is equivalent to 0.028 m^3/s. Write a script titled *flowrate* that will prompt the user for flow in cubic meters per second and will

print the equivalent flow rate in cubic feet per second. Here is an example of running the script. Your script must produce output in exactly the same *format* as this:

```
>> flowrate
Enter the flow in m^3/s: 15.2
A flow rate of 15.200 meters per sec
is equivalent to 542.857 feet per sec
```

13. On average, people in a region spend 8% to 10% of their income on food. Write a script that will prompt the user for annual income. It will then print the range that would typically be spent on food annually. Also, print a monthly range.

14. Wing loading, which is airplane weight divided by wing area, is an important design factor in aeronautical engineering. Write a script that will prompt the user for the weight of the aircraft in kilograms, and the wing area in meters squared, and then calculate and print the wing loading of the aircraft in kilograms per square meter.

15. Write a script that assigns values for the x coordinate and then y coordinate of a point, and then plots this using a green +.

16. Plot **exp(x)** for values of x ranging from -2 to 2 in steps of 0.1. Place an appropriate title on the plot, and label the axes.

17. Create a vector x with values ranging from 1 to 100 in steps of 5. Create a vector y, which is the square root of each value in x. Plot these points. Next, use the **bar** function instead of **plot** to get a bar chart instead.

18. Create a y vector that stores random integers in the 1 to 100 range. Create an x vector that iterates from 1 to the length of the y vector. Experiment with the **plot** function using different colors, line types, and plot symbols.

19. Plot **sin(x)** for x values ranging from 0 to π (in separate Figure Windows)
    - using 10 points in this range
    - using 100 points in this range

20. Atmospheric properties, such as temperature, air density, and air pressure, are important in aviation. Create a file that stores temperatures in degrees Kelvin at various altitudes. The altitudes are in the first column and the temperatures in the second. For example, it may look like this:

1000	288
2000	281
3000	269
5000	256
10000	223

Write a script that will load these data into a matrix, separate it into vectors, and then plot the data with appropriate axis labels and a title.

21. Create a 3×6 matrix of random integers, each in the range from 50 to 100. Write this to a file called *randfile.dat*. Then, create a new matrix of random integers, but this time make it a 2×6 matrix of random integers, each in the range from 50 to 100. Append this matrix to the original file. Then, read the file in (which will be to a variable called *randfile*) just to make sure that it worked!

22. Create a file called *testtan.dat* comprised of two lines with three real numbers on each line (some negative, some positive, in the −1 to 3 range). The file can be created from the Editor, or saved from a matrix. Then, **load** the file into a matrix and calculate the tangent of every element in the resulting matrix.

23. Write a function *calcrectarea* that will calculate and return the area of a rectangle. Pass the length and width to the function as input arguments.

24. Write a function called *fn* that will calculate y as a function of x, as follows:

$$y = x^3 - 4x^2 + \sin(x)$$

Here are two examples of using the function:

```
>> help fn
    Calculates y as a function of x

>> y = fn(7)
y =
    147.6570
```

Renewable energy sources, such as biomass, are gaining increasing attention. Biomass energy units include megawatt hours (MWh) and gigajoules (GJ). One MWh is equivalent to 3.6 GJ. For example, one cubic meter of wood chips produces 1 MWh.

25. Write a function *mwh_to_gj* that will convert from MWh to GJ. Here are some examples of using the function:

```
>> gj = mwh_to_gj(mwh)
gj =
    11.8800

>> disp(mwh_to_gj(1.1))
    3.9600

>> help mwh_to_gj
    Converts from MWh to GJ
```

26. The velocity of an aircraft is typically given in either miles/hour or meters/second. Write a function that will receive one input argument, the velocity of an airplane in miles per hour, and will return the velocity in meters per second. The relevant conversion factors are one hour = 3600 seconds, one mile = 5280 feet, and one foot = 0.3048 meters.

27. If a certain amount of money (called the principal P) is invested in a bank account, earning an interest rate i compounded annually, the total amount of money T_n that will be in the account after n years is given by:

$$T_n = P(1 + i)^n$$

Write a function that will receive input arguments for P, i, and n, and will return the total amount of money T_n. Also, give an example of calling the function.

28. List some differences between a script and a function.

29. The velocity of a moving fluid can be found by calculating the difference between the total and static pressure P_t and P_s. For water, this is given by

$$V = 1.016\sqrt{P_t - P_s}$$

Write a function that will receive as input arguments the total and static pressures and will return the velocity of the water.

30. For a project, some biomedical engineering students are designing a device that will monitor a person's heart rate while on a treadmill. The device will let the subject know when the target heart rate has been reached. A simple calculation of the target heart rate (THR) for a moderately active person is

$$THR = (220 - A) * .6$$

where A is the person's age. Write a function that will calculate and return the THR.

31. An approximation for a factorial can be found using Stirling's formula:

$$n! \approx \sqrt{2\pi n}\left(\frac{n}{e}\right)^n$$

Write a function to implement this, passing the value of n as an argument.

32. The cost of manufacturing n units (where n is an integer) of a particular product at a factory is given by the equation

$$C(n) = 5n^2 - 44n + 11$$

Write a script *mfgcost* that will
- prompt the user for the number of units n
- call a function *costn* that will calculate and return the cost of manufacturing n units
- print the result (the format must be exactly as shown next)

Next, write the function *costn*, which simply receives the value of n as an input argument, and calculates and returns the cost of manufacturing n units. Here is an example of executing the script:

```
>> mfgcost
Enter the number of units: 100
The cost for 100 units will be $45611.00
>>
```

33. The desired conversion depends on temperature and other factors, but an approximation is that 1 inch of rain is equivalent to 6.5 inches of snow. Write a script that prompts the user for the number of inches of rain, calls a function to return the equivalent amount of snow, and prints this result. Write the function as well!

34. The volume V of a regular tetrahedron is given by $V = \frac{1}{12}\sqrt{2}s^3$, where s is the length of the sides of the equilateral triangles that form the faces of the tetrahedron. Write a program to calculate such a volume. The program will consist of one script and one function. The function will receive one input argument, which is the length of the sides, and will return the volume of the tetrahedron. The script will prompt the user for the length of the sides, call the function to calculate the volume, and print the result in a nice sentence format. For simplicity, we will ignore units.

35. Write a function that is called *pickone*, which will receive one input argument *x*, which is a vector, and will return one random element from the vector. For example,

```
>> pickone(4:7)
ans =
      5

>> disp(pickone(-2:0))
-1

>> help pickone
pickone(x) returns a random element from vector x
```

36. A function can return a vector as a result. Write a function *vecout* that will receive one integer argument and will return a vector that increments from the value of the input argument to its value plus 5, using the colon operator. For example,

```
>> vecout(4)
ans =
    4    5    6    7    8    9
```

37. If the lengths of two sides of a triangle and the angle between them are known, the length of the third side can be calculated. Given the lengths of two sides (*b* and *c*) of a triangle, and the angle between them α in degrees, the third side a is calculated as follows:

$$a^2 = b^2 + c^2 - 2\,b\,c\,\cos(\alpha)$$

Write a script *thirdside* that will prompt the user and read in values for *b, c,* and α (in degrees), and then calculate and print the value of *a* with three decimal places. (Note: To convert an angle from degrees to radians, multiply the angle by $\pi/180$.) The format of the output from the script should look exactly like this:

```
>> thirdside
Enter the first side: 2.2
Enter the second side: 4.4
Enter the angle between them: 50

The third side is 3.429
```

For more practice, write a function to calculate the third side, so that the script will call this function.

38. A part is being turned on a lathe. The diameter of the part is supposed to be 20,000 mm. The diameter is measured every 10 minutes and the results are stored in a file called *partdiam.dat.* Create a data file to simulate this. The file will store the time in minutes and the diameter at each time. Plot the data.

39. A file *floatnums.dat* has been created for use in an experiment. However, it contains float (real) numbers and what is desired instead is integers. Also, the file is not exactly in the correct format; the values are stored columnwise rather than rowwise. For example, if the file contains the following:

```
90.5792   27.8498   97.0593
12.6987   54.6882   95.7167
91.3376   95.7507   48.5376
63.2359   96.4889   80.0280
 9.7540   15.7613   14.1886
```

what is really desired is:

```
91   13   91   63   10
28   55   96   96   16
97   96   49   80   14
```

Create the data file in the specified format. Write a script that would read from the file *floatnums.dat* into a matrix, round the numbers, and write the matrix in the desired format to a new file called "intnums.dat."

40. A file named *costssales.dat* stores a company's cost and sales figures for the last *n* quarters (*n* is not defined ahead of time). The costs are in the first column, and the sales are in the second column. For example, *if* five quarters were represented, there would be five lines in the file, and it might look like this:

```
1100   800
1233   650
1111   1001
1222   1300
 999   1221
```

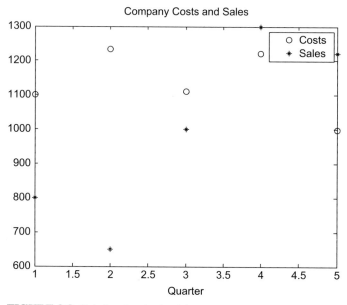

FIGURE 2.8 Plot of cost and sales data

Write a script called *salescosts* that will read the data from this file into a matrix. When the script is executed, it will do three things. First, it will print how many quarters were represented in the file, such as:

```
>> salescosts
There were 5 quarters in
the file
```

Next, it will plot the costs using black circles and sales using black stars (*) in a Figure Window with a legend (using default axes) as seen in Figure 2.8.

Finally, the script will write the data to a new file called *newfile.dat* in a different order. The sales will be in the first row, and the costs in the second row. For example, if the file is as shown before, the resulting file will store the following:

Note

It should not be assumed that the number of lines in the file is known.

```
 800    650    1001   1300   1221
1100   1233   1111   1222    999
```

Selection Statements

KEY TERMS

selection statements	logical expression	temporary variable
branching statements	relational operators	error-checking
condition	logical operators	nesting statements
relational expression	truth table	cascading if-else
Boolean expression	action	"is" functions

In the scripts and functions we've seen thus far, every statement was executed in sequence. That is not always desirable, and in this section we'll see how to make choices as to whether statements are executed or not, and how to choose between or among statements. The statements that accomplish this are called *selection* or *branching* statements.

The MATLAB® software has two basic statements that allow choices: the **if** statement and the **switch** statement. The **if** statement has optional **else** and **elseif** clauses for branching. The **if** statement uses expressions that are logically **true** or **false**. These expressions use relational and logical operators. MATLAB also has "is" functions that ask whether something is **true** or **false**; these functions are covered at the end of this chapter.

3.1 RELATIONAL EXPRESSIONS

Conditions in **if** statements use expressions that are conceptually either true or false. These expressions are called *relational expressions*; they are also sometimes called *Boolean expressions* or *logical expressions*. These expressions can use both *relational operators*, which relate two expressions of compatible types, and *logical operators*, which operate on **logical** operands.

The relational operators in MATLAB are:

Operator	Meaning
>	greater than
<	less than
>=	greater than or equals
<=	less than or equals
==	equality
~=	inequality

All concepts should be familiar, although the operators used may be different from those used in other programming languages or in mathematics classes. In particular, it is important to note that the operator for equality is two consecutive equal signs, not a single equal sign (since the single equal sign is already used as the assignment operator).

For numerical operands, the use of these operators is straightforward. For example, $3 < 5$ means "3 less than 5," which is conceptually a true expression. In MATLAB, as in many programming languages, "true" is represented by the **logical** value 1, and "false" is represented by the **logical** value 0. So, the expression $3 < 5$ actually displays in the Command Window the value 1 **(logical)** in MATLAB. Displaying the result of expressions like this in the Command Window demonstrates the values of the expressions.

```
>> 3 < 5
ans =
    1

>> 2 > 9
ans =
    0
```

The result type is **logical**. MATLAB also has built-in **true** and **false**. In other words, **true** is equivalent to **logical(1)** and **false** is equivalent to **logical(0)**. (In some versions of MATLAB, the value shown for the result of these expressions is **true** or **false** in the Workspace Window.) Although these are **logical** values, mathematical operations could be performed on the resulting 1 or 0.

```
>> 5 < 7
ans =
    1

>> ans + 3
ans =
    4
```

Comparing characters (e.g., 'a' < 'c') is also possible. Characters are compared using their ASCII equivalent values in the character encoding. So, 'a' < 'c' is a **true** expression, because the character 'a' comes before the character 'c'.

```
>> 'a' < 'c'
ans =
    1
```

The logical operators are:

Operator	Meaning
||	or (for scalars)
&&	and (for scalars)
~	not

Note
The logical operators for matrices will be covered in Chapter 5.

All logical operators operate on **logical** or Boolean operands. The **not** operator is a unary operator; the others are binary. The **not** operator will take a **logical** expression, which is **true** or **false**, and give the opposite value. For example, $\sim(3 < 5)$ is **false** since $(3 < 5)$ is **true**. The **or** operator has two **logical** expressions as operands. The result is **true** if either or both of the operands are **true**, and **false** only if both operands are **false**. The **and** operator also operates on two **logical** operands. The result of an **and** expression is **true** only if both operands are **true**; it is **false** if either or both are **false**.

In addition to these logical operators, MATLAB also has a function **xor**, which is the exclusive-or function. It returns **logical true** if one (and only one) of the arguments is **true**. For example, in the following only the first argument is **true**, so the result is **true**:

```
>> xor(3 < 5, 'a' > 'c')
ans =
    1
```

In this example, both arguments are **true** so the result is **false**:

```
>> xor(3 < 5, 'a' < 'c')
ans =
    0
```

Given the **logical** values of **true** and **false** in variables x and y, the *truth table* (see Table 3.1) shows how the logical operators work for all combinations. Note that the logical operators are commutative (e.g., x || y is the same as y || x).

Table 3.1 Truth Table for Logical Operators

x	y	~x	x || y	x && y	xor(x,y)
true	true	false	true	true	false
true	false	false	true	false	true
false	false	true	false	false	false

As with the numerical operators, it is important to know the operator precedence rules. Table 3.2 shows the rules for the operators that have been covered thus far in order of precedence.

Table 3.2 Operator Precedence Rules

Operators	Precedence
parentheses: ( )	highest
transpose and power ', ^	
unary: negation (−), not (~)	
multiplication, division *,/,\	
addition, subtraction +, −	
colon operator :	
relational <, <=, >, >=, ==, ~=	
and &&	
or ||	
assignment =	lowest

QUICK QUESTION!

Assume that there is a variable *x* that has been initialized. What would be the value of the expression

3 < x < 5

if the value of *x* is 4? What if the value of *x* is 7?

Answer: The value of this expression will always be **logical true**, or 1, regardless of the value of the variable *x*. Expressions are evaluated from left to right. So, first the expression 3 < x will be evaluated. There are only two possibilities: either this will be **true** or **false**, which means that either the

expression will have the value 1 or 0. Then, the rest of the expression will be evaluated, which will be either 1 < 5, or 0 < 5. Both of these expressions are **true**.

So, the value of *x* does not matter: The expression 3 < x < 5 would be **true** regardless of the value of the variable *x*. This is a logical error; it would not enforce the desired range. If we wanted an expression that was **logical true** only if *x* was in the range from 3 to 5, we could write 3 < x && x < 5.

PRACTICE 3.1

Think about what would be produced by the following expressions, and then type them in to verify your answers.

```
4 > 3 + 1

'e' == 'd' + 1

3 < 9 - 2

(3 < 9) - 2

4 == 3 + 1 && 'd' > 'c'

3 >= 2 || 'x' == 'y'

xor(3 >= 2, 'x' == 'y')

xor(3 >= 2, 'x' ~= 'y')
```

3.2 THE IF STATEMENT

The **if** statement chooses whether another statement, or group of statements, is executed or not. The general form of the **if** statement follows:

```
if condition
     action
end
```

A *condition* is a relational expression that is conceptually, or logically, **true** or **false**. The *action* is a statement, or a group of statements, that will be executed if the condition is **true**. When the **if** statement is executed, first the condition is evaluated. If the value of the condition is **true**, the action will be executed; if not, the action will not be executed. The action can be any number of statements until the reserved word **end**; the action is naturally bracketed by the reserved words **if** and **end**. (Note: This is different from the **end** that is used as an index into a vector or matrix.) The action is usually indented to make it easier to see.

For example, the following **if** statement checks to see whether the value of a variable is negative. If it is, the value is changed to a positive number by using the absolute value function; otherwise, nothing is changed.

```
if num < 0
     num = abs(num)
end
```

If statements can be entered in the Command Window, although they generally make more sense in scripts or functions. In the Command Window, the **if** line would be entered, followed by the Enter key, the action, the Enter key, and finally **end** and Enter. The results immediately follow. For example, the preceding **if** statement is shown twice here.

```
>> num = -4;
>> if num < 0
      num = abs(num)
   end
num =
      4

>> num = 5;
>> if num < 0
      num = abs(num)
   end
>>
```

Note that the output from the assignment is not suppressed, so the result of the action will be shown if the action is executed. The first time the value of the variable is negative so the action is executed and the variable is modified, but in the second case the variable is positive so the action is skipped.

This may be used, for example, to make sure that the square root function is not used on a negative number. The following script prompts the user for a number, and prints the square root. If the user enters a negative number, the **if** statement changes it to positive before taking the square root.

sqrtifexamp.m

```
% Prompt the user for a number and print its sqrt
num = input('Please enter a number: ');
% If the user entered a negative number, change it
if num < 0
      num = abs(num);
end
fprintf('The sqrt of %.1f is %.1f\n',num,sqrt(num))
```

Here are two examples of running this script:

```
>> sqrtifexamp
Please enter a number: -4.2
The sqrt of 4.2 is 2.0

>> sqrtifexamp
Please enter a number: 1.44
The sqrt of 1.4 is 1.2
```

Note that in the script the output from the assignment statement is suppressed. In this case, the action of the **if** statement was a single assignment statement. The action can be any number of valid statements. For example, we may wish to print a note to the user to say that the number entered was being changed.

```
sqrtifexampii.m
% Prompt the user for a number and print its sqrt

num = input('Please enter a number: ');

% If the user entered a negative number, tell
% the user and change it
if num < 0
    disp('OK, we''ll use the absolute value')
    num = abs(num);
end
fprintf('The sqrt of %.1f is %.1f\n', num, sqrt(num))
```

Note
As seen here, two single quotes in the **disp** statement are used to print one single quote.

```
>> sqrtifexampii
Please enter a number: −25
OK, we'll use the absolute value
The sqrt of 25.0 is 5.0
```

PRACTICE 3.2

Write an **if** statement that would print "No overtime for you!" if the value of a variable *hours* is less than 40. Test the **if** statement for values of *hours* both less than and greater than 40.

QUICK QUESTION!

Assume that we want to create a vector of increasing integer values from *mymin* to *mymax*. We will write a function *createvec* that receives two input arguments, *mymin* and *mymax*, and returns a vector with values from *mymin* to *mymax* in steps of one. First, we would make sure that the value of *mymin* is less than the value of *mymax*. If not, we would need to exchange their values before creating the vector. How would we accomplish this?

Answer: To exchange values, a temporary variable is required. For example, let's say that we have two variables *a* and *b*, storing the values:

```
a = 3;
b = 5;
```

To exchange values, we could *not* just assign the value of *b* to *a*:

```
a = b;
```

If that were done, then the value of *a*(3) is lost! Instead, we need to assign the value of *a* first to a ***temporary variable*** so that the value is not lost. The algorithm would be:

1. Assign the value of *a* to *temp*.
2. Assign the value of *b* to *a*.
3. Assign the value of *temp* to *b*.

```
>> temp = a;
>> a = b
a =
    5

>> b = temp
b =
    3
```

Continued

Now, for the function. An **if** statement is used to determine whether or not the exchange is necessary.

`createvec.m`

```
function outvec = createvec(mymin, mymax)
% createvec creates a vector that iterates from a
% specified minimum to a maximum
% Format of call: createvec(minimum, maximum)
% Returns a vector

% If the "minimum" isn't smaller than the "maximum",
% exchange the values using a temporary variable
if mymin > mymax
     temp = mymin;
     mymin = mymax;
     mymax = temp;
end

% Use the colon operator to create the vector
outvec = mymin:mymax;
end
```

Examples of calling the function are:

```
>> createvec(4,6)
ans =
     4    5    6

>> createvec(7,3)
ans =
     3    4    5    6    7
```

3.2.1 Representing logical true and false

It has been stated that conceptually true expressions have the **logical** value of 1, and expressions that are conceptually false have the **logical** value of 0. Representing the concepts of **logical true** and **false** in MATLAB is slightly different: The concept of false is represented by the value of 0, but the concept of true can be represented by *any nonzero value* (not just 1). This can lead to some strange **logical** expressions. For example, consider the following **if** statement:

```
>> if 5
     disp('Yes, this is true!')
   end
Yes, this is true!
```

Since 5 is a nonzero value, the condition is **true**. Therefore, when this **logical** expression is evaluated, it will be **true**, so the **disp** function will be executed and "Yes, this is true!" is displayed. Of course, this is a pretty bizarre **if** statement, and one that hopefully would never be encountered!

However, a simple mistake in an expression can lead to a similar result. For example, let's say that the user is prompted for a choice of 'Y' or 'N' for a yes/no question.

```
letter = input('Choice (Y/N): ','s');
```

In a script we might want to execute a particular action if the user responded with 'Y'. Most scripts would allow the user to enter either lowercase or uppercase; for example, either 'y' or 'Y' to indicate "yes." The proper expression that would return **true** if the value of "letter" was 'y' or 'Y' would be

```
letter == 'y' || letter == 'Y'
```

However, if by mistake this was written as

```
letter == 'y' || 'Y'
```

this expression would ALWAYS be **true**, regardless of the value of the variable *letter*. This is because 'Y' is a nonzero value, so it is a **true** expression. The first part of the expression may be **false**, but since the second expression is **true** the entire expression would be **true**.

3.3 THE IF-ELSE STATEMENT

The **if** statement chooses whether or not an action is executed. Choosing between two actions, or choosing from several actions, is accomplished using **if-else**, nested **if**, and **switch** statements.

The **if-else** statement is used to choose between two statements, or sets of statements. The general form is:

```
if condition
    action1
else
    action2
end
```

First, the condition is evaluated. If it is **true**, then the set of statements designated as "action1" is executed, and that is the end of the **if-else** statement. If instead the condition is **false**, the second set of statements designated as "action2" is executed, and that is the end of the **if-else** statement. The first set of statements ("action1") is called the action of the **if** clause; it is what will be executed if the expression is **true**. The second set of statements ("action2") is called the action of the **else** clause; it is what will be executed if the expression is **false**. One of these actions, and only one, will be executed—which one depends on the value of the condition.

For example, to determine and print whether or not a random number in the range from 0 to 1 is less than 0.5, an **if-else** statement could be used:

```
if rand < 0.5
    disp('It was less than .5!')
else
    disp('It was not less than .5!')
end
```

PRACTICE 3.3

Write a script *printsindegorrad* that will:

1. Prompt the user for an angle.
2. Prompt the user for (r)adians or (d)egrees, with radians as the default.
3. If the user enters 'd', the **sind** function will be used to get the sine of the angle in degrees; otherwise, the **sin** function will be used. Which sine function to use will be based solely on whether the user enters a 'd' or not. A 'd' means degrees, so **sind** is used; otherwise, for any other character the default of radians is assumed so **sin** is used.
4. Print the result.

Here are examples of running the script:

```
>> printsindegorrad
Enter the angle: 45
(r)adians (the default) or (d)egrees: d
The sin is 0.71

>> printsindegorrad
Enter the angle: pi
(r)adians (the default) or (d)egrees: r
The sin is 0.00
```

One application of an **if-else** statement is to check for errors in the inputs to a script (this is called *error-checking*). For example, an earlier script prompted the user for a radius, and then used that to calculate the area of a circle. However, it did not check to make sure that the radius was valid (e.g., a positive number). Here is a modified script that checks the radius:

checkradius.m

```
% This script calculates the area of a circle
% It error-checks the user's radius
radius = input('Please enter the radius: ');
if radius <= 0
    fprintf('Sorry; %.2f is not a valid radius\n',radius)
else
    area = calcarea(radius);
    fprintf('For a circle with a radius of %.2f,',radius)
    fprintf(' the area is %.2f\n',area)
end
```

Examples of running this script when the user enters invalid and then valid radii are shown as follows:

```
>> checkradius
Please enter the radius: -4
Sorry; -4.00 is not a valid radius

>> checkradius
Please enter the radius: 5.5
For a circle with a radius of 5.50, the area is 95.03
```

The **if-else** statement in this example chooses between two actions: printing an error message, or using the radius to calculate the area and then printing out the result. Note that the action of the **if** clause is a single statement, whereas the action of the **else** clause is a group of three statements.

3.4 NESTED IF-ELSE STATEMENTS

The **if-else** statement is used to choose between two actions. To choose from more than two actions the **if-else** statements can be *nested*, meaning one statement inside of another. For example, consider implementing the following continuous mathematical function $y = f(x)$:

$$y = 1 \quad \text{if } x < -1$$
$$y = x^2 \text{ if } -1 \leq x \leq 2$$
$$y = 4 \quad \text{if } x > 2$$

The value of y is based on the value of x, which could be in one of three possible ranges. Choosing which range could be accomplished with three separate **if** statements, as follows:

```
if x < -1
    y = 1;
end
if x >= -1 && x <=2
    y = x^2;
end
if x > 2
    y = 4;
end
```

Since the three possibilities are mutually exclusive, the value of y can be determined by using three separate **if** statements. However, this is not very efficient code: all three **logical** expressions must be evaluated, regardless of the range in which x falls. For example, if x is less than -1, the first expression is **true** and 1 would be assigned to y. However, the two expressions in the next two **if** statements are still evaluated. Instead of writing it in the preceding way, the expressions can be *nested* so that the statement ends when an expression is found to be **true**:

```
if x < -1
    y = 1;
else
    % If we are here, x must be >= -1
    % Use an if-else statement to choose
    % between the two remaining ranges
    if x <= 2
       y = x^2;
    else
       % No need to check
       % If we are here, x must be > 2
       y = 4;
    end
end
```

By using a nested **if-else** to choose from among the three possibilities, not all conditions must be tested as they were in the previous example. In this case, if x is less than –1, the statement to assign 1 to y is executed, and the **if-else** statement is completed so no other conditions are tested. If, however, x is not less than –1, then the **else** clause is executed. If the **else** clause is executed, then we already know that x is greater than or equal to –1 so that part does not need to be tested.

Instead, there are only two remaining possibilities: either x is less than or equal to 2, or it is greater than 2. An **if-else** statement is used to choose between those two possibilities. So, the action of the **else** clause was another **if-else** statement. Although it is long, this is one **if-else** statement, a nested **if-else** statement. The actions are indented to show the structure of the statement. Nesting **if-else** statements in this way can be used to choose from among 3, 4, 5, 6, ... The possibilities are practically endless!

This is actually an example of a particular kind of nested **if-else** statement called a *cascading* **if-else** statement. This is a type of nested **if-else** statement in which the conditions and actions cascade in a stairlike pattern (similar to a waterfall).

3.4.1 The elseif clause

THE PROGRAMMING CONCEPT

In most programming languages, choosing from multiple options means using nested if-else statements. However, MATLAB has another method of accomplishing this using the elseif clause.

THE EFFICIENT METHOD

To choose from among more than two actions, the **elseif** clause is used. For example, if there are *n* choices (where *n* > 3 in this example), the following general form would be used:

```
if condition1
    action1
elseif condition2
    action2
elseif condition3
    action3
% etc: there can be many of these
else
    actionn % the nth action
end
```

The actions of the **if**, **elseif**, and **else** clauses are naturally bracketed by the reserved words **if**, **elseif**, **else**, and **end**. For example, the previous example could be written using the **elseif** clause rather than nesting **if-else** statements:

```
if x < -1
    y = 1;
elseif x <= 2
    y = x^2;
else
    y = 4;
end
```

Note that in this example we only need one **end**. So, there are three ways of accomplishing this task: using three separate **if** statements, using nested **if-else** statements, and using an **if** statement with **elseif** clauses, which is the simplest. This could be implemented in a function that receives a value of *x* and returns the corresponding value of *y*:

```
calcy.m
```

```
function y = calcy(x)
% calcy calculates y as a function of x
% Format of call: calcy(x)
% y = 1      if    x < -1
% y = x^2    if    -1 <= x <= 2
% y = 4      if    x > 2

if x < -1
    y = 1;
elseif x <= 2
    y = x^2;
else
    y = 4;
end
end
```

```
>> x = 1.1;
>> y = calcy(x)
y =
    1.2100
```

QUICK QUESTION!

How could you write a function to determine whether an input argument is a scalar, vector, or matrix?

Answer: To do this, the **size** function can be used to find the dimensions of the input argument. If both the number of rows and columns is equal to 1, then the input argument is a scalar. If, on the other hand, only one dimension is 1, the input argument is a vector (either a row or column vector). If neither dimension is 1, the input argument is a matrix. These three options can be tested using a nested **if-else** statement. In this example, the word 'scalar', 'vector', or 'matrix' is returned from the function.

findargtype.m

```
function outtype = findargtype(inputarg)
% findargtype determines whether the input
% argument is a scalar, vector, or matrix
% Format of call: findargtype(inputArgument)
% Returns a string

[r c] = size(inputarg);
if r == 1 && c == 1
    outtype = 'scalar';
elseif r == 1 || c == 1
    outtype = 'vector';
else
    outtype = 'matrix';
end
end
```

Examples of calling this function are:

```
>> findargtype(33)
ans =
scalar
```

```
>> findargtype(2:5)
ans =
vector
```

```
>> findargtype(zeros(2,3))
ans =
matrix
```

Note: There is no need to check for the last case: If the input argument Isn't a scalar or a vector, it must be a matrix!

PRACTICE 3.4

Modify the function *findargtype* to return 'scalar', 'row vector', 'column vector', or 'matrix' depending on the input argument.

Another example demonstrates choosing from more than just a few options. The following function receives an integer quiz grade, which should be in the range from 0 to 10. The function then returns a corresponding letter grade, according to the following scheme: a 9 or 10 is an 'A', an 8 is a 'B', a 7 is a 'C', a 6 is a 'D', and anything below that is an 'F'. Since the possibilities are mutually exclusive, we could implement the grading scheme using separate **if** statements. However, it is more efficient to have one **if-else** statement with multiple **elseif** clauses. Also, the function returns the value 'X' if the quiz grade is not valid. The function assumes that the input is an integer.

letgrade.m

```
function grade = letgrade(quiz)
% letgrade returns the letter grade corresponding
% to the integer quiz grade argument
% Format of call: letgrade(integerQuiz)
% Returns a character

% First, error-check
if quiz < 0 || quiz > 10
    grade = 'X';

% If here, it is valid so figure out the
% corresponding letter grade
elseif quiz == 9 || quiz == 10
    grade = 'A';
elseif quiz == 8
    grade = 'B';
elseif quiz == 7
    grade = 'C';
elseif quiz == 6
    grade = 'D';
else
    grade = 'F';
end
end
```

Three examples of calling this function are:

```
>> quiz = 8;
>> lettergrade = letgrade(quiz)
lettergrade =
B

>> quiz = 4;
>> letgrade(quiz)
ans =
F
```

```
>> lg = letgrade(22)
lg =
X
```

In the part of this **if** statement that chooses the appropriate letter grade to return, all of the **logical** expressions are testing the value of the variable *quiz* to see if it is equal to several possible values, in sequence (first 9 or 10, then 8, then 7, etc.). This part can be replaced by a **switch** statement.

3.5 THE SWITCH STATEMENT

A **switch** statement can often be used in place of a nested **if-else** or an **if** statement with many **elseif** clauses. **Switch** statements are used when an expression is tested to see whether it is *equal to* one of several possible values.

The general form of the **switch** statement is:

```
switch switch_expression
   case caseexp1
      action1
   case caseexp2
      action2
   case caseexp3
      action3
   % etc: there can be many of these
   otherwise
      actionn
end
```

The **switch** statement starts with the reserved word **switch**, and ends with the reserved word **end**. The *switch_expression* is compared, in sequence, to the **case** expressions (*caseexp1*, *caseexp2*, etc.). If the value of the *switch_expression* matches *caseexp1*, for example, then *action1* is executed and the **switch** statement ends. If the value matches *caseexp3*, then *action3* is executed, and in general if the value matches *caseexpi* where *i* can be any integer from 1 to *n*, then the *actioni* is executed. If the value of the *switch_expression* does not match any of the **case** expressions, the action after the word **otherwise** is executed (the *n*th action, *actionn*).

Note

It is assumed that the user will enter an integer value. If the user does not, an error message will be printed or an incorrect result will be returned. Methods for remedying this are discussed in Chapter 4.

For the previous example, the **switch** statement can be used as follows:

switchletgrade.m

```
function grade = switchletgrade(quiz)
% switchletgrade returns the letter grade corresponding
% to the integer quiz grade argument using switch
% Format of call: switchletgrade(integerQuiz)
% Returns a character
```

```
% First, error-check
if quiz < 0 || quiz > 10
    grade = 'X';
else
    % If here, it is valid so figure out the
    % corresponding letter grade using a switch
    switch quiz
        case 10
            grade = 'A';
        case 9
            grade = 'A';
        case 8
            grade = 'B';
        case 7
            grade = 'C';
        case 6
            grade = 'D';
        otherwise
            grade = 'F';
    end
end
end
```

Here are two examples of calling this function:

```
>> quiz = 22;
>> lg = switchletgrade(quiz)
lg =
X

>> switchletgrade(9)
ans =
A
```

Since the same action of printing 'A' is desired for more than one grade, these can be combined as follows:

```
switch quiz
    case {10, 9}
        grade = 'A';
    case 8
        grade = 'B';
        % etc.
```

The curly braces around the **case** expressions 10 and 9 are necessary.

In this example, we error-checked first using an **if-else** statement. Then if the grade was in the valid range, a **switch** statement was used to find the corresponding letter grade.

Sometimes the **otherwise** clause is used for the error message. For example, if the user is supposed to enter only a 1, 3, or 5, the script might be organized as follows:

switcherror.m

```
% Example of otherwise for error message
choice = input('Enter a 1, 3, or 5: ');

switch choice
    case 1
        disp('It' 's a one!!')
    case 3
        disp('It''s a three!!')
    case 5
        disp('It''s a five!!')
    otherwise
        disp('Follow directions next time!!')
end
```

In this example, actions are taken if the user correctly enters one of the valid options. If the user does not, the **otherwise** clause handles printing an error message. Note the use of two single quotes within the string to print a single quote.

```
>> switcherror
Enter a 1, 3, or 5: 4
Follow directions next time!!
```

3.6 THE MENU FUNCTION

MATLAB has a built-in function called **menu** that will display a Figure Window with pushbuttons for the options. The first string passed to the **menu** function is the heading (an instruction), and the rest are labels that appear on the pushbuttons. The function returns the number of the button that is pushed. For example,

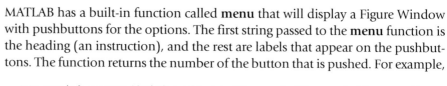

```
>> mypick = menu('Pick a pizza','Cheese','Shroom','Sausage');
```

will display the Figure Window seen in Figure 3.1 and store the result of the user's button push in the variable *mypick*.

FIGURE 3.1 Menu in Figure Window

There are three buttons, the equivalent values of which are 1, 2, and 3. For example, if the user pushes the "Sausage" button, *mypick* would have the value 3:

```
>> mypick
mypick =
    3
```

Note that the strings 'Cheese', 'Shroom', and 'Sausage' are just labels on the buttons. The actual value of the button push in this example would be 1, 2, or 3.

A script that uses this **menu** function would then use either an **if-else** statement or a **switch** statement to take an appropriate action based on the button pushed. For example, the following script simply prints which pizza to order, using a **switch** statement.

pickpizza.m

```
%This script asks the user for a type of pizza
% and prints which type to order using a switch

mypick = menu('Pick a pizza','Cheese','Shroom','Sausage');
switch mypick
    case 1
        disp('Order a cheese pizza')
    case 2
        disp('Order a mushroom pizza')
    case 3
        disp('Order a sausage pizza')
    otherwise
        disp('No pizza for us today')
end
```

This is an example of running this script and clicking on the "Sausage" button:

```
>> pickpizza
Order a sausage pizza
```

QUICK QUESTION!

How could the **otherwise** action get executed in this **switch** statement?

Answer: If the user clicks on the red "X" on the top right of the menu box instead of on one of the three buttons, the value returned from the **menu** function will be 0, which will cause the **otherwise** clause to be executed. This could also have been accomplished using a **case** 0 label instead of **otherwise**.

Instead of using a **switch** statement in this script, an alternative method would be to use an **if-else** statement with **elseif** clauses.

pickpizzaifelse.m

```
%This script asks the user for a type of pizza
% and prints which type to order using if-else
mypick = menu('Pick a pizza', 'Cheese', 'Shroom', 'Sausage');
if mypick == 1
    disp('Order a cheese pizza')
```

Continued

```
elseif mypick == 2
    disp('Order a mushroom pizza')
elseif mypick == 3
    disp('Order a sausage pizza')
else
    disp('No pizza for us today')
end
```

PRACTICE 3.5

Write a function that will receive one number as an input argument. It will use the **menu** function to display 'Choose a function' and will have buttons labeled 'fix', 'floor', and 'abs'. Using a <u>switch</u> statement, the function will then calculate and return the requested function (e.g., if 'abs' is chosen, the function will return the absolute value of the input argument). Choose a fourth function to return if the user clicks on the red 'X' instead of pushing a button.

3.7 THE "IS" FUNCTIONS IN MATLAB

There are a lot of functions that are built into MATLAB that test whether or not something is **true**; these functions have names that begin with the word "is." Since these functions are frequently used in <u>if</u> statements, they are introduced in this chapter. For example, the function called **isletter** returns **logical** 1 if the character argument is a letter of the alphabet, or 0 if it is not:

```
>> isletter('h')
ans =
    1

>> isletter('4')
ans =
    0
```

The **isletter** function will return **logical true** or **false** so that it can be used in a condition in an <u>if</u> statement. For example, here is code that would prompt the user for a character, and then print whether or not it is a letter:

```
mychar = input('Please enter a char: ','s');
if isletter(mychar)
    disp('Is a letter')
else
    disp('Not a letter')
end
```

When used in an **if** statement, it is not necessary to test the value to see whether the result from **isletter** is equal to 1 or 0; this is redundant. In other words, in the condition of the **if** statement,

```
isletter(mychar)
```

and

```
isletter(mychar) == 1
```

would produce the same results.

QUICK QUESTION!

How can we write our own function *myisletter* to accomplish the same result as **isletter**?

Answer: The function would compare the character's position within the character encoding.

```
myisletter.m

function outlog = myisletter(inchar)
% myisletter returns true if the input argument
% is a letter of the alphabet or false if not
% Format of call: myisletter(inputCharacter)
% Returns logical 1 or 0

if inchar >= 'a' && inchar <= 'z' ...
        || inchar >= 'A' && inchar <= 'Z'
    outlog = logical(1);
else
    outlog = logical(0);
end
end
```

Remember that it is necessary to check for both lowercase and uppercase letters. Also, the function must return **logical** 1 or 0.

The function **isempty** returns **logical true** if a variable is empty, **logical false** if it has a value, or an error message if the variable does not exist. Therefore, it can be used to determine whether a variable has a value yet or not. For example,

```
>> clear
>> isempty(evec)
??? Undefined function or variable 'evec'.
```

```
>> evec = [];
>> isempty(evec)
ans =
     1

>> evec = [evec 11];
>> isempty(evec)
ans =
     0
```

The **isempty** function will also determine whether or not a string variable is empty. For example, the following can be used to determine whether the user entered a string in an **input** function:

```
>> istr = input('Please enter a string: ','s');
Please enter a string:
>> isempty(istr)
ans =
     1
```

PRACTICE 3.6

Prompt the user for a string, and then print either the string that the user entered or an error message if the user failed to enter a string.

The function **iskeyword** will determine whether or not a name is a key word in MATLAB, and therefore something that cannot be used as an identifier name. By itself (with no arguments), it will return the list of all keywords. Note that the names of functions like "sin" are not key words, so their values can be overwritten if used as an identifier name.

```
>> iskeyword('sin')
ans =
     0

>> iskeyword('switch')
ans =
     1

>> iskeyword
ans =
    'break'
    'case'
    'catch'

    % etc.
```

There are many other "is" functions; the list of them can be found in the Help browser.

SUMMARY

Common Pitfalls

- Using = instead of == for equality
- Not using quotes when comparing a string variable to a string, such as in

  ```
  letter == y
  ```

  instead of

  ```
  letter == 'y'
  ```

- Confusing && and ||
- Confusing || and **xor**
- Putting a space in two-character operators (e.g., typing "< =" instead of "<=").
- Not spelling out an entire **logical** expression. An example is typing

  ```
  radius || height <= 0
  ```

  instead of

  ```
  radius <= 0 || height <= 0
  ```

  or typing

  ```
  letter == 'y' || 'Y'
  ```

  instead of

  ```
  letter == 'y' || letter == 'Y'
  ```

  Note that these are logically incorrect, but would not result in error messages. Note also that the expression "letter == 'y' || 'Y'" will ALWAYS be **true**, regardless of the value of the variable *letter*, since 'Y' is a nonzero value and therefore a **true** expression.
- Writing conditions that are more complicated than necessary, such as

  ```
  if (x < 5) == 1
  ```

  instead of just

  ```
  if (x < 5)
  ```

- Using an **if** statement instead of an **if-else** statement for error-checking; for example,

  ```
  if error occurs
      print error message
  end

  continue rest of code
  ```

instead of

```
if error occurs
    print error message
else
    continue rest of code
end
```

In the first example, the error message would be printed but then the program would continue anyway.

Programming Style Guidelines

- Use indentation to show the structure of a script or function. In particular, the actions in an **if** statement should be indented.
- When the **else** clause is not needed, use an **if** statement rather than an **if-else** statement. The following is an example:

```
if unit == 'i'
    len = len * 2.54;
else
    len = len; % this does nothing so skip it!
end
```

Instead, just use:

```
if unit == 'i'
    len = len * 2.54;
end
```

- Do not put unnecessary conditions on **else** or **elseif** clauses. For example, the following prints one thing if the value of a variable "number" is equal to 5, and something else if it is not.

```
if number == 5
    disp('It is a 5')
elseif number ~= 5
    disp('It is not a 5')
end
```

The second condition, however, is not necessary. Either the value is 5 or not, so just the **else** would handle this:

```
if number == 5
    disp('It is a 5')
else
    disp('It is not a 5')
end
```

- When using the **menu** function, ensure that the program handles the situation when the user clicks on the red 'X' on the menu box rather than pushing one of the buttons.

MATLAB Reserved Words	
if	case
else	otherwise
switch	elseif

MATLAB Functions and Commands	
true	menu
false	isletter
xor	isempty
sind	iskeyword

MATLAB Operators	
less than <	inequality ~=
greater than >	or for scalars ||
less than or equals <=	and for scalars &&
greater than or equals >=	not ~
equality ==	

Exercises

1. What would be the result of the following expressions?

   ```
   'b' >= 'c' - 1

   3 == 2 + 1

   (3 == 2) + 1

   xor(5 < 6, 8 > 4)
   ```

2. Write a script that tests whether the user can follow instructions. It prompts the user to enter an 'x'. If the user enters anything other than an 'x', it prints an error message; otherwise, the script does nothing.

3. Write a function *nexthour* that receives one integer argument, which is an hour of the day, and returns the next hour. This assumes a 12-hour clock; so, for example, the next hour after 12 would be 1. Here are two examples of calling this function:

   ```
   >> fprintf('The next hour will be %d.\n',nexthour(3))
   The next hour will be 4.
   >> fprintf('The next hour will be %d.\n',nexthour(12))
   The next hour will be 1.
   ```

4. Write a script to calculate the volume of a pyramid, which is 1/3 * base * height, where the base is length * width. Prompt the user to enter values for the length, width, and height, and then calculate the volume of the pyramid. When the user enters each value, he or she will then also be prompted for 'i' for inches or 'c' for centimeters. (*Note:* 2.54 cm = 1 inch.) The script should print the volume in cubic inches with three decimal places. As an example, the output format will be:

```
This program will calculate the volume of a pyramid.
Enter the length of the base: 50
Is that i or c? i
Enter the width of the base: 6
Is that i or c? c
Enter the height: 4
Is that i or c? i

The volume of the pyramid is xxx.xxx cubic inches.
```

5. Write a script to prompt the user for a character, and then print either that it is a letter of the alphabet or that it is not.

6. Write a script that will prompt the user for a numerator and a denominator for a fraction. If the denominator is 0, it will print an error message saying that division by 0 is not possible. If the denominator is not 0, it will print the result of the fraction.

7. The eccentricity of an ellipse is defined as

$$\sqrt{1 - \frac{b^2}{a}}$$

where a is the semimajor axis and b is the semiminor axis of the ellipse. A script prompts the user for the values of a and b. Since division by 0 is not possible, the script prints an error message if the value of a is 0 (it ignores any other errors, however). If a is not 0, the script calls a function to calculate and return the eccentricity, and then the script prints the result. Write the script and the function.

8. The systolic and diastolic blood pressure readings are found when the heart is pumping and the heart is at rest, respectively. A biomedical experiment is being conducted only on subjects whose blood pressure is optimal. This is defined as a systolic blood pressure less than 120 and a diastolic blood pressure less than 80. Write a script that will prompt for the systolic and diastolic blood pressures of a person, and will print whether or not that person is a candidate for this experiment, or not.

9. The continuity equation in fluid dynamics for steady fluid flow through a stream tube equates the product of the density, velocity, and area at two points that have varying cross-sectional areas. For incompressible flow, the densities are constant so that the equation is $A_1 V_1 = A_2 V_2$. If the areas and V_1 are known, V_2 can be found as $\frac{A_1}{A_2} V_1$. Therefore, whether the velocity at the second point increases or decreases depends on the areas at the two points. Write a script that will prompt the user for the two areas in square feet, and will print whether the velocity at the second point will increase, decrease, or remain the same as at the first point.

10. In chemistry, the pH of an aqueous solution is a measure of its acidity. The pH scale ranges from 0 to 14, inclusive. A solution with a pH of 7 is said to be *neutral*, a solution with a pH greater than 7 is *basic*, and a solution with a pH less than 7 is *acidic*. Write a script that will prompt the user for the pH of a solution, and will print whether it is neutral, basic, or acidic. If the user enters an invalid pH, an error message will be printed.

11. Write a function *createvecMToN* that will create and return a vector of integers from *m* to *n* (where *m* is the first input argument and *n* is the second), regardless of whether *m* is less than *n* or greater than *n*. If *m* is equal to *n*, the "vector" will just be *1 × 1* or a scalar.

12. Write a function *flipvec* that will receive one input argument. If the input argument is a row vector, the function will reverse the order and return a new row vector. If the input argument is a column vector, the function will reverse the order and return a new column vector. If the input argument is a matrix or a scalar, the function will return the input argument unchanged.

13. In a script, the user is supposed to enter either a 'y' or 'n' in response to a prompt. The user's input is read into a character variable called "letter." The script will print "OK, continuing" if the user enters either a 'y' or 'Y'; "OK, halting" if the user enters an 'n' or 'N'; or "Error" if the user enters anything else. Put this statement in the script first:

```
letter = input('Enter your answer: ', 's');
```

Write the script using a single nested **if-else** statement (**elseif** clause is permitted).

14. Write the script from the previous exercise using a **switch** statement instead.

15. In aerodynamics, the Mach number is a critical quantity. It is defined as the ratio of the speed of an object (e.g., an aircraft) to the speed of sound. If the Mach number is less than 1, the flow is subsonic; if the Mach number is equal to 1, the flow is transonic; and if the Mach number is greater than 1, the flow is supersonic. Write a script that will prompt the user for the speed of an aircraft and the speed of sound at the aircraft's current altitude and will print whether the condition is subsonic, transonic, or supersonic.

16. Write a script that will prompt the user for a temperature in degrees Celsius, and then an 'F' for Fahrenheit or 'K' for Kelvin. The script will print the corresponding temperature in the scale specified by the user. For example, the output might look like this:

```
Enter the temp in degrees C: 29.3
Do you want K or F? F
The temp in degrees F is 84.7
```

The format of the output should be exactly as specified. The conversions follow:

$$F = \frac{9}{5}C + 32$$

$$K = C + 273.15$$

17. Write a script that will generate one random integer, and will print whether the random integer is an even or an odd number. (Hint: An even number is divisible by 2, whereas an odd number is not; so check the remainder after dividing by 2.)

18. Write a function *isdivby4* that will receive an integer input argument, and will return **logical** 1 for **true** if the input argument is divisible by 4, or **logical false** if it is not.

19. Write a function *isint* that will receive a number input argument *innum*, and will return 1 for **true** if this number is an integer, or 0 for **false** if not. Use the fact that *innum* should be equal to **int32(innum)** if it is an integer. Unfortunately, due to round-off errors, it should be noted that it is possible to get **logical** 1 for **true** if the input argument is close to an integer. Therefore, the output may not be what you might expect, as shown here.

```
>> isint(4)
ans =
     1
>> isint(4.9999)
ans =
     0

>> isint(4.9999999999999999999999999999)
ans =
     1
```

20. A Pythagorean triple is a set of positive integers (a,b,c) such that $a^2 + b^2 = c^2$. Write a function *ispythag* that will receive three positive integers (a, b, c in that order) and will return **logical** 1 for **true** if they form a Pythagorean triple, or 0 for **false** if not.

21. In fluid dynamics, the Reynolds number Re is a dimensionless number used to determine the nature of a fluid flow. For an internal flow (e.g., water flow through a pipe), the flow can be categorized as follows:

$Re \leq 2300$	Laminar region
$2300 < Re \leq 4000$	Transition region
$Re > 4000$	Turbulent region

Write a script that will prompt the user for the Reynolds number of a flow and will print the region the flow is in. An example of running the script follows:

```
>> Reynolds
Enter a Reynolds number: 3500
The flow is in transition region
```

Would it be a good idea to write the selection statements using **switch**? Why or why not?

22. The area A of a rhombus is defined as $A = \dfrac{d_1 d_2}{2}$, where d_1 and d_2 are the lengths of the two diagonals. Write a script *rhomb* that first prompts the user for the lengths of the two diagonals. If either is a negative number or zero, the script prints an error message. Otherwise, if they are both positive, it calls a function *rhombarea* to return the area of the rhombus, and prints the result. Write the function, also! The lengths of the diagonals, which you can assume are in inches, are passed to the *rhombarea* function.

Global temperature changes have resulted in new patterns of storms in many parts of the world. Tracking wind speeds and a variety of categories of storms is important in understanding the ramifications of these temperature variations. Programs that work with storm data will use selection statements to determine the severity of storms and also to make decisions based on the data.

23. Whether a storm is a tropical depression, tropical storm, or hurricane is determined by the average sustained wind speed. In miles per hour, a storm is a tropical depression if the winds are less than 38 mph. It is a tropical storm if the winds are between 39 and 73 mph, and it is a hurricane if the wind speeds are ≥74 mph. Write a script that will prompt the user for the wind speed of the storm, and will print which type of storm it is.

24. Hurricanes are categorized based on wind speeds. The following table shows the category number for hurricanes with varying wind ranges and what the storm surge is (in feet above normal).

Cat	Wind speed	Storm surge
1	74–95	4–5
2	96–110	6–8
3	111–130	9–12
4	131–155	13–18
5	>155	>18

Write a script that will prompt the user for the wind speed, and will print the hurricane category number and the typical storm surge.

25. The Beaufort Wind Scale is used to characterize the strength of winds. The scale uses integer values and goes from a force of 0, which is no wind, up to 12, which is a hurricane. The following script first generates a random force value. Then, it prints a message regarding what type of wind that force represents, using a **switch** statement. You are to rewrite this **switch** statement as one nested **if-else** statement that accomplishes exactly the same thing. You may use **else** and/or **elseif** clauses.

```
ranforce = round(rand* 12);

switch ranforce
    case 0
        disp('There is no wind')
    case { 1,2,3,4,5,6}
        disp('There is a breeze')
    case { 7,8,9}
        disp('This is a gale')
    case { 10,11}
        disp('It is a storm')
    case 12
        disp('Hello, Hurricane!')
end
```

26. Clouds are generally classified as high, middle, or low level. The height of the cloud is the determining factor, but the ranges vary depending on the temperatue. For example, in tropical regions the classifications may be based on the following height ranges (given in feet):

Low	0-6500
Middle	6500-20,000
High	>20,000

Write a script that will prompt the user for the height of the cloud in feet, and print the classification.

27. Rewrite the following **switch** statement as one nested **if-else** statement (**elseif** clauses may be used). Assume that there is a variable *letter* and that it has been initialized.

```
switch letter
    case 'x'
        disp('Hello')
    case {'y', 'Y'}
        disp('Yes')
    case 'Q'
        disp('Quit')
    otherwise
        disp('Error')
end
```

28. Rewrite the following nested **if-else** statement as a **switch** statement that accomplishes exactly the same thing. Assume that *num* is an integer variable that has been initialized, and that there are functions *f1*, *f2*, *f3*, and *f4*. Do not use any **if** or **if-else** statements in the actions in the **switch** statement; use only calls to the four functions.

```
if num < -2 || num > 4
    f1(num)
else
    if num <= 2
        if num >= 0
            f2(num)
        else
            f3(num)
        end
    else
        f4(num)
    end
end
```

0 1 2 → f2

-1 -2 +3

3 4 → f4

29. Write a script *areaMenu* that will print a list consisting of "cylinder," "circle," and "rectangle." It prompts the user to choose one, and then prompts the user for the appropriate quantities (e.g., the radius of the circle) and then prints its area. If the user enters an invalid choice, the script simply prints an error message. The script should use a nested **if-else** statement to accomplish this. Here are two examples of running it (units are assumed to be inches).

```
>> areaMenu
Menu
1. Cylinder
2. Circle
3. Rectangle

Please choose one: 2
Enter the radius of the circle: 4.1
The area is 52.81
>> areaMenu
Menu
1. Cylinder
2. Circle
3. Rectangle
Please choose one: 3
Enter the length: 4
Enter the width: 6
The area is 24.00
```

30. Modify the *areaMenu* script to use a **switch** statement to decide which area to calculate.

31. Modify the *areaMenu* script (either version) to use the built-in **menu** function instead of printing the menu choices.

32. Write a script that prompts the user for a value of a variable *x*. Then, it uses the **menu** function to present choices between 'sin(x)', 'cos(x)', and 'tan(x)'. The script will print whichever function of x the user chooses. Use an **if-else** statement to accomplish this.

33. Modify the above script to use a **switch** statement instead.

34. Write a function that will receive one number as an input argument. It will use the **menu** function that will display 'Choose a function' and will have buttons labeled 'ceil', 'round', and 'sign'. Using a **switch** statement, the function will then calculate and return the requested function (e.g., if 'round' is chosen, the function will return the rounded value of the input argument).

35. Modify the function in Question 34 to use a nested **if-else** statement instead.

36. Simplify this statement:

```
if num < 0
    num = abs(num);
else
    num = num;
end
```

37. Simplify this statement:

```
if val >= 4
    disp('OK')
elseif val < 4
    disp('smaller')
end
```

Loop Statements

Consider the problem of calculating the area of a circle with a radius of 0.3 centimeters. A MATLAB® program certainly is not needed to do that; you'd use your calculator instead, and punch in $\pi * 0.3^2$. However, if a table of circle areas is desired, for radii ranging from 0.1 centimeters to 100 centimeters in steps of 0.05 (e.g., 0.1, 0.15, 0.2, etc.), it would be very tedious to use a calculator and write it all down. One of the great uses of a computer program such as MATLAB is the ability to repeat a process such as this.

This chapter will cover statements in MATLAB that allow other statement(s) to be repeated. The statements that do this are called *looping statements*, or *loops*. There are two basic kinds of loops in programming: *counted loops* and *conditional loops*. A counted loop is a loop that repeats statements a specified number of times (so, ahead of time it is known how many times the statements are to be repeated). In a counted loop, for example, you might say "repeat these statements 10 times." A conditional loop also repeats statements, but ahead of time it is not known *how many* times the statements will need to be repeated. With a conditional loop, for example, you might say "repeat these statements until this condition becomes false." The statement(s) that are repeated in any loop are called the *action* of the loop.

MATLAB®: A Practical Introduction to Programming and Problem Solving
© 2012 Elsevier Inc. All rights reserved.

There are two different loop statements in MATLAB: the **for** statement and the **while** statement. In practice, the **for** statement is usually used as the counted loop, and the **while** is used as the conditional loop. To keep it simple, that is how they will be presented here.

4.1 THE FOR LOOP

The **for** statement, or the **for** loop, is used when it is necessary to repeat statement(s) in a script or function, and when it *is* known ahead of time how many times the statements will be repeated. The statements that are repeated are called the action of the loop. For example, it may be known that the action of the loop will be repeated five times. The terminology used is that we *iterate* through the action of the loop five times.

The variable that is used to iterate through values is called a *loop variable*, or an *iterator variable*. For example, the variable might iterate through the integers 1 through 5 (e.g., 1, 2, 3, 4, and then 5). Although variable names in general should be mnemonic, it is common for an iterator variable to be given the name i (and if more than one iterator variable is needed, i, j, k, l, etc.). This is historical, and is because of the way integer variables were named in Fortran. However, in MATLAB both **i** and **j** are built-in values for $\sqrt{-1}$, so using either as a loop variable will override that value. If that is not an issue, then it is okay to use i as a loop variable.

The general form of the **for** loop is

```
for loopvar = range
    action
end
```

where *loopvar* is the loop variable, "range" is the range of values through which the loop variable is to iterate, and the action of the loop consists of all statements up to the **end**. Just like with **if** statements, the action is indented to make it easier to see. The range can be specified using any vector, but normally the easiest way to specify the range of values is to use the colon operator.

For instance, the following prints a column of numbers from 1 to 5:

```
for i = 1:5
    fprintf('%d\n',i)
end
```

This loop could be entered in the Command Window, although like **if** and **switch** statements, loops will make more sense in scripts and functions. In the Command Window, the results would appear after the **for** loop:

```
>> for i = 1:5
    fprintf('%d\n',i)
    end
```

```
1
2
3
4
5
```

What the **for** statement accomplished was to print the value of i and then the newline character for every value of i, from 1 through 5 in steps of 1. The first thing that happens is that i is initialized to have the value 1. Then, the action of the loop is executed, which is the **fprintf** statement that prints the value of i (1), and then the newline character to move the cursor down. Then, i is incremented to have the value of 2.

Next, the action of the loop is executed, which prints 2 and the newline. Then, i is incremented to 3 and that is printed, then i is incremented to 4 and that is printed, and then finally i is incremented to 5 and that is printed. The final value of i is 5; this value can be used once the loop has finished.

QUICK QUESTION!

How could you print the following column of integers?

```
  0
 50
100
150
200
```

Answer: In a loop, you could print these values starting at 0, incrementing by 50, and ending at 200. Each is printed using a field width of 3.

```
>> for i = 0:50:200
       fprintf('%3d\n',i)
   end
```

4.1.1 Finding sums and products

A very common application of a **for** loop is to calculate sums and products. For example, instead of just printing the integers 1 through 5, we could calculate the sum of the integers 1 through 5 (or, in general, 1 through n where n is any positive integer). Basically, we want to implement

$$\sum_{i=1}^{n} i$$

or calculate the sum $1 + 2 + 3 + \ldots + n$.

To do this, we need to add each value to a *running sum*. A running sum keeps changing, as we keep adding to it. First the sum has to be initialized to 0, then in this case it will be 1 $(0+1)$, then 3 $(0+1+2)$, then 6 $(0+1+2+3)$, and so forth.

In a function to calculate the sum, we need a loop or iterator variable i, as before, and also a variable to store the running sum. In this case we will use the output argument *runsum* as the running sum. Every time through the loop, the next value of *i* is added to the value of *runsum*. This function will return the end result, which is the sum of all integers from 1 to the input argument *n* stored in the output argument *runsum*.

```
sum1ToN.m
```

```
function runsum = sum1ToN(n)
% sum1ToN returns the sum of integers from 1 to n
% Format of call: sum1ToN(n)

runsum = 0;
for i = 1:n
    runsum = runsum + i;
end
end
```

Note
The output was suppressed when initializing the sum to 0 and when adding to it during the loop.

As an example, if 5 is passed to be the value of the input argument *n*, the function will calculate and return $1 + 2 + 3 + 4 + 5$, or 15:

```
>> sum1ToN(5)
ans =
    15
```

PRACTICE 4.1

Write a function *sumMToN* that is similar to the preceding function but will calculate the sum of the integers from *m* to *n*. For example, if the integers 4 and 7 are passed to the function, it will calculate $4 + 5 + 6 + 7$:

```
>> sumMToN(4,7)
ans =
    22
```

Another very common application of a **for** loop is to find a ***running product***. For example, instead of finding the sum of the integers 1 through *n*, we could find the product of the integers 1 through *n*. Basically, we want to implement

$$\prod_{i=1}^{n} i$$

or calculate the product $1 * 2 * 3 * 4 * \ldots * n$, which is called the *factorial* of *n* and is written *n*!.

THE PROGRAMMING CONCEPT

The basic algorithm is similar to finding a sum, except that we need to multiply each value of the loop variable to a running product. The difference is that while a running sum variable is initialized to 0, a running product variable must be initialized to 1. Thus, the first time a value is multiplied by it, it does not change the original value.

```
myfact.m

function runprod = myfact(n)
% myfact returns n!
% Format of call: myfact(n)

runprod = 1;
for i = 1:n
    runprod = runprod * i;
end
end
```

Any positive integer argument could be passed to this function, and it will calculate the factorial of that number. For example, if 5 is passed, the function will calculate and return 1 * 2 * 3 * 4 * 5, or 120:

```
>> myfact(5)
ans =
    120
```

THE EFFICIENT METHOD

MATLAB has a built-in function, **factorial**, that will find the factorial of an integer n:

```
>> factorial(5)
ans =
    120
```

Sums and products with vectors

The previous examples found either the sum or product of values at regular intervals, for example, from 1 to an integer n. Frequently, however, we wish to find the sum and/or product of the elements in a vector, regardless of what those values might be. For example, we will write a function to sum all of the elements in a vector.

THE PROGRAMMING CONCEPT

The vector is passed as an argument to the function. The function loops through all of the vector's elements, from 1 to the length of the vector, to add them all to the running sum.

```
myvecsum.m

function outarg = myvecsum(vec)
% myvecsum returns the sum of the elements in a
```

```
%     vector
% Format of call: myvecsum(vector)

outarg = 0;
for i = 1:length(vec)
    outarg = outarg + vec(i);
end
end
```

Any vector could be passed to this function; for example, we could just specify values for the elements in square brackets:

```
>> myvecsum([5 9 4])
ans =
    18
```

THE EFFICIENT METHOD

MATLAB has a built-in function, **sum**, that will sum all values in a vector. Again, any vector can be passed to the **sum** function:

```
>> sum([5 9 4])
ans =
    18
```

The function *myvecsum* illustrates a very important concept: looping through all of the elements in a vector to do something with each one. In this case, we are adding every element in the vector to a running sum, which is stored in the output argument *outarg*. Notice that the loop variable, *i*, is used as the index into the vector.

The first time through the loop, when *i* has the value 1, the value of vec(1) which is 5, is added to the value of *outarg* (so it is $0 + 5$, or 5). The second time through the loop, when *i* has the value 2, vec(2) or 9 is added to *outarg* (so it now stores $0 + 5 + 9$, or 14). Then, the third and final time through the loop vec (3) or 4 is added, so *outarg* now stores $0 + 5 + 9 + 4$, or 18.

This is in fact one reason to store values in a vector. Values in a vector typically represent "the same thing," so in a program typically the same operation would be performed on every element. The general form of a **for** loop to accomplish this is

```
for i = 1:length(vectorvariable)
    do something with vectorvariable(i)
end
```

The loop variable iterates through all elements in the vector, from 1 through the end (given by length(vectorvariable)), doing something with each element, specified as vectorvariable(i).

As another example, we will write a function to find the product of all the elements in a vector.

THE PROGRAMMING CONCEPT

The vector is passed as an argument to the function. The function loops through all of the vector's elements, from 1 to the length of the vector, to multiply them all by the running product.

```
myvecprod.m
```

```
function outarg = myvecprod(vec)
% myvecprod returns the product of
% the elements in a vector
% Format of call: myvecprod(vector)

outarg = 1;
for i = 1:length(vec)
    outarg = outarg * vec(i);
end
end
```

```
>> myvecprod([5 9 4])
ans =
    180
```

THE EFFICIENT METHOD

MATLAB has a built-in function, **prod**, that will return the product of all values in a vector.

```
>> prod([5 9 4])
ans =
    180
```

QUICK QUESTION!

How could we write a function *prodMToN* to calculate and return the product of the integers *m* to *n* without assuming a specific order of the arguments? In other words, both the function calls *prodMToN(3,6)* and *prodMToN(6,3)* would return the result of 3*4*5*6 or 360.

The answer is in the following boxes.

THE PROGRAMMING CONCEPT

To loop from the smaller value to the larger, we would first have to compare their values and exchange them if necessary.

prodMToN.m

```
function runprod = prodMToN(m,n)
% prodMToN returns the product of m:n
%      using a for loop
% Format: prodMToN(m,n) or prodMToN(n,m)

% Make sure m is less than n
if m > n
    temp = m;
    m = n;
    n = temp;
end

% Loop to calculate the running product
runprod = 1;
for i = m:n
    runprod = runprod * i;
end
end
```

THE EFFICIENT METHOD

Instead of exchanging the values of *m* and *n*, we could use the colon operator with steps of either +1 or –1 to create a vector. Also, instead of looping to calculate a running product, we could use the **prod** function.

prodMToNii.m

```
function outprod = prodMToNii(m,n)
% prodMToNii returns the product of m:n
%   using : and prod
% Format: prodMToNii(m,n) or profMToNii(n,m)

if m < n
    outprod = prod(m:n);
else
    outprod = prod(m:-1:n);
end
end
```

Additionally, MATLAB has the functions **cumsum** and **cumprod** that return a vector of the running sums or products. For example, for the following vector the first value is 5, so that is the first value in the vector returned by **cumsum**. Then the next value is 9 so 5 + 9 is 14, and finally 14 + 4 is 18.

```
>> vec = [5 9 4];
>> cumsum(vec)
ans =

     5    14    18
```

Since *vec* is a row vector, the result is also a row vector; passing a column vector to **cumsum** would result in a column vector. The cumulative product function **cumprod** instead calculates 5, then 5 * 9, and finally 5 * 9 * 4:

```
>> cumprod(vec)
ans =
     5    45   180
```

The **cumsum** and **cumprod** functions will return a vector with the same dimensions as the input vector.

Preallocating a vector

There are essentially two programming methods that could be used to simulate the **cumsum** function. One method is to start with an empty vector and concatenate each running sum value to the vector. Extending a vector, however, is very inefficient. A better method is to *preallocate* the vector to the correct size and then change the value of each element to be successive running sums. Both methods will be shown here.

In the following function, the output argument is initialized to the empty vector [ ]. Then, every time the next element in the vector is added to the running sum, this new sum is concatenated to the end of the vector.

myveccumsum.m

```
function outvec = myveccumsum(vec)
% myveccumsum simulates cumsum for a vector
% Format: myveccumsum(vector)

outvec = [];
runsum = 0;
for i = 1:length(vec)
    runsum = runsum + vec(i);
    outvec = [ outvec runsum] ;
end
end
```

An example of calling the function follows:

```
>> myveccumsum([5 9 4])
ans =
      5    14    18
```

The first time in the loop, *outvec* will be [5]. Then, the second time *runsum* will be 14 and *outvec* will store [5 14]. Finally, *runsum* will be 18 and *outvec* will store [5 14 18].

Although the previous method works, it is inefficient. What happens is that every time a vector is extended, a new "chunk" of memory must be found that is large enough for the new vector, and all the values must be copied from the original location in memory to the new one. This can take a long time to execute.

A better method involves referring to each index in the output vector, and placing each partial sum into the next element in the output vector. As each value of *vec(i)* is added to the running sum, this new sum is stored in *outvec(i)*.

myveccumsumii.m

```
function outvec = myveccumsumii(vec)
% myveccumsumii imitates cumsum for a vector
% It preallocates the output vector
% Format: myveccumsumii(vector)

outvec = zeros(size(vec));
runsum = 0;
for i = 1:length(vec)
    runsum = runsum + vec(i);
    outvec(i) = runsum;
end
end
```

Although initializing the output vector *outvec* to all zeros is not strictly necessary, it greatly improves the efficiency of the function. Initializing this vector to all zeros with the same size as the input argument **preallocates** that much memory for *outvec*. Then, each element is changed in the loop to its correct value.

MATLAB has many other functions that work with vectors. Many of these functions, which are statistical in nature, will be seen in Chapter 13.

PRACTICE 4.2

Write a function that imitates the **cumprod** function. Use the method of preallocating the output vector.

4.1.2 Combining for loops with if statements

Another example of a common application on a vector is to find the minimum and/or maximum value in the vector.

THE PROGRAMMING CONCEPT

For instance, the algorithm to find the minimum value in a vector follows:

1. The working minimum (the minimum that has been found so far) is the first element in the vector to begin with.
2. Loop through the rest of the vector (from the second element to the end).

   ■ If any element is less than the working minimum, then that element is the new minimum so far.

The following function implements this algorithm, and returns the minimum value found in the vector.

```
myminvec.m
```

```
function outmin = myminvec(vec)
% myminvec returns the minimum value in a vector
% Format: myminvec(vector)

outmin = vec(1);
for i = 2:length(vec)
    if vec(i) < outmin
        outmin = vec(i);
    end
end
end
```

```
>> vec = [3 8 99 −1];
>> myminvec(vec)
ans =
    −1

>> vec = [3 8 99 11];
>> myminvec(vec)
ans =
    3
```

Note
An **if** statement is used in the loop rather than an **if-else** statement. If the value of the next element in the vector is less than *outmin*, then the value of *outmin* is changed; otherwise, no action is necessary.

THE EFFICIENT METHOD

MATLAB has functions **min** and **max** that find the minimum and maximum values in a vector.

```
>> vec = [5 9 4];
>> min(vec)
ans =
    4
```

PRACTICE 4.3

Write a function to find and return the maximum value in a vector.

4.1.3 For loops that do not use the iterator variable in the action

In all examples discussed thus far, the value of the loop variable has been used in some way in the action of the **for** loop: we have printed the value of *i*, or added it to a sum, or multiplied it by a running product, or used it as an index into a vector. It is not always necessary to actually use the value of the loop variable, however. Sometimes the variable is simply used to iterate, or repeat, an action a specified number of times. For example,

```
for i = 1:3
    fprintf('I will not chew gum\n')
end
```

produces the output:

```
I will not chew gum
I will not chew gum
I will not chew gum
```

The variable *i* is necessary to repeat the action three times, even though the value of *i* is not used in the action of the loop.

QUICK QUESTION!

What would be the result of the following **for** loop?

```
for i = 4:2:8
    fprintf('I will not chew gum\n')
end
```

Answer: Exactly the same output as above! It doesn't matter that the loop variable iterates through the values 4, then 6,

then 8, instead of 1, 2, 3. Since the loop variable is not used in the action, this is just another way of specifying that the action should be repeated three times. Of course, using 1:3 makes more sense!

4.1.4 Input in a for loop

The following script repeats the process of prompting the user for a number, and *echo printing* the number (which means simply printing it back out). A **for** loop specifies how many times this is to occur. This is another example in which the loop variable is not used in the action, but instead just specifies how many times to repeat the action.

```
forecho.m
% This script loops to repeat the action of
% prompting the user for a number and echo printing it

for iv = 1:3
    inputnum = input('Enter a number: ');
    fprintf('You entered %.1f\n',inputnum)
end
```

```
>> forecho
Enter a number: 33
You entered 33.0
Enter a number: 1.1
You entered 1.1
Enter a number: 55
You entered 55.0
```

In this example, the loop variable *iv* iterates through the values 1 through 3, so the action is repeated three times. The action consists of prompting the user for a number and echo printing it with one decimal place.

PRACTICE 4.4

Modify the *forecho* script to sum the numbers that the user enters and print the result.

Instead of simply echo printing the numbers, it is often necessary to store them in a vector. One way of accomplishing this is to start by preallocating the vector and then putting values in each element, as we saw in a previous example. The following is a function that does this, and returns the resulting vector. The function receives an input argument *n*, and repeats the process *n* times.

```
forinputvec.m
function numvec = forinputvec(n)
% forinputvec returns a vector of length n
% It Prompts the user and puts n numbers into a vector
% Format: forinputvec(n)

numvec = zeros(1,n);
for iv = 1:n
    inputnum = input('Enter a number: ');
    numvec(iv) = inputnum;
end
end
```

Next is an example of calling this function and storing the resulting vector in a variable called *myvec*.

```
>> myvec = forinputvec(3)
Enter a number: 44
Enter a number: 2.3
Enter a number: 11

myvec =
    44.0000    2.3000    11.0000
```

QUICK QUESTION!

If you need to just print the sum or average of the numbers that the user enters, would you need to store them in a vector variable?

Answer: No. You could just add each to a running sum as you read them in a loop.

QUICK QUESTION!

What if you wanted to calculate how many of the numbers that the user entered were greater than the average?

Answer: Yes, then you would need to store them in a vector, because you would have to go back through them to count how many were greater than the average (or, alternatively, you could go back and ask the user to enter them again!!).

4.2 NESTED FOR LOOPS

The action of a loop can be any valid statement(s). When the action of a loop is another loop, this is called a *nested loop*.

The general form of a nested **for** loop is as follows:

```
for loopvarone = rangeone    ←   outer loop

    % actionone includes the inner loop

    for loopvartwo = rangetwo    ←    inner loop
        actiontwo
    end
end
```

The first **for** loop is called the *outer loop*; the second **for** loop is called the *inner loop*. The action of the outer loop consists (in part; there could be other statements) of the entire inner loop.

As an example, a nested **for** loop will be demonstrated in a script that will print a box of stars (*). Variables in the script will specify how many rows and columns to print. For example, if *rows* has the value 3, and *columns* has the value 5, the following would be the output:

```
* * * * *
* * * * *
* * * * *
```

Since lines of output are controlled by printing the newline character, the basic algorithm is:

- For every row of output,
  - Print the required number of stars
  - Move the cursor down to the next line (print '\n')

printstars.m

```
% Prints a box of stars
% How many will be specified by 2 variables
%    for the number of rows and columns

rows = 3;
columns = 5;
% loop over the rows
for i = 1:rows
    % for every row loop to print * 's and then one \n
    for j = 1:columns
        fprintf('*')
    end
    fprintf('\n')
end
```

Running the script displays the output:

```
>> printstars
* * * * *
* * * * *
* * * * *
```

The variable *rows* specifies the number of rows to print, and the variable *columns* specifies how many stars to print in each row. There are two loop variables: i is the loop variable for the rows, and j is the loop variable for the columns. Since the number of rows is known and the number of columns is known (given by the variables *rows* and *columns*), **for** loops are used. There is one **for** loop to loop over the rows, and another to print the required number of stars.

The values of the loop variables are not used within the loops, but are used simply to iterate the correct number of times. The first **for** loop specifies that the action will be repeated "rows" times. The action of this loop is to print stars and then the newline character. Specifically, the action is to loop to print *columns* stars (e.g., five stars) across on one line. Then, the newline character is printed after all five stars to move the cursor down to the next line.

In this case, the outer loop is over the rows, and the inner loop is over the columns. The outer loop must be over the rows because the script is printing a certain number of rows of output. For each row, a loop is necessary to print the required number of stars; this is the inner **for** loop.

When this script is executed, first the outer loop variable *i* is initialized to 1. Then, the action is executed. The action consists of the inner loop, and then printing the newline character. So, while the outer loop variable has the value 1, the inner loop variable *j* iterates through all of its values. Since the value of *columns* is 5, the inner loop will print a star five times. Then, the newline character is printed and then the outer loop variable *i* is incremented to 2. The action of the outer loop is then executed again, meaning the inner loop will print five stars, and then the newline character will be printed. This continues, and in all the action of the outer loop will be executed *rows* times.

Notice the action of the outer loop consists of two statements (the **for** loop and an **fprintf** statement). The action of the inner loop, however, is only a single **fprintf** statement.

The **fprintf** statement to print the newline character must be separate from the other **fprintf** statement that prints the star character. If we simply had

```
fprintf('*\n')
```

as the action of the inner loop, this would print a long column of 15 stars, not a box.

QUICK QUESTION!

How could this script be modified to print a triangle of stars instead of a box, such as the following?

```
*
**
***
```

printtristars.m

```
% Prints a triangle of stars
% How many will be specified by a variable
%    for the number of rows
rows = 3;
for i=1:rows
    % inner loop just iterates to the value of i
    for j=1:i
        fprintf('*')
    end
    fprintf('\n')
end
```

```
>> printtristars
*
**
***
```

Answer: In this case, the number of stars to print in each row is the same as the row number (e.g., one star is printed in row 1, two stars in row 2, and so on). The inner **for** loop does not loop to *columns*, but to the value of the row loop variable (so we do not need the variable *columns*):

In the previous examples, the loop variables were just used to specify the number of times the action is to be repeated. In the next example, the actual values of the loop variables will be printed.

```
printloopvars.m
```
```
% Displays the loop variables
for i = 1:3
    for j = 1:2
        fprintf('i=%d, j=%d\n',i,j)
    end
    fprintf('\n')
end
```

Executing this script would print the values of both *i* and *j* on one line every time the action of the inner loop is executed. The action of the outer loop consists of the inner loop and printing a newline character, so there is a separation between the actions of the outer loop:

```
>> printloopvars
i=1, j=1
i=1, j=2

i=2, j=1
i=2, j=2

i=3, j=1
i=3, j=2
```

Now, instead of just printing the loop variables, we can use them to produce a multiplication table, by multiplying the values of the loop variables.

The following function *multtable* calculates and returns a matrix that is a multiplication table. Two arguments are passed to the function, which are the number of rows and columns for this matrix.

```
multtable.m
```
```
function outmat = multtable(rows, columns)
% multtable returns a matrix which is a
% multiplication table
% Format: multtable(nRows, nColumns)

% Preallocate the matrix
outmat = zeros(rows,columns);
for i = 1:rows
    for j = 1:columns
        outmat(i,j) = i * j;
    end
end
end
```

In the following example of calling this function, the resulting matrix has three rows and five columns:

```
>> multtable(3,5)
ans =
     1     2     3     4     5
     2     4     6     8    10
     3     6     9    12    15
```

Note that this is a function that returns a matrix; it does not print anything. It preallocates the matrix to zeros, and then replaces each element. Since the number of rows and columns are known, **for** loops are used. The outer loop loops over the rows, and the inner loop loops over the columns. The action of the nested loop calculates $i * j$ for all values of i and j.

First, when i has the value 1, j iterates through the values 1 through 5, so first we are calculating $1*1$, then $1 * 2$, then $1 * 3$, then $1 * 4$, and finally $1 * 5$. These are the values in the first row (first in element $(1,1)$, then $(1,2)$, then $(1,3)$, then $(1,4)$, and finally $(1,5)$). Then, when i has the value 2, the elements in the second row of the output matrix are calculated, as j again iterates through the values from 1 through 5. Finally, when i has the value 3, the values in the third row are calculated $(3 * 1, 3 * 2, 3 * 3, 3 * 4,$ and $3 * 5)$.

This function could be used in a script that prompts the user for the number of rows and columns, calls this function to return a multiplication table, and writes the resulting matrix to a file:

createmulttab.m

```
% Prompt the user for rows and columns and
%    create a multiplication table to store in
%    a file "mymulttable.dat"

num_rows = input('Enter the number of rows: ');
num_cols = input('Enter the number of columns: ');
multmatrix = multtable(num_rows, num_cols);
save mymulttable.dat multmatrix -ascii
```

Following is an example of running this script, and then loading from the file into a matrix to verify that the file was created:

```
>> createmulttab
Enter the number of rows: 6
Enter the number of columns: 4

>> load mymulttable.dat
>> mymulttable
mymulttable =
```

```
    1     2     3     4
    2     4     6     8
    3     6     9    12
    4     8    12    16
    5    10    15    20
    6    12    18    24
```

PRACTICE 4.5

For each of the following (they are separate), determine what would be printed. Then, check your answers by trying them in MATLAB.

```
mat = [ 2:4; 8 2 5] ;
[ r c]  = size(mat);
for i = 1:r
    fprintf('The sum is %d\n', sum(mat(i,:)))
end
--------------------------------------
   for i = 1:3
    fprintf('%d ', i)
    for j= 1:2
        fprintf('%d ', j)
    end
    fprintf('\n')
end
```

4.2.1 Nested loops and matrices

Nested loops are often used when it is necessary to loop through all of the elements of a matrix. As an example, we will calculate the overall sum of the elements in a matrix.

THE PROGRAMMING CONCEPT

The matrix is passed as an input argument to the function. The function then uses the **size** function to determine the number of rows and columns in the matrix. It then loops over all elements in the matrix by using a nested loop: one loop over the rows, and another loop over the columns, adding each element to the running sum. Note that the loop variables i and j are used as the indices into the matrix: first the row index and then the column index.

```
mymatsum.m
```
```
function outsum = mymatsum(mat)
% mymatsum returns the overall sum of the elements
```

```
% in a matrix
% Format: mymatsum(matrix)

[row col] = size(mat);
outsum = 0;

% The outer loop is over the rows
for i = 1:row
    for j = 1:col
        outsum = outsum + mat(i,j);
    end
end
end
```

```
>> mat = [3:5; 2 5 7]
mat =
     3     4     5
     2     5     7

>> mymatsum(mat)
ans =
    26
```

THE EFFICIENT METHOD

MATLAB has a built-in function **sum**, as we have seen. For matrices, like many built-in functions, the **sum** function operates columnwise, meaning that it will return the sum of each column.

```
>> mat
mat =
     3     4     5
     2     5     7

>> sum(mat)
ans =
     5     9    12
```

So, to get the overall sum, it is necessary to sum the column sums!

```
>> sum(sum(mat))
ans =
    26
```

In the current example, the outer loop was over the rows, and the inner loop was over the columns. This order could easily be switched, however, so that the outer loop is over the columns and the inner loop is over the rows:

mymatsumb.m

```
function outsum = mymatsumb(mat)
% mymatsumb returns the overall sum of the elements
% in a matrix, with outer loop over columns
% Format: mymatsumb(matrix)

[ row col] = size(mat);
outsum = 0;

% The outer loop is over the columns
for i = 1:col
    for j = 1:row
        outsum = outsum + mat(j,i);
    end
end
end
```

Note
When referring to an element in the matrix, the row index is always given first, and then the column index, regardless of the order of the loops.

```
>> mat = [3:5; 2 5 7]
mat =

     3     4     5
     2     5     7

>> mymatsumb(mat)
ans =
    26
```

The order of the loops does not matter in this example, since all that is required is to add each element to the overall sum.

QUICK QUESTION!

How would we sum each individual column, rather than just getting an overall sum?

Answer: The programming method would require a nested loop in which the outer loop is over the columns. For example, we will modify the previous function to sum each column and return a row vector containing the results, as shown on the next page.

QUICK QUESTION!—CONT'D

matcolsum.m

```
function outsum = matcolsum(mat)
% matcolsum finds the sum of every column in a matrix
% Returns a vector of the column sums
% Format: matcolsum(matrix)
[ row col] = size(mat);

% Preallocate the vector to the number of columns
outsum = zeros(1,col);

% Every column is being summed so the outer loop
% has to be over the columns
for i = 1:col
    % Initialize the running sum to 0 for every column
    runsum = 0;
    for j = 1:row
        runsum = runsum + mat(j,i);
    end
    outsum(i) = runsum;
end
end
```

Note that the output argument will be a row vector containing the same number of columns as the input argument matrix. Also, since the function is calculating a sum for each column, the *runsum* variable must be initialized to 0 for every column, so it is initialized inside of the outer loop.

```
>> mat = [3:5; 2 5 7]
mat =
     3     4     5
     2     5     7
```

```
>> matcolsum(mat)
ans =
     5     9    12
```

Of course, the built-in **sum** function in MATLAB would accomplish the same thing, as we have already seen.

PRACTICE 4.6

Modify the function *matcolsum*. Create a function *matrowsum* to calculate and return a vector of all row sums, instead of column sums. For example, calling it and passing the *mat* variable would result in the following:

```
>> matrowsum(mat)
ans =
    12    14
```

Note that since the built-in **sum** function sums each column, one way of using **sum** to find the sum of each row is to transpose the matrix, as follows:

```
>> sum(mat')
ans =
    12    14
```

For matrices, the **cumsum** function returns a matrix consisting of the first row of the matrix argument, then the sum of the values in the first and second rows (for a matrix that has two rows).

```
>> cumsum(mat)
ans =
    3    4    5
    5    9   12
```

The functions **min** and **max** also operate columnwise; these functions find the minimum or maximum values in each column.

```
>> mat
mat =
    3    4    5
    2    5    7

>> max(mat)
ans =
    3    5    7
```

4.2.2 Combining nested for loops and if statements

The statements inside of a nested loop can be any valid statement, including any selection statement. For example, there could be an **if** or **if-else** statement as the action, or part of the action, in a loop.

As an example, assume that there is a file called *datavals.dat* containing results recorded from an experiment. However, some were erroneously recorded. The numbers are all supposed to be positive. The script below reads from this file into a matrix. It prints the sum from each row of only the positive numbers. We will assume that the file contains integers but will not assume how many lines are in the file nor how many numbers per line.

sumonlypos.m

```
% Sums only positive numbers from file
% Reads from the file into a matrix and then
%    calculates and prints the sum of only the
%    positive numbers from each row
```

```
load datavals.dat
[r c] = size(datavals);

for i = 1:r
    sumrow = 0;
    for j = 1:c
        if datavals(i,j) >= 0
            sumrow = sumrow + datavals(i,j);
        end
    end
    fprintf('The sum for row %d is %d\n',i,sumrow)
end
```

For example, *if* the file contains:

33	−11	2
4	5	9
22	5	−7
2	11	3

the output from the program would look like this:

```
>> sumonlypos
The sum for row 1 is 35
The sum for row 2 is 18
The sum for row 3 is 27
The sum for row 4 is 16
```

The file is loaded and the data are stored in a matrix variable. The script finds the dimensions of the matrix, and then loops through all elements in the matrix by using a nested loop; the outer loop iterates through the rows and the inner loop iterates through the columns. For each element, an **if-else** statement determines whether the element is positive or not. It only adds the positive values to the row sum. Since the sum is found for each row, the *sumrow* variable is initialized to 0 for every row, meaning inside of the outer loop.

QUICK QUESTION!

Would it matter if the order of the loops was reversed in this example, so that the outer loop iterates over the columns and the inner loop over the rows?

Answer: Yes, since we want a sum for every row the outer loop must be over the rows.

PRACTICE 4.7

Write a function *mymatmin* that finds the minimum value in each column of a matrix argument and returns a vector of the column minimums. An example of calling the function follows:

```
>> mat = round(rand(3,4)*19+1)
mat =
    15    19    17     5
     6    14    13    13
     9     5     3    13

>> mymatmin(mat)
ans =
     6     5     3     5
```

QUICK QUESTION!

Would the function *mymatmin* in the Practice 4.7 also work for a vector argument?

Answer: Yes, it should, since a vector is just a subset of a matrix. In this case, one of the loop actions would be executed only one time (for the rows if it is a row vector, or for the columns if it is a column vector).

4.3 WHILE LOOPS

The **while** statement is used as the conditional loop in MATLAB; it is used to repeat an action when ahead of time it is *not known how many* times the action will be repeated. The general form of the **while** statement is:

```
while condition
      action
end
```

The action, which consists of any number of statement(s), is executed as long as the condition is **true**.

The way it works is that first the condition is evaluated. If it is logically **true**, the action is executed. So, to begin with, the **while** statement is just like an **if** statement. However, at that point the condition is evaluated again. If it is still **true**, the action is executed again. Then, the condition is evaluated again. If it is still **true**, the action is executed again. Then, the condition is. . . . eventually, this has to stop! Eventually something in the action has to change something in the condition so that it becomes **false**. The condition must eventually become **false** to avoid an *infinite loop*. (If this happens, Ctrl-C will exit the loop.)

As an example of a conditional loop, we will write a function that will find the first factorial that is greater than the input argument *high*. Previously, we wrote a function to calculate a particular factorial. For example, to calculate 5! we found the product 1 * 2 * 3 * 4 * 5. In that case a **for** loop was used, since it was known that the loop would be repeated five times. Now, we do not know how many times the loop will be repeated.

The basic algorithm is to have two variables, one that iterates through the values 1, 2, 3, and so on, and one that stores the factorial of the iterator at each step. We start with 1, and 1 factorial, which is 1. Then, we check the factorial. If it is not greater than *high*, the iterator variable will then increment to 2, and find its factorial (2). If this is not greater than *high*, the iterator will then increment to 3, and the function will find its factorial (6). This continues until we get to the first factorial that is greater than *high*.

So, the process of incrementing a variable and finding its factorial is repeated until we get to the first value greater than **high**. This is implemented using a **while** loop:

factgthigh.m

```
function facgt = factgthigh(high)
% factgthigh returns the first factorial > input
% Format: factgthigh(inputInteger)

i=0;
fac=1;
while fac <= high
    i=i+1;
    fac = fac * i;
end
facgt = fac;
end
```

An example of calling the function, passing 5000 for the value of the input argument *high*, follows:

```
>> factgthigh(5000)
ans =
        5040
```

The iterator variable i is initialized to 0, and the running product variable *fac*, which will store the factorial of each value of i, is initialized to 1. The first time the **while** loop is executed, the condition is **true**: 1 is less than or equal to 5000. So, the action of the loop is executed, which is to increment i to 1 and *fac* becomes 1 (1 * 1).

After the execution of the action of the loop, the condition is evaluated again. Since it will still be **true**, the action is executed: i is incremented to 2, and *fac* will get the value 2 (1 * 2). The value 2 is still $\leq$5000, so the action will be executed again: i will be incremented to 3, and *fac* will get the value 6 (2 * 3). This continues, until the first value of *fac* is found that is greater than 5000. As soon as *fac* gets to this value, the condition will be **false** and the **while** loop will end. At that point the factorial is assigned to the output argument, which returns the value.

The reason that i is initialized to 0 rather than 1 is that the first time the loop action is executed, i becomes 1 and *fac* becomes 1 so that we have 1 and 1!, which is 1.

4.3.1 Multiple conditions in a while loop

In the *factgthigh* function, the condition in the **while** loop consisted of one expression, which tested whether or not the variable *fac* was less than or equal to the variable *high*. In many cases, however, the condition will be more complicated than that and could use either the **or** operator || or the **and** operator &&. For example, it may be that it is desired to stay in a **while** loop as long as a variable x is in a particular range:

```
while x >= 0 && x <= 100
```

As another example, continuing the action of a loop may be desired as long as at least one of two variables is in a specified range:

```
while x < 50 || y < 100
```

4.3.2 Reading from a file in a while loop

The following example illustrates reading from a data file using a **while** loop. Data from an experiment has been recorded in a file called *experd.dat*. The file has some numbers followed by a –99, then more numbers, all on the same line. The only data values that we are interested in, however, are those before −99.

The algorithm for the script will be:

1. Read the data from the file into a vector.
2. Create a new vector variable *newvec* that only has the data values up to but not including the −99.
3. Plot the new vector values, using black circles.

For example, *if* the file contains the following:

```
3.1  11  5.2  8.9  -99  4.4  62
```

the plot produced would look like Figure 4.1.

For simplicity, we will assume that the file format is as specified. Using **load** will create a vector with the name *experd*, which contains the values from the file. Also, since this is generic data we will omit the plot labels for simplicity.

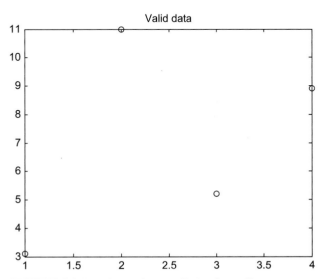

FIGURE 4.1 Plot of some (but not all) data from a file

THE PROGRAMMING CONCEPT

Using the programming method, we would loop through the vector until the −99 is found, creating the new vector by storing each element from *experd* in the vector *newvec*.

findvalwhile.m

```
% Reads data from a file, but only plots the numbers
% up to a flag of -99. Uses a while loop.

load experd.dat

i = 1;
while experd(i) ~= -99
    newvec(i) = experd(i);
    i = i + 1;
end

plot(newvec,'ko')
title('Valid data')
```

THE EFFICIENT METHOD

A more efficient method using built-in functions will be shown in the next chapter.

4.3.3 Input in a while loop

Sometimes a **while** loop is used to process input from the user as long as the user is entering data in a correct format. The following script repeats the process of prompting the user, reading in a positive number, and echo printing it, as long as the user correctly enters positive numbers when prompted. As soon as the user types in a negative number, the program will print "OK" and end.

whileposnum.m

```
% Prompts the user and echo prints the numbers entered
% until the user enters a negative number

inputnum=input('Enter a positive number: ');
while inputnum >= 0
    fprintf('You entered a %d.\n\n',inputnum)
    inputnum = input('Enter a positive number: ');
end
fprintf('OK!\n')
```

When the script is executed, the input/output might look like this:

```
>> whileposnum
Enter a positive number: 6
You entered a 6.

Enter a positive number: -2
OK!
```

Note that the prompt is repeated in the script: once before the loop, and then again at the end of the action. This is done so that every time the condition is evaluated, there is a new value of *inputnum* to check. If the user enters a negative number the first time, no values would be echo printed:

```
>> whileposnum
Enter a positive number: -33
OK!
```

Note
This example illustrates a very important feature of **while** loops: It is possible that the action will not be executed at all, if the value of the condition is false the first time it is evaluated.

As we have seen previously, MATLAB will give an error message if a character is entered rather than a number.

```
>> whileposnum
Enter a positive number: a
??? Error using ==> input
Undefined function or variable 'a'.

Enter a positive number: -4
OK!
```

However, if the character is actually the name of a variable, it will use the value of that variable as the input. For example:

```
>> a=5;
>> whileposnum
Enter a positive number: a
You entered a 5.

Enter a positive number: -4
OK!
```

4.3.4 Counting in a while loop

Although **while** loops are used when the number of times the action will be repeated is not known ahead of time, it is frequently useful to know how many times the action was in fact repeated. In that case, it is necessary to *count* the number of times that the action is executed. The following variation on the previous script counts the number of positive numbers that the user successfully enters.

countposnum.m

```
% Prompts the user for positive numbers and echo prints as
% long as the user enters positive numbers
% Counts the positive numbers entered by the user
counter=0;
inputnum=input('Enter a positive number: ');
while inputnum >= 0
    fprintf('You entered a %d.\n\n',inputnum)
    counter = counter + 1;
    inputnum = input('Enter a positive number: ');
end
fprintf('Thanks, you entered %d positive numbers.\n',counter)
```

The script initializes a variable *counter* to 0. Then, in the **while** loop action, every time the user successfully enters a number, the script increments the counter variable. At the end of the script, it prints the number of positive numbers that were entered.

```
>> countposnum
Enter a positive number: 4
You entered a 4.

Enter a positive number: 8
You entered a 8.

Enter a positive number: 11
You entered a 11.

Enter a positive number: −4
Thanks, you entered 3 positive numbers.
```

PRACTICE 4.8

Write a script *aveposnum* that will repeat the process of prompting the user for positive numbers, until the user enters a negative number, as just shown. Instead of echo printing them, however, the script will print the average (of just the positive numbers). If no positive numbers are entered, the script will print an error message instead of the average. Examples of executing this script follow:

```
>> aveposnum
Enter a positive number: −5
No positive numbers to average.

>> aveposnum
Enter a positive number: 8
Enter a positive number: 3
Enter a positive number: 4
Enter a positive number: −6
The average was 5.00.
```

4.3.5 Error-checking user input in a while loop

In most applications, when the user is prompted to enter something, there is a valid range of values. If the user enters an incorrect value, rather than having the program carry on with an incorrect value, or just printing an error message, the program should repeat the prompt. The program should keep prompting the user, reading the value, and checking it, until the user enters a value that is in the correct range. This is a very common application of a conditional loop: looping until the user correctly enters a value in a program. This is called *error-checking*.

For example, the following script prompts the user to enter a positive number, and loops to print an error message and repeat the prompt until the user finally enters a positive number.

readonenum.m

```
% Loop until the user enters a positive number

inputnum=input('Enter a positive number: ');
while inputnum < 0
      inputnum = input('Invalid! Enter a positive number: ');
end
fprintf('Thanks, you entered a %.1f \n',inputnum)
```

An example of running this script follows:

```
>> readonenum
Enter a positive number: -5
Invalid! Enter a positive number: -2.2
Invalid! Enter a positive number: c
??? Error using ==> input
Undefined function or variable 'c'.

Invalid! Enter a positive number: 44
Thanks, you entered a 44.0
```

Note
MATLAB itself catches the character input and prints an error message and repeats the prompt when the *c* was entered.

QUICK QUESTION!

How could we vary the previous example, so that the script asks the user to enter positive numbers *n* times, where *n* is an integer defined to be 3?

Answer: Every time that the user enters a value, the script checks and in a <u>while</u> loop keeps telling the user that it's

invalid until a valid positive number is entered. By putting the error-check in a <u>for</u> loop that repeats *n* times, the user is forced eventually to enter three positive numbers, as shown in the following.

QUICK QUESTION!—CONT'D

```
readnnums.m
% Loop until the user enters n positive numbers
n=3;
fprintf('Please enter %d positive numbers\n\n',n)
for i=1:n
    inputnum=input('Enter a positive number: ');
    while inputnum < 0
        inputnum = input('Invalid! Enter a positive number: ');
    end
    fprintf('Thanks, you entered a %.1f \n',inputnum)
end
```

```
>> readnnums
Please enter 3 positive numbers

Enter a positive number: 5.2
Thanks, you entered a 5.2
```

```
Enter a positive number: 6
Thanks, you entered a 6.0
Enter a positive number: −7.7
Invalid! Enter a positive number: 5
Thanks, you entered a 5.0
```

Error-checking for integers

Since MATLAB uses the type **double** by default for all values, to check to make sure that the user has entered an integer, the program has to convert the input value to an integer type (e.g., **int32**) and then checks to see whether that is equal to the original input. The following examples illustrate the concept.

If the value of the variable *num* is a real number, converting it to the type **int32** will round it, so the result is not the same as the original value.

```
>> num = 3.3;
>> inum = int32(num)
inum =
        3

>> num == inum
ans =
      0
```

If, on the other hand, the value of the variable *num* is an integer, converting it to an integer type will not change the value.

```
>> num = 4;
>> inum = int32(num)
inum =
        4
```

```
>> num == inum
ans =
     1
```

The following script uses this idea to error-check for integer data; it loops until the user correctly enters an integer.

readoneint.m

```
% Error-check until the user enters an integer
inputnum = input('Enter an integer: ');
num2 = int32(inputnum);
while num2 ~= inputnum
    inputnum = input('Invalid! Enter an integer: ');
    num2 = int32(inputnum);
end
fprintf('Thanks, you entered a %d \n',inputnum)
```

Examples of running this script are:

```
>> readoneint
Enter an integer: 9.5
Invalid! Enter an integer: 3.6
Invalid! Enter an integer: −11
Thanks, you entered a −11
```

```
>> readoneint
Enter an integer: 5
Thanks, you entered a 5
```

Putting these ideas together, the following script loops until the user correctly enters a positive integer. There are two parts to the condition, since the value must be positive and must be an integer.

readoneposint.m

```
% Error-checks until the user enters a positive integer
inputnum = input('Enter a positive integer: ');
num2 = int32(inputnum);
while num2 ~= inputnum || num2 < 0
    inputnum = input('Invalid! Enter a positive integer: ');
    num2 = int32(inputnum);
end
fprintf('Thanks, you entered a %d \n',inputnum)
```

```
>> readoneposint
Enter a positive integer: 5.5
Invalid! Enter a positive integer: −4
Invalid! Enter a positive integer: 11
Thanks, you entered a 11
```

PRACTICE 4.9

Modify the previous script to read n positive integers, instead of just one.

SUMMARY

Common Pitfalls

- Forgetting to initialize a running sum or count variable to 0
- Forgetting to initialize a running product variable to 1
- In cases where loops are necessary, not realizing that if an action is required for every row in a matrix, the outer loop must be over the rows (and if an action is required for every column, the outer loop must be over the columns)
- Not realizing that it is possible that the action of a **while** loop will never be executed
- Failing to error-check input into a program

Programming Style Guidelines

- Use loops for repetition only when necessary:
  - **for** statements as counted loops
  - **while** statements as conditional loops
- Do not use i or j for iterator variable names if the use of the built-in constants **i** and **j** is desired
- Indent the action of loops
- If the loop variable is just being used to specify how many times the action of the loop is to be executed, use the colon operator $1{:}n$ where n is the number of times the action is to be executed
- Preallocate vectors and matrices whenever possible (when the size is known ahead of time)
- When data are read in a loop, only store them in an array if it will be necessary to access the individual data values again

MATLAB Reserved Words	
while	end
for	

MATLAB Functions and Commands	
factorial	cumprod
sum	min
prod	max
cumsum	

Exercises

1. Write a **for** loop that will print the column of real numbers from 1.5 to 3.1 in steps of 0.2.

2. Write a function *sumsteps2* that calculates and returns the sum of 1 to *n* in steps of 2, where *n* is an argument passed to the function. For example, if 11 is passed, it will return $1 + 3 + 5 + 7 + 9 + 11$. Do this using a **for** loop. Calling the function will look like this:

```
>> sumsteps2(11)
ans =
    36
```

3. Write a function *prodby2* that will receive a value of a positive integer *n* and will calculate and return the product of the odd integers from 1 to *n* (or from 1 to *n* – 1 if *n* is even).

4. Prompt the user for an integer *n* and print "I love this stuff!" *n* times.

5. In the Command Window, write a **for** loop that will iterate through the integers from 32 to 255. For each, show the corresponding character from the character encoding.

6. In the Command Window, write a **for** loop that will print the elements from a vector variable in sentence format. For example, if the following is the vector:

```
>> vec = [5.5 11 3.45];
```

this would be the result:

```
Element 1 is 5.50.
Element 2 is 11.00.
Element 3 is 3.45.
```

The **for** loop should work regardless of how many elements are in the vector.

7. Write a script that will:
   - Generate a random integer in the range from 2 to 5
   - Loop that many times to
   - Prompt the user for a number
   - Print the sum of the numbers entered thus far with one decimal place

   There are many signal processing applications. Voltages, currents, and sounds are all examples of signals studied in a diverse range of disciplines such as biomedical engineering, acoustics, and telecommunications. Sampling discrete data points from a continuous signal is an important concept.

8. A sound engineer has recorded a sound signal from a microphone. The sound signal was "sampled," meaning that values at discrete intervals were recorded (rather than a continuous sound signal). The units of each data sample are volts. The microphone was not on at all times, however, so the data samples that are below a certain threshold are considered to be data values that were samples when the microphone

was not on, and therefore not valid data samples. The sound engineer would like to know the average voltage of the sound signal.

Write a script that will ask the user for the threshold and the number of data samples, and then for the individual data samples. The program will then print the average and a count of the *valid* data samples, or an error message if there were no valid data samples. An example of what the input and output would look like in the Command Window is shown here.

```
Please enter the threshold below which samples will be
considered to be invalid: 3.0
Please enter the number of data samples to enter: 7
```

Note

In the absence of valid data samples, the program would print an error message instead of the last line shown on the right.

```
Please enter a data sample: 0.4
Please enter a data sample: 5.5
Please enter a data sample: 5.0
Please enter a data sample: 2.1
Please enter a data sample: 6.2
Please enter a data sample: 0.3
Please enter a data sample: 5.4
```

```
The average of the 4 valid data samples is 5.53 volts.
```

9. Write a script that will load data from a file into a matrix. Create the data file first, and make sure that there is the same number of values on every line in the file so that it can be loaded into a matrix. Using a **for** loop, it will then create as many Figure Windows as there are rows in the matrix, and will plot the numbers from each row in a separate Figure Window.

10. A machine cuts *N* pieces of a pipe. After each cut, each piece of pipe is weighed and its length is measured; these two values are then stored in a file called *pipe.dat* (first the weight and then the length on each line of the file). Ignoring units, the weight is supposed to be between 2.1 and 2.3, inclusive, and the length is supposed to be between 10.3 and 10.4, inclusive. The following is just the beginning of what will be a long script to work with these data. For now, the script will just count how many rejects there are. A reject is any piece of pipe that has an invalid weight and/or length. For a simple example—if *N* is 3 (meaning three lines in the file) and the file stores

```
2.14   10.30
2.32   10.36
2.20   10.35
```

there is only one reject, the second one, as it weighs too much. The script would print:

```
There were 1 rejects.
```

11. Improve the output from the previous problem. If there is only one reject, it should print "There was 1 reject."; otherwise, for *n* rejects it should print "There were *n* rejects."

12. When would it matter if a **for** loop contained `for i = 1:4` vs. `for i =[ 3 5 2 6]`, and when would it not matter?

13. Create a vector of five random integers, each in the range from –10 to 10. Perform each of the following using loops (with **if** statements if necessary):
    - Subtract 3 from each element
    - Count how many are positive
    - Get the absolute value of each element
    - Find the maximum

14. Write a function that will receive a matrix as an input argument, and will calculate and return the overall average of all numbers in the matrix. Use loops, not built-in functions, to calculate the average.

15. We have seen that by default, when using built-in functions such as **sum** and **prod** on matrices, MATLAB will perform the function on each column. A dimension can also be specified when calling these functions. MATLAB refers to the columns as dimension 1 and the rows as dimension 2, such as the following:

```
>> sum(mat,1)
>> sum(mat,2)
```

Create a matrix and find the product of each row and column using **prod**.

16. Create a 3×5 matrix. Perform each of the following using loops (with **if** statements if necessary):
    - Find the maximum value in each column
    - Find the maximum value in each row
    - Find the maximum value in the entire matrix

17. With a matrix, when would:
    - Your outer loop be over the rows?
    - Your outer loop be over the columns?
    - It not matter which is the outer and which is the inner loop?

18. Assume that you have a matrix of integers *mat*. Fill in the rest of the **fprintf** statement so that this would print the product of every row in the matrix in the following format:

```
The product of row 1 is 162
The product of row 2 is 320
        etc.

[ row col] = size(mat);
for i = 1:row
    fprintf('The product of row %d is %d\n',      )
end
```

Note
The value of *col* is not needed.

19. Write a script *beautyofmath* that produces the following output. The script should iterate from 1 to 9 to produce the expressions on the left, perform the specified operation to get the results shown on the right, and print exactly in the format shown on the next page.

```
>> beautyofmath
1 x 8 + 1 = 9
12 x 8 + 2 = 98
123 x 8 + 3 = 987
1234 x 8 + 4 = 9876
12345 x 8 + 5 = 98765
123456 x 8 + 6 = 987654
1234567 x 8 + 7 = 9876543
12345678 x 8 + 8 = 98765432
123456789 x 8 + 9 = 987654321
```

20. Write a script that will print the following multiplication table:

```
1
2   4
3   6   9
4   8   12   16
5   10  15   20   25
```

21. The wind chill factor (WCF) measures how cold it feels with a given air temperature T (in degrees Fahrenheit) and wind speed V (in miles per hour). One formula for WCF is

$$\text{WCF} = 35.7 + 0.6\,\text{T} - 35.7(\text{V}^{0.16}) + 0.43\,\text{T}(\text{V}^{0.16})$$

Write a function to receive the temperature and wind speed as input arguments, and return the WCF. Using loops, print a table showing wind chill factors for temperatures ranging from –20 to 55 in steps of 5, and wind speeds ranging from 0 to 55 in steps of 5. Call the function to calculate each wind speed.

22. Instead of printing the WCFs in the previous problem, create a matrix of WCFs and write them to a file.

23. The inverse of the mathematical constant e can be approximated as follows:

$$\frac{1}{e} \approx \left(1 - \frac{1}{n}\right)^{n}$$

Write a script that will loop through values of n until the difference between the approximation and the actual value is less than 0.0001. The script should then print out the built-in value of e^{-1} and the approximation to four decimal places, and also print the value of n required for such accuracy.

24. Given the following loop:

```
while x < 10
    action
end
```

for what values of the variable x would the action of the loop be skipped entirely? If the variable x is initialized to have the value of 5 before the loop, what would the action have to include for this to not be an infinite loop?

25. Write a script that will prompt the user for the radius r and height of a cone, error-check the user's input for the radius and the height, and then calculate and print the volume of the cone (volume $= \pi/3 \; r^2 h$).

26. Write a script (e.g., called *findmine*) that will prompt the user for minimum and maximum integers, and then another integer that is the user's choice in the range from the minimum to the maximum. The script will then generate random integers in the range from the minimum to the maximum, until a match for the user's choice is generated. The script will print how many random integers had to be generated until a match for the user's choice was found. For example, running this script might result in this output:

```
>> findmine
Please enter your minimum value: -2
Please enter your maximum value: 3
Now enter your choice in this range: 0
It took 3 tries to generate your number
```

27. Write a script that will prompt the user for N integers, and then write the positive numbers (≥ 0) to an ASCII file called *pos.dat* and the negative numbers to an ASCII file called *neg.dat*. Error-check to ensure that the user enters N integers.

28. In thermodynamics, the Carnot efficiency is the maximum possible efficiency of a heat engine operating between two reservoirs at different temperatures. The Carnot efficiency is given as

$$\eta = 1 - \frac{T_C}{T_H}$$

where T_C and T_H are the absolute temperatures at the cold and hot reservoirs, respectively. Write a script that will prompt the user for the two reservoir temperatures in Kelvin and print the return of the corresponding Carnot efficiency to three decimal places. The script should error-check the user's input since absolute temperature cannot be less than or equal to zero. The script should also swap the temperature values if T_H is less than T_C.

29. Write a script that will continue prompting the user for positive numbers, and storing them in a vector variable, until the user types in a negative number.

30. Write a script *echoletters* that will prompt the user for letters of the alphabet and echo-print them until the user enters a character that is not a letter of the alphabet. At that point, the script will print the nonletter, and a count of how many letters were entered. Here are examples of running this script:

```
>> echoletters
Enter a letter: T
Thanks, you entered a T
Enter a letter: a
Thanks, you entered a a
Enter a letter: 8
8 is not a letter
You entered 2 letters

>> echoletters
Enter a letter: !
! is not a letter
You entered 0 letters
```

Note
The format must be exactly as shown in this code.

31. Write a script that will use the **menu** function to present the user with choices for functions "fix," "floor," and "ceil." Error-check by looping to display the menu until the user pushes one of the buttons (an error could occur if the user clicks on the "X" on the menu box rather than pushing one of the buttons). Then, generate a random number and print the result of the user's function choice of that number (e.g., **fix(5)**).

32. Write a script called *prtemps* that will prompt the user for a maximum Celsius value in the range from –16 to 20; error-check to make sure it's in that range. Then, print a table showing degrees Fahrenheit and degrees Celsius until this maximum is reached. The first value that exceeds the maximum should not be printed. The table should start at 0 degrees Fahrenheit, and increment by 5 degrees Fahrenheit until the max (in Celsius) is reached. Both temperatures should be printed with a field width of 6 and one decimal place. The formula is $C = 5/9 (F - 32)$. For example, the execution of the script might look like this (the format should be exactly like this):

```
>> prtemps
When prompted, enter a temp in degrees C in range -16 to 20.
Enter a maximum temp: 30
Error! Enter a maximum temp: 9

    F      C
   0.0   -17.8
   5.0   -15.0
          .
          .
          .
  40.0     4.4
  45.0     7.2
```

33. Create an *x* vector that has integers 1 through 10, and set a *y* vector equal to *x*. Plot this straight line. Now, add noise to the data points by creating a new *y2* vector that stores the values of *y* plus or minus 0.25. Plot the straight line and also these noisy points.

34. A blizzard is a massive snowstorm. Definitions vary, but for our purposes we will assume that a blizzard is characterized by both winds of 30 mph or higher and blowing snow that leads to visibility of 0.5 mile or less, sustained for at least four hours. Data from a storm one day has been stored in a file *stormtrack.dat*. There are 24 lines in the file, one for each hour of the day. Each line in the file has the wind speed and visibility at a location. Create a sample data file. Read these data from the file and determine whether blizzard conditions were met during this day or not.

Vectorized Code

CONTENTS

Although loops are extremely useful in most programming applications and necessary in many languages, in the MATLAB® software they are frequently not necessary especially when dealing with vectors or matrices. In this chapter, the concept of *vectorizing* will be introduced, which is the term used in MATLAB for rewriting code that was written using constructs such as loops in a traditional programming language and instead taking advantage of array operations in MATLAB. The vectorized code is faster and easier for the programmer to write, and in many cases, is also faster for MATLAB to execute.

5.1 LOOPS WITH VECTORS AND MATRICES

In most programming languages when performing an operation on a vector, a **for** loop is used to loop through the entire vector, using the loop variable as the index into the vector. In general, in MATLAB, assuming there is a vector variable *vec*, the indices range from 1 to the length of the vector:

```
for i = 1:length(vec)
    % do something with vec(i)
end
```

For example, let's say that we want to multiply every element of a vector v by 3, and store the result back in v, where v is initialized as follows:

```
>> v = [3 7 2 1];
```

MATLAB®: A Practical Introduction to Programming and Problem Solving

155

THE PROGRAMMING CONCEPT

To accomplish this, we can loop through all of the elements in the vector and multiply each element by 3. In the following, the output is suppressed in the loop, and then the resulting vector is shown:

```
>> for i = 1:length(v)
        v(i) = v(i) * 3;
   end
>> v
v =
     9   21   6   3
```

Similarly, for an operation on a matrix, a nested loop would be required, and the loop variables over the rows and columns are used as the subscripts into the matrix. In general, assuming a matrix variable *mat*, we use **size** to return separately the number of rows and columns and use these variables in the **for** loops:

```
[r c] = size(mat);
for row = 1:r
   for col = 1:c
      % do something with mat(row,col)
   end
end
```

Typically, this is not necessary in MATLAB!!

5.2 OPERATIONS ON VECTORS AND MATRICES

Numerical operations can be done on entire vectors or matrices. For example, let's say that we want to multiply every element of a vector v by 3 as in the previous section.

THE EFFICIENT METHOD

In MATLAB, we can simply multiply v by 3 and store the result back in v in an assignment statement:

```
>> v = v*3
v =
     9   21   6   3
```

As another example, we can divide every element by 2:

```
>> v= [3 7 2 1];
>> v/2
ans =
    1.5000    3.5000    1.0000    0.5000
```

For a matrix, numerical operations can also be performed on every element. For example, to multiply every element in a matrix by 2 with most languages would involve a nested loop, but in MATLAB it is automatic.

```
>> mat = [4:6; 3:-1:1]
mat =
      4      5      6
      3      2      1
>> mat * 2
ans =
      8     10     12
      6      4      2
```

These are *scalar operations*; we are multiplying every element in a vector or matrix by a scalar, or dividing every element in a vector or a matrix by a scalar.

Array operations are operations that are performed on vectors or matrices term by term, or element by element. This means that the two arrays (vectors or matrices) must be the same size to begin with. The following examples demonstrate the array addition and subtraction operators.

```
>> v1 = 2:5
v1 =
      2      3      4      5
>> v2 = [33 11 5 1]
v2 =
     33     11      5      1
>> v1 + v2
ans =
     35     14      9      6

>> mata = [5:8; 9:-2:3]
mata =
      5      6      7      8
      9      7      5      3
>> matb = reshape(1:8,2,4)
matb =
      1      3      5      7
      2      4      6      8
>> mata - matb
ans =
      4      3      2      1
      7      3     -1     -5
```

However, for any operation that is based on multiplication (which means multiplication, division, and exponentiation), a dot must be placed in front of the operator for array operations. For example, the exponentiation operator, .^, must be used when working with vectors and matrices, rather than just the ^ operator. Squaring a vector, for example, means multiplying each element by itself, so the .^ operator must be used.

```
>> v = [3 7 2 1];
>> v ^ 2
??? Error using ==> mpower
Inputs must be a scalar and a square matrix.
To compute elementwise POWER, use POWER (.^) instead.

>> v .^ 2
ans =
      9     49      4      1
```

Similarly, the operator .* must be used for array multiplication and ./ or .\ for array division. The following examples demonstrate array multiplication and array division.

```
>> v1 = 2:5
v1 =
      2      3      4      5
>> v2 = [33 11 5 1]
v2 =
     33     11      5      1

>> v1 .* v2
ans =
     66     33     20      5

>> mata = [5:8; 9:-2:3]
mata =
      5      6      7      8
      9      7      5      3

>> matb = reshape (1:8, 2,4)
matb =
      1      3      5      7
      2      4      6      8

>> mata ./ matb
ans =
     5.0000    2.0000    1.4000    1.1429
     4.5000    1.7500    0.8333    0.3750
```

The operators .^, .*, ./, and .\ are called **array operators** and are used when multiplying or dividing vectors or matrices of the same size term by term.

Note

Matrix multiplication is a very different operation, and will be covered in Chapter 12.

PRACTICE 5.1

1. Create a vector variable and add 2 to every element in it.
2. Create a matrix variable and divide every element by 3.
3. Create a matrix variable and square every element.

5.3 VECTORS AND MATRICES AS FUNCTION ARGUMENTS

Using most programming languages, if it is desired to evaluate a function on every element in a vector or a matrix, loop(s) would be necessary to accomplish this. However, as we have already seen, in MATLAB an entire vector or matrix can be passed as an argument to a function; the function will be evaluated on every element. This means that the result will be the same size as the input argument.

For example, let us find the sine in radians of every element of a vector *vec*. The **sin** function will automatically return the sine of each individual element and the result will be a vector with the same length as the input vector.

```
>> vec = −2:1
vec =
     −2    −1     0     1

>> sinvec = sin(vec)
sinvec =
    −0.9093  −0.8415        0    0.8415
```

For a matrix, the resulting matrix will have the same size as the input argument matrix. For example, the **sign** function will find the sign of each element in a matrix:

```
>> mat = [0 4 -3; -1 0 2]
mat =
      0     4    −3
     −1     0     2

>> sign(mat)
ans =
      0     1    −1
     −1     0     1
```

Vectors or matrices can be passed to user-defined functions, as well, as long as the operators used in the function are correct. For example, we previously defined a function that calculates the area of a circle:

```
>> type calcarea

function area = calcarea(rad)
% calcarea calculates the area of a circle
% Format of call: calcarea(radius)
% Returns the area

area = pi * rad * rad;
end
```

The previous function was written assuming that the argument was a scalar, so calling it with a vector instead would produce an error message:

```
>> calcarea(1:3)
??? Error using ==> mtimes
Inner matrix dimensions must agree.
Error in ==> calcarea at 6
     area = pi * rad * rad;
```

This is because the * was used for multiplication in the function, but .* must be used when multiplying vectors term by term. Changing this in the function would allow either scalars or vectors to be passed to this function:

calcareaii.m

```
function area = calcareaii(rad)
% calcareaii returns the area of a circle
% The input argument can be a vector of radii
% Format: calcareaii(radiiVector)

area = pi * rad .* rad;
end
```

```
>> calcareaii(1:3)
ans =
    3.1416    12.5664    28.2743

>> calcareaii(4)
ans =
50.2655
```

Note the * operator is only necessary when multiplying the radius vector by itself. Multiplying by **pi** is scalar multiplication, so the .* operator is not needed there. We could have also used

```
area = pi * rad .^ 2;
```

5.4 LOGICAL VECTORS

The relational operators can also be used with vectors and matrices. For example, let's say that there is a vector *vec*, and we want to compare every element in the vector to 5 to determine whether it is greater than 5 or not. The result would be a vector (with the same length as the original) with **logical true** or **false** values.

```
>> vec = [5  9  3  4  6  11];
```

THE PROGRAMMING CONCEPT

To accomplish this using the programming method, we would have to loop through all of the vector's elements and compare each element with 5 to determine whether the corresponding value in the result would be **logical true** or **false**.

THE EFFICIENT METHOD

In MATLAB, this can be accomplished automatically by simply using the relational operator >.

```
>> isg = vec > 5
isg =
      0   1   0   0   1   1
```

Note that using the relational operator creates a vector consisting of all **logical true** or **false** values. Although the current example is a vector of ones and zeros, and numerical operations can be done on the vector *isg*—its type is **logical** rather than **double**.

```
>> doubres = isg + 5
ans =
      5   6   5   5   6   6
```

```
>> whos
  Name        Size          Bytes  Class

  doubres     1x6              48  double array
  isg         1x6               6  logical array
  vec         1x6              48  double array
```

To determine how many of the elements in the vector *vec* were greater than 5, the **sum** function could be used on the resulting vector *isg*:

```
>> sum(isg)
ans =
      3
```

What we have done is to create a *logical vector isg*. This logical vector can be used to index into the original vector. For example, if only the elements from the vector that are greater than 5 are desired:

```
>> vec(isg)
ans =
      9   6   11
```

This is called *logical indexing*. Only the elements from *vec* for which the corresponding element in the logical vector *isg* is **logical true** are returned.

QUICK QUESTION!

Why doesn't the following work?

```
>> vec([0 1 0 0 1 1])
??? Subscript indices must either be real
positive integers or logicals.
```

Answer: The difference between the vector in this example and *isg* is that *isg* is a vector of logicals (**logical** 1s and 0s), whereas [0 1 0 0 1 1] by default is a vector of **double** values. *Only logical 1s and 0s can be used to index into a vector.*

To preallocate a vector or matrix of all **logical** 1s or 0s, the functions **ones** and **zeros** can be used, and then cast to the type **logical**:

```
>> logical(zeros(2))
ans =
      0    0
      0    0

>> logical(ones(1,5))
ans =
      1    1    1    1    1
```

However, MATLAB also has the functions **true** and **false** that accomplish this, and are faster and manage memory more efficiently than using **logical** with **zeros** or **ones**:

```
>> false(2)
ans =
      0    0
      0    0

>> true(1,5)
ans =
      1    1    1    1    1
```

How can we write a function that will receive a vector and an integer and will return a logical vector, storing **logical true** only for elements of the vector that are greater than the integer?

THE PROGRAMMING CONCEPT

The function receives two input arguments: the vector, and an integer n with which to compare (so it is somewhat more general than using 5). It loops through every element in the input vector, and stores in the result vector either a 1 or 0 depending on whether vec(i) > n is **true** or **false**.

testvecgtn.m

```
function outvec = testvecgtn(vec,n)
% testvecgtn tests whether elements in vector
% are greater than n or not
% Format: testvecgtn(vector, n)

% Preallocate the vector to logical false
outvec = false(size(vec));
for i = 1:length(vec)
    % Each element in the output vector stores 1 or 0
    if vec(i) > n
        outvec(i) = 1;
```

```
      else
            outvec(i) = 0;
      end
end
end
```

THE EFFICIENT METHOD

```
testvecgtnii.m
```

```
function outvec = testvecgtnii(vec,n)
% testvecgtnii tests whether elements in vector
% are greater than n or not with no loop
% Format: testvecgtnii(vector, n)

outvec = vec > n;
```

5.4.1 Logical built-in functions

There are built-in functions in MATLAB that are useful in conjunction with **logical** vectors or matrices; two of these are the functions **any** and **all**. The function **any** returns **logical true** if any element in a vector is nonzero, and **false** if not. The function **all** returns **logical true** only if all elements are nonzero. Here are some examples. For the variable *vec1*, all elements are nonzero, so both **any** and **all** return **true**. (Recall that any nonzero value can be used to represent the concept of **true**, not just **logical** 1.)

```
>> vec1 = [1 3 1 1 2];
>> any(vec1)
ans =
      1

>> all(vec1)
ans =
      1
```

For *vec2*, some but not all elements are nonzero; consequently, **any** returns **true** but **all** returns **false**.

```
>> vec2 = [1 1 0 1]
vec2 =
      1     1     0     1
```

```
>> any(vec2)
ans =
      1

>> all(vec2)
ans =
      0
```

The function **find** returns the indices of a vector that meet given criteria. For example, to find all of the elements in a vector that are greater than 5:

```
>> vec = [5 3 6 7 2]
vec =
      5     3     6     7     2

>> find(vec > 5)
ans =
      3     4
```

As an example of using this, in the previous chapter we solved the following problem using a **while** loop but now we can solve it more efficiently using **find**. Data from an experiment has been recorded in a file called *experd.dat*. The file has some numbers followed by a –99 and then more numbers, all on the same line. The only data values that we are interested in, however, are those before the –99. The algorithm for the script follows:

- Read the data from the file into a vector.
- Create a new vector variable *newvec* that only has the data values up to but not including the –99.
- Plot the new vector values, using black circles.

THE EFFICIENT METHOD

Using the **find** function, we can locate the index of the element that stores the –99. Then, the new vector comprises all of the original vector from the first element to the index *before* the index of the element that stores the –99.

findval.m

```
% Reads data from a file, but only plots the numbers
% up to a flag of -99. Uses find and the colon operator

load experd.dat

where = find(experd == -99);
newvec = experd(1:where-1);

plot(newvec,'ko')
```

For matrices, the **find** function will use linear indexing when returning the indices that meet the specified criteria. For example:

```
>> mata
mata =
        5       6       7       8
        9       7       5       3

>> find(mata == 5)
ans =
        1
        6
```

For both vectors and matrices, an empty vector will be returned if no elements match the criterion.

The function **isequal** is useful in comparing vectors. In MATLAB, using the equality operator with arrays will return 1 or 0 for each element; the **all** function could then be used on the resulting array to determine whether all elements were equal or not. The built-in function **isequal** also accomplishes this:

```
>> vec1 = [1 3 -4 2 99];
>> vec2 = [1 2 -4 3 99];
>> vec1 == vec2
ans =
        1       0       1       0       1

>> all(vec1 == vec2)
ans =
        0

>> isequal(vec1,vec2)
ans =
        0
```

MATLAB also has *or* and *and* operators that work elementwise for matrices:

Operator	Meaning
|	elementwise or for matrices
&	elementwise and for matrices

These operators will compare any two vectors or matrices, as long as they are the same size, element by element and return a vector or matrix of the same size of **logical** 1s and 0s. The operators || and && are only used with scalars, not matrices. For example:

```
>> v1 = [3 0 5 1];
>> v2 = [0 0 2 0];
```

```
>> v1 & v2
ans =
     0     0     1     0

>> v1 | v2
ans =
     1     0     1     1

>> v1 && v2
??? Operands to the || and && operators must be convertible to
logical scalar values.
```

As with the numerical operators, it is important to know the operator precedence rules. Table 5.1 shows the rules for the operators that have been covered so far, in the order of precedence.

Table 5.1 Operator Precedence Rules	
Operators	**Precedence**
parentheses ( )	highest
transpose and power ', ^, .^	
unary: negation (−), not (∼)	
multiplication, division *, /, \, .*, ./, .\	
addition, subtraction +, −	
colon operator :	
relational <, <=, >, >=, ==, ∼=	
elementwise and &	
elementwise or		
and &&	
or ||	
assignment =	lowest

5.5 VECTORIZING CODE

The term *vectorizing code* means essentially rewriting code that has been written inefficiently, perhaps in another language, to make use of the built-in functions and operations in MATLAB. In many cases this means using these built-in functions and operators instead of writing loops and selection statements.

For example, a function *signum* follows:

```
signum.m
```

```
function outmat = signum(mat)
% signum imitates the sign function
% Format: signum(matrix)
```

```
[r c] = size(mat);
for i = 1:r
    for j = 1:c
        if mat(i,j) > 0
            outmat(i,j) = 1;
        elseif mat(i,j) == 0
            outmat(i,j) = 0;
        else
            outmat(i,j) = -1;
        end
    end
end
end
```

To test this function, we will create a matrix and pass it to the function. Here is an example of using this function:

```
>> mat = [0 4 -3; -1 0 2]
mat =
      0     4    -3
     -1     0     2

>> signum(mat)
ans =
      0     1    -1
     -1     0     1
```

Close inspection reveals that the function accomplishes the same task as the built-in **sign** function! Therefore, a function such as *signum* is not necessary in MATLAB.

QUICK QUESTION!

Determine what the following function accomplishes:

xxx.m

```
function logresult = xxx(vec)
% QQ for you — what does this do?

logresult = logical(0);
i = 1;
while i <= length(vec) && logresult == 0
    if vec(i) ~= 0
        logresult = logical(1);
    end
    i = i + 1;
end
end
```

Answer: The output produced by this function is the same as the **any** function.

QUICK QUESTION!

Determine what the following function accomplishes:

yyy.m

```
function logresult = yyy(vec)
% QQ for you - what does this do?

count = 0;
for i = 1:length(vec)
    if vec(i) ~= 0
        count = count + 1;
    end
end

if count == length(vec)
    logresult = logical(1);
else
    logresult = logical(0);
end
end
```

```
>> yyy([1 1 1])
ans =
    1
>> vec1 = 1:5;
>> yyy(vec1)
ans =
    1
>> vec2 = [1 0 1 1];
>> yyy(vec2)
ans =
    0
```

Answer: The output produced by this function is the same as the **all** function.

To be able to vectorize code in MATLAB, there are several important features to keep in mind:

- Scalar and array operations
- Logical vectors
- Built-in functions
- Preallocation of vectors

There are many functions in MATLAB that can be utilized instead of code that uses loops and selection statements. These functions have been demonstrated already but it is worth repeating them to emphasize their utility:

- **sum** and **prod**: Find the sum or product of every element in a vector, or column in a matrix.
- **cumsum** and **cumprod**: Return a vector or matrix of the cumulative (running) sums or products.
- **min** and **max**: Find the minimum value in a vector, or in every column of a matrix.
- **any, all, find**: Work with logical expressions.
- "is" functions such as **isletter** and **isequal**: return **logical** values.

PRACTICE 5.2

Vectorize the following:

```
i = 0;
for inc = 0: 0.5: 3
    i = i + 1;
    myvec(i) = sqrt(inc);
end
-----------------------------------
[r c] = size(mat);
newmat = zeros(r,c);
for i = 1:r
    for j = 1:c
        newmat(i,j) = sign(mat(i,j));
    end
end
```

QUICK QUESTION!

If we have a vector *vec* that erroneously stores negative values, how can we eliminate those negative values?

Answer: One method is to determine where they are and delete these elements:

```
>> vec = [11 -5 33 2 8 -4 25];
>> neg = find(vec < 0)
neg =
     2    6
```

```
>> vec(neg) = []
vec =
     11    33    2    8    25
```

Alternatively, we can just use a logical vector rather than **find**:

```
>> vec = [11 -5 33 2 8 -4 25];
>> vec(vec < 0) = []
vec =
     11    33    2    8    25
```

PRACTICE 5.3

Modify the result seen in the previous Quick Question. Instead of deleting the "bad" elements, retain only the "good" ones. (Hint: Do it two ways, using **find** and using a logical vector with the expression *vec >= 0*.)

There are several other functions that can be useful in vectorizing code, including **diff** and **meshgrid**. The function **diff** returns the differences between consecutive elements in a vector. An example follows:

```
>> diff([4  7  15  32])
ans =
     3    8    17
```

```
>> diff([4 7 2 32])
ans =
     3    -5    30
```

For a vector v with a length of n, the length of **diff(v)** will be $n - 1$.

As an example, a vector that stores a signal can contain both positive and negative values. (For simplicity, we will assume no zeroes, however.) For many applications it is useful to find the zero crossings, or where the signal goes from being positive to negative or vice versa. This can be accomplished with vectorized code, using the functions **sign**, **diff**, and **find**.

```
>> vec = [0.2 -0.1 -0.2 -0.1 0.1 0.3 -0.2];
>> sv = sign(vec)
sv =
     1   -1   -1   -1    1    1   -1

>> dsv = diff(sv)
dsv =
    -2    0    0    2    0   -2

>> find(dsv)
ans =
     1    4    6
```

This shows that the signal crossings are between elements 1 and 2, 4 and 5, and 6 and 7.

The **meshgrid** function receives as input arguments two vectors, and returns as output arguments two matrices that can specify the x and y coordinates of points in images, or can be used to calculate functions on two variables x and y. For example, the x and y coordinates of a 2×3 image would be specified by the coordinates:

```
(1,1)   (2,1)   (3,1)
(1,2)   (2,2)   (3,2)
```

The matrices that separately specify the coordinates are created by the **meshgrid** function:

```
>> [x y] = meshgrid(1:3,1:2)
x =
     1    2    3
     1    2    3
y =
     1    1    1
     2    2    2
```

As an example, let's say we want to evaluate a function f of two variables x and y:

```
f(x,y) = 2*x + y
```

where x ranges from 1 to 4 and y ranges from 1 to 3. We could use nested **for** loops to accomplish this. Using **meshgrid**, however, we can use vectorized code instead of nested loops:

```
>> [x y] = meshgrid(1:4,1:3)
x =
        1       2       3       4
        1       2       3       4
        1       2       3       4
y =
        1       1       1       1
        2       2       2       2
        3       3       3       3

>> f = 2*x + y
f =
        3       5       7       9
        4       6       8      10
        5       7       9      11
```

5.6 TIMING

MATLAB has built-in functions that determine how long it takes code to execute. One set of related functions is **tic/toc**. These functions are placed around code, and will print the time it took for the code to execute. The function **tic** essentially turns a timer on, and then **toc** evaluates the timer and prints the result. The following is a script that illustrates these functions.

fortictoc.m

```
tic
mysum = 0;
for i = 1:20000000
    mysum = mysum + i;
end
toc
```

```
>> fortictoc
Elapsed time is 0.088591 seconds.
```

Here is an example of a script that demonstrates how much preallocating a vector speeds up the code.

tictocprealloc.m

```
% This shows the timing difference between
% preallocating a vector vs. not

clear
disp('No preallocation')
tic
for i = 1:10000
    x(i) = sqrt(i);
end
toc

disp('Preallocation')
tic
y = zeros(1,10000);
for i = 1:10000
    y(i) = sqrt(i);
end
toc
```

```
>> tictocprealloc
No preallocation
Elapsed time is 0.070526 seconds.
Preallocation
Elapsed time is 0.001177 seconds.
```

QUICK QUESTION!

Preallocation can speed up code, but to preallocate it is necessary to know the desired size. What if you do not know the eventual size of a vector (or matrix)? Does that mean that you have to extend it rather than preallocating?

Answer: If you know the maximum size that it could possibly be, you can preallocate to a size that is larger than necessary, then delete the "unused" elements. To do that, you would

have to count the number of elements that are actually used. For example, if you have a vector *vec* that has been preallocated, and a variable *count* that stores the number of elements that were actually used, this will trim the unnecessary elements:

```
vec = vec(1:count)
```

SUMMARY

Common Pitfalls

- Attempting to use an array of **double** 1s and 0s to index into an array (must be **logical**, instead).
- Forgetting that for array operations based on multiplication, the dot must be used in the operator. In other words, for multiplying, dividing by, dividing into, or raising to an exponent term by term, the operators are .*, ./, .\, and .^.
- Attempting to use || or && with arrays. Always use | and & when working with arrays; || and && are only used with scalars.

Programming Style Guidelines

- Vectorize code whenever possible. If it is not necessary to use loops in MATLAB, don't!
- Use the array operators .*, ./, .\, and .^ in functions so that the input arguments can be arrays and not just scalars.
- Use **true** instead of **logical(1)** and **false** instead of **logical(0)**, especially when creating vectors or matrices.

MATLAB Functions and Commands		
any	isequal	tic
all	diff	toc
find	meshgrid	

MATLAB Operators
array operators .^, .*, ./, .\
elementwise or for matrices |
elementwise and for matrices &

Exercises

1. The following code was written by somebody who does not know how to use MATLAB efficiently. Rewrite this as a single statement that will accomplish exactly the same thing for a matrix variable *mat* (e.g., vectorize this code):

```
[r c] = size(mat);
for i = 1:r
    for j = 1:c
        mat(i,j) = mat(i,j) * 2;
    end
end
```

2. Vectorize this code! Write *one* assignment statement that will accomplish exactly the same thing as the given code (assume that the variable *vec* has been initialized):

```
result = 0;
for i = 1:length(vec)
    result = result + vec(i);
end
```

3. Vectorize this code! Write *one* assignment statement that will accomplish exactly the same thing as the given code (assume that the variable *vec* has been initialized):

```
newv = zeros(size(vec));
myprod = 1;
for i = 1:length(vec)
    myprod = myprod * vec(i);
    newv(i) = myprod;
end
newv % Note: this is just to display the value
```

4. Create a *1 × 6* vector of random integers, each in the range from 1 to 20. Use built-in functions to find the minimum and maximum values in the vector. Also, create a vector of cumulative sums using **cumsum**.

5. Write a relational expression for a vector variable that will verify that the last value in a vector created by **cumsum** is the same as the result returned by **sum**.

6. Create a vector of five random integers, each in the range from −10 to 10. Perform each of the following using only vectorized code:
   - Subtract 3 from each element
   - Count how many are positive
   - Get the absolute value of each element
   - Find the maximum.

7. Create a *3 × 5* matrix. Perform each of the following using only vectorized code:
   - Find the maximum value in each column
   - Find the maximum value in each row
   - Find the maximum value in the entire matrix

8. Write a function called **geomser** that will receive values of r and n, and will calculate and return the sum of the geometric series:

   $$1 + r + r^2 + r^3 + r^4 + \ldots + r^n$$

   The following examples of calls to this function illustrate what the result should be:

   ```
   >> geomser(1,5)
   ans =

            6

   >> disp(geomser(2,4))
            31
   ```

9. Generate a random integer n, create a vector of the integers one through n in steps of 2, square them, and plot the squares.

10. A vector v stores for several employees of the Green Fuel Cells Corporation their hours worked for one week followed for each by the hourly pay rate. For example, if the variable stores

    ```
    >> v
    v =
       10.5000   40.0000   18.0000   20.0000    7.5000
    ```

    that means the first employee worked 33 hours at $10.50 per hour, the second worked 40 hours at $18 an hour, and so on. Write code that will separate this into two vectors, one that stores the hours worked and another that stores the hourly rates. Then, use the array multiplication operator to create a vector, storing in the new vector the total pay for every employee.

11. Write a function **repvec** that receives a vector and the number of times each element is to be duplicated. The function should then return the resulting vector. Do this problem using built-in functions only. Here are some examples of calling the function:

```
>> repvec(5:-1:1,2)
ans =
     5     5     4     4     3     3     2     2     1     1

>> repvec([ 0 1 0] ,3)
ans =
     0     0     0     1     1     1     0     0     0
```

12. The mathematician Euler proved the following:

$$\frac{\pi^2}{6} = 1 + \frac{1}{4} + \frac{1}{9} + \frac{1}{16} + \cdots$$

Rather than finding a mathematical proof for this, try to verify whether the conjecture seems to be true or not. (*Note:* Two basic ways to approach this are choosing a number of terms to add or loop until the sum is close to $\pi^2/6$.)

When working with images, it is often necessary to "crop" one image to match the size of another. Images are represented very simply in MATLAB as matrices of numbers. However, these matrices are quite large. Depending on the resolution, the number of rows and columns could easily be in the thousands. It is therefore extremely important when working with image matrices to vectorize the code.

13. Write a script that will read from a file *oldfile.dat* into a matrix. It will create a square matrix (same number of rows and columns) by deleting rows or columns as necessary, and then write this new square matrix to a new file called *squarefile.dat*. For example, if the original matrix is *4 × 6*, the new matrix would be created by deleting the fifth and sixth columns to result in a *4 × 4* matrix. Another example: If the original matrix is *3 × 2*, the third row would be deleted to result in a *2 × 2* matrix. The script should be general and work regardless of the size of the original file and should not use any loops or **if** statements. Create the data file first.

14. A file called *hightemp.dat* was created some time ago that stores, on every line, a year followed by the high temperature at a specific site for each month of that year. For example, the file might look like this:

```
89     42     49     55     72     63     68     77     82     76     67
90     45     50     56     59     62     68     75     77     75     66
91     44     43     60     60     60     65     69     74     70     70
etc.
```

As can be seen, only two digits were used for the year (which was common in the last century). Write a script that will read this file into a matrix, create a new matrix that stores the years correctly as 19xx, and then write this to a new file called *y2ktemp. dat.* (*Hint:* Add 1900 to the entire first column of the matrix.) Such a file, for example, would look like this:

```
1989  42    49    55    72    63    68    77    82    76    67
1990  45    50    56    59    62    68    75    77    75    66
1991  44    43    60    60    60    65    69    74    70    70
etc.
```

15. Write a script that will prompt the user for a quiz grade and error-check until the user enters a valid quiz grade. The script will then echo print the grade. For this case, valid grades are in the range from 0 to 10 in steps of 0.5. Do this by creating a vector of valid grades and then use **any** or **all** in the condition in the **while** loop.

16. Which is faster—using **false** or using **logical**(0) to preallocate a matrix to all **logical** zeros? Write a script to test this.

17. Which is faster—using a **switch** statement or using a nested **if-else**? Write a script to test this.

18. Vectorize the following code:

```
n = 3;
x = zeros(n);
y = x;
for i = 1:n
    x(:,i) = i;
    y(i,:) = i;
end
```

19. A company is calibrating some measuring instrumentation and has measured the radius and height of one cylinder 10 separate times; the measurements are in vector variables r and h. Use vectorized code to find the volume from each trial, which is given by $\pi r^2 h$. Also, use logical indexing first to make sure that all measurements were valid (> 0).

MATLAB Programs

KEY TERMS

functions that return more than one value

functions that do not return any values

side effects

call-by-value

modular programs

main program

primary function

subfunction

menu-driven program

variable scope

base workspace

local variable

global variable

persistent variable

bug

debugging

syntax errors

runtime errors

logical errors

tracing

breakpoints

function stubs

Chapter 2 introduced scripts and user-defined functions. In that chapter, we saw how to write script files, which are sequences of statements that are stored in an M-file and then executed. We also saw how to write user-defined functions, also stored in M-files, that calculate and return a single value. In this chapter, we will expand on these concepts, and introduce other kinds of user-defined functions. We will show how MATLAB® programs consist of combinations of scripts and user-defined functions. The mechanisms for interactions of variables in M-files and the Command Window will be explored. Finally, techniques for finding and fixing mistakes in programs will be reviewed.

6.1 MORE TYPES OF USER-DEFINED FUNCTIONS

We have already seen how to write a user-defined function, stored in an M-file, that calculates and returns one value. This is just one type of function. It is also possible for a function to return multiple values, and it is possible for a function to return nothing. We will categorize functions as follows:

- Functions that calculate and return one value
- Functions that calculate and return more than one value
- Functions that just accomplish a task, such as printing, without returning any values

Thus, although many functions calculate and return values, some do not. Some functions instead just accomplish a task. Categorizing the functions as such is somewhat arbitrary, but there are differences between these three types of functions, including the format of the function headers and also the way in which the functions are called. Regardless of what kind of function it is, all functions must be defined, and all function definitions consist of the header and the body. Also, the function must be called for it to be utilized.

In general, any function in MATLAB consists of the following:

- The function *header* (the first line), which contains
  - The reserved word **function**. (If the function *returns* values, the name(s) of the output argument(s), followed by the assignment operator =.)
  - The name of the function. (*IMPORTANT:* This should be the same as the name of the M-file in which this function is stored to avoid confusion.)
- The input arguments in parentheses, if there are any (separated by commas if there is more than one).
- A comment that describes what the function does (this is printed if **help** is used)
- The *body* of the function, which includes all statements, including putting values in all output arguments, if there are any
- **end** at the end of the function

6.1.1 Functions that return more than one value

Functions that return one value have one output argument, as can be seen in Section 2.7. Functions that return more than one value must instead have more than one output argument in the function header in square brackets. This means that in the body of the function, values must be put in all output arguments listed in the function header. The general form of a function definition for a function that calculates and *returns more than one value* looks like the following:

```
functionname.m
```
```
function [ output arguments] = functionname (input arguments)
% Comment describing the function

Statements here; these must include putting values in all of the
output arguments listed in the header

end
```

In the vector of output arguments, the output argument names are by convention separated by commas. In more recent versions of MATLAB, choosing File, New, then Function (rather than File, New, then Script) brings up a template in the Editor that can then be filled in:

```
function [ output_args ] = Untitled2 (input_args )
%UNTITLED2 Summary of this function goes here
% Detailed explanation goes here

end
```

For example, here is a function that calculates two values, both the area and the circumference of a circle; this is stored in a file called *areacirc.m*:

```
areacirc.m
function [area, circum] = areacirc(rad)
% areacirc returns the area and
% the circumference of a circle
% Format: areacirc(radius)

area = pi * rad .* rad;
circum = 2 * pi * rad;
end
```

Since this function is calculating two values, there are two output arguments in the function header (*area* and *circum*) that are placed in square brackets [ ] . Therefore, somewhere in the body of the function, values have to be put in both. Since the function is returning two values, it is important to capture and store these values in separate variables when the function is called. In this case, the first value returned, the area of the circle, is stored in a variable *a* and the second value returned is stored in a variable *c*.

```
>> [a c] = areacirc(4)
a =
    50.2655
c =
    25.1327
```

If this is not done, only the first value returned is retained—in this case, the area:

```
>> disp(areacirc(4))
    50.2655
```

Note that in capturing the values the order matters. In this case, the function returns first the area and then the circumference of the circle. The order in which values are assigned to the output arguments within the function, however, does not matter.

What would happen if a vector of radii was passed to the function?

Answer: Since the .* operator is used in the function to square *rad*, a vector can be passed to the input argument *rad*. Therefore, the results will also be vectors, so the variables on the left side of the assignment operator would become vectors of areas and circumferences.

```
>> [a c] = areacirc (1:4)
a =
    3.1416   12.5664   28.2743   50.2655
c =
    6.2832   12.5664   18.8496   25.1327
```

The **help** function shows the comment listed under the function header:

```
>> help areacirc
   areacirc calculates the area and
   the circumference of a circle
   Format: areacirc(radius)
```

The *areacirc* function could be called from the Command Window as shown here, or from a script. Here is a script that will prompt the user for the radius of just one circle, call the *areacirc* function to calculate and return the area and circumference of the circle, and print the results:

calcareacirc.m

```
% This script prompts the user for the radius of a circle,
% calls a function to calculate and return both the area
% and the circumference, and prints the results
% It ignores units and error-checking for simplicity

radius = input ('Please enter the radius of the circle: ');
[area circ] = areacirc(radius);
fprintf('For a circle with a radius of %.1f,\n', radius)
fprintf('the area is %.1f and the circumference is %.1f\n',...
    area, circ)
```

```
>> calcareacirc
Please enter the radius of the circle: 5.2
For a circle with a radius of 5.2,
the area is 84.9 and the circumference is 32.7
```

PRACTICE 6.1

Write a function *perimarea* that calculates and returns the perimeter and area of a rectangle. Pass the length and width of the rectangle as input arguments. For example, this function might be called from the following script:

calcareaperim.m

```
% Prompt the user for the length and width of a rectangle,
% call a function to calculate and return the perimeter
```

```
% and area, and print the result
% For simplicity it ignores units and error-checking

length = input('Please enter the length of the rectangle: ');
width = input('Please enter the width of the rectangle: ');
[perim area] = perimarea(length, width);
fprintf('For a rectangle with a length of %.1f and a', length)
fprintf(' width of %.1f,\nthe perimeter is %.1f,', width, perim)
fprintf(' and the area is %.1f\n', area)
```

As another example, consider a function that calculates and returns three output arguments. The function will receive one input argument representing a total number of seconds, and returns the number of hours, minutes, and remaining seconds that it represents. For example, 7515 total seconds is 2 hours, 5 minutes, and 15 seconds because $7515 = 3600 * 2 + 60 * 5 + 15$.

The algorithm follows:

- Divide the total seconds by 3600, which is the number of seconds in an hour. For example, 7515/3600 is 2.0875. The integer part is the number of hours (e.g., 2).
- The remainder of the total seconds divided by 3600 is the remaining number of seconds; it is useful to store this in a local variable.
- The number of minutes is the remaining number of seconds divided by 60 (again, the integer part).
- The number of seconds is the remainder of the division.

breaktime.m

```
function [ hours, minutes, secs] = breaktime(totseconds)
% breaktime breaks a total number of seconds into
% hours, minutes, and remaining seconds
% Format: breaktime(totalSeconds)

hours = floor(totseconds/3600);
remsecs = rem(totseconds, 3600);
minutes = floor(remsecs/60);
secs = rem(remsecs, 60);
end
```

An example of calling this function is

```
>> [h m s] = breaktime(7515)
h =
     2
m =
     5
s =
    15
```

As before, it is important to store all values that the function returns in separate variables.

6.1.2 Functions that accomplish a task without returning values

Many functions do not calculate values, but rather accomplish a task such as printing formatted output. Since these functions do not return any values, there are no output arguments in the function header.

The general form of a function definition for a *function that does not return any values* looks like this:

```
functionname.m
function functionname(input arguments)
% Comment describing the function

Statements here
end
```

Note what is missing in the function header: there are no output arguments, and no assignment operator.

For example, the following function just prints the number arguments passed to it in a sentence format:

```
printem.m
function printem(a,b)
% printem prints two numbers in a sentence format
% Format: printem(num1, num2)

fprintf('The first number is %.1f and the second is %.1f\n',a,b)
end
```

Since this function performs no calculations, there are no output arguments in the function header and no equal symbol (=). An example of a call to the *printem* function is

```
>> printem(3.3, 2)
The first number is 3.3 and the second is 2.0
```

Note that since the function does not return a value, it cannot be called from an assignment statement. Any attempt to do this would result in an error, such as the following:

```
>> x = printem(3, 5) % Error!!
??? Error using ==> printem
Too many output arguments.
```

We can therefore think of the call to a function that does not return values as a statement by itself, in that the function call cannot be imbedded in another statement such as an assignment statement or an output statement.

The tasks that are accomplished by functions that do not return any values (e.g., output from an **fprintf** statement or a **plot**) are sometimes referred to as *side effects*. Some standards for commenting functions include putting the side effects in the block comment.

PRACTICE 6.2

Write a function that receives a vector as an input argument and prints the elements from the vector in a sentence format.

```
>> printvecelems([5.9 33 11])
Element 1 is 5.9
Element 2 is 33.0
Element 3 is 11.0
```

6.1.3 Functions that return values versus printing

A function that calculates and *returns* values (through the output arguments) does not normally also print them; that is left to the calling script or function. It is good programming practice to separate these tasks.

If a function just prints a value, rather than returning it, the value cannot be used later in other calculations. For example, here is a function that just prints the circumference of a circle:

calccircum1.m
```
function calccircum1(radius)
% calccircum1 displays the circumference of a circle
% but does not return the value
% Format: calccircum1(radius)

disp(2 * pi * radius)
end
```

Calling this function prints the circumference, but there is no way to store the value so that it can be used in subsequent calculations:

```
>> calccircum1(3.3)
   20.7345
```

Since no value is returned by the function, attempting to store the value in a variable would be an error:

```
>> c = calccircum1(3.3)
??? Error using ==> calccircum1
Too many output arguments.
```

By contrast, the following function calculates and returns the circumference, so that it can be stored and used in other calculations. For example, if the circle is the base of a cylinder, and we wish to calculate the surface area of the cylinder, we would need to multiply the result from the *calccircum2* function by the height of the cylinder.

calccircum2.m

```
function circle_circum = calccircum2(radius)
% calccircum2 calculates and returns the
% circumference of a circle
% Format: calccircum2(radius)

circle_circum = 2 * pi * radius;
end
```

```
>> circumference = calccircum2(3.3)
circumference =
    20.7345
```

```
>> height = 4;
>> surf_area = circumference * height
surf_area =
    82.9380
```

6.1.4 Passing arguments to functions

In all function examples presented thus far, at least one argument was passed in the function call to be the value(s) of the corresponding input argument(s) in the function header. The *call-by-value* method is the term for this method of passing the values of the arguments to the input arguments in the functions.

In some cases, however, it is not necessary to pass any arguments to the function. Consider, for example, a function that simply prints a random real number with two decimal places:

printrand.m

```
function printrand()
% printrand prints one random number
% Format: printrand or printrand()

fprintf('The random # is %.2f\n', rand)
end
```

Here is an example of calling this function:

```
>> printrand()
The random # is 0.94
```

Since nothing is passed to the function, there are no arguments in the parentheses in the function call, and none in the function header, either. In fact, the parentheses are not even needed in either the function or the function call. The following works as well:

printrandnp.m

```
function printrandnp
% printrandnp prints one random number
% Format: printrandnp or printrandnp()

fprintf('The random # is %.2f\n',rand)
end
```

```
>> printrandnp
The random # is 0.52
```

In fact, the function can be called with or without empty parentheses in the function header. This was an example of a function that did not receive any input arguments nor did it return any output arguments; it simply accomplished a task.

The following is another example of a function that does not receive any input arguments, but in this case it does return a value. The function prompts the user for a string and returns the value entered.

stringprompt.m

```
function outstr = stringprompt
% stringprompt prompts for a string and returns it
% Format stringprompt or stringprompt()

disp('When prompted, enter a string of any length.')
outstr = input('Enter the string here: ', 's');
end
```

```
>> mystring = stringprompt
When prompted, enter a string of any length.
Enter the string here: Hi there

mystring =
Hi there
```

PRACTICE 6.3

Write a function that will prompt the user for a positive number, loop to error-check to make sure that the number is positive, and return the positive number.

QUICK QUESTION!

It is important that the number of arguments in the call to a function must be the same as the number of input arguments in the function header, even if that number is zero. Also, if a function returns more than one value, it is important to "capture" all values by having an equivalent number of variables in a vector on the left side of an assignment statement. Although it is not an error if there aren't enough variables, some of the values returned will be lost. The following question is posed to highlight this.

Given the following function header (note that this is just the function header, not the entire function definition):

```
function[ outa, outb] = qq1(x, y, z)
```

which of the following proposed calls to this function would be valid?

a) [var1 var2] = qq1(a, b, c);

b) answer = qq1(3, y, q);

c) [a b] = myfun(x, y, z);

d) [outa outb] = qq1(x, z);

Answer: The first proposed function call, (a), is valid. There are three arguments that are passed to the three input arguments in the function header, the name of the function is *qq1*, and there are two variables in the assignment statement to store the two values returned from the function. Function call (b) is valid, although only the first value returned from the function would be stored in *answer*; the second value would be lost. Function call (c) is invalid because the name of the function is given incorrectly. Function call (d) is invalid because only two arguments are passed to the function, but there are three input arguments in the function header.

6.2 MATLAB PROGRAM ORGANIZATION

A MATLAB program typically consists of a script that calls functions to do the actual work.

6.2.1 Modular programs

A *modular program* is a program in which the solution is broken down into modules, and each is implemented as a function. The script that calls these functions is typically called the *main program*.

To demonstrate the concept, we will use the very simple example of calculating the area of a circle. In Section 6.3 a much longer and more realistic example will be given. For this example, there are three steps in the algorithm to calculate the area of a circle:

- Get the input (the radius)
- Calculate the area
- Display the results

In a modular program, there would be one main script that calls three separate functions to accomplish these tasks:

- A function to prompt the user and read in the radius
- A function to calculate and return the area of the circle
- A function to display the results

Since scripts and functions are stored in M-files, there would therefore be four separate M-files altogether for this program; one M-file script, and three M-file functions, as follows:

calcandprintarea.m

```
% This is the main script to calculate the
% area of a circle
% It calls 3 functions to accomplish this
radius = readradius;
area = calcarea(radius);
printarea(radius,area)
```

readradius.m

```
function radius = readradius
% readradius prompts the user and reads the radius
% Format: readradius or readradius()

disp('When prompted, please enter the radius in inches.')
radius = input('Enter the radius: ');
end
```

calcarea.m

```
function area = calcarea(rad)
% calcarea returns the area of a circle
% Format: calcarea(radius)

area = pi * rad .* rad;
end
```

printarea.m

```
function printarea(rad,area)
% printarea prints the radius and area
% Format: printarea(radius, area)

fprintf('For a circle with a radius of %.2f inches,\n',rad)
fprintf('the area is %.2f inches squared.\n',area)
end
```

When the program is executed, the following steps will take place:

- The script *calcandprintarea* begins executing.
- *calcandprintarea* calls the *readradius* function.
    - *readradius* executes and returns the radius.
- *calcandprintarea* resumes executing and calls the *calcarea* function, passing the radius to it.
    - *calcarea* executes and returns the area.

- *calcandprintarea* resumes executing and calls the *printarea* function, passing both the radius and the area to it.
  - *printarea* executes and prints.
- The script finishes executing.

Running the program would be accomplished by typing the name of the script; this would call the other functions:

```
>> calcandprintarea
When prompted, please enter the radius in inches.
Enter the radius: 5.3
For a circle with a radius of 5.30 inches,
the area is 88.25 inches squared.
```

Note how the function calls and the function headers match up. For example:

readradius function:

```
function call: radius = readradius;
function header: function radius = readradius
```

In the function call, no arguments are passed so there are no input arguments in the function header. The function returns one output argument so that is stored in one variable.

calcarea function:

```
function call: area = calcarea(radius);
function header: function area = calcarea(rad)
```

In the function call, one argument is passed in parentheses so there is one input argument in the function header. The function returns one output argument so that is stored in one variable.

printarea function:

```
function call: printarea(radius,area)
function header: function printarea(rad,area)
```

In the function call, there are two arguments passed, so there are two input arguments in the function header. The function does not return anything, so the call to the function is a statement by itself; it is not in an assignment or output statement.

Of course, the *readradius* function should error-check the user's input.

PRACTICE 6.4

Modify the *readradius* function to error-check the user's input to make sure that the radius is valid. The function should ensure that the radius is a positive number by looping to print an error message until the user enters a valid radius.

6.2.2 Subfunctions

Thus far, every function has been stored in a separate M-file. However, it is possible to have more than one function in a given M-file. For example, if one function calls another, the first (calling) function would be the *primary function*, and the function that is called is a *subfunction*. These functions would both be stored in the same M-file, first the primary function and then the subfunction. The name of the M-file would be the same as the name of the primary function, to avoid confusion.

To demonstrate this, a program that is similar to the previous one, but calculates and prints the area of a rectangle, is shown here. The script, or main program, first calls a function that reads the length and width of the rectangle, and then calls a function to print the results. This function calls a subfunction to calculate the area.

rectarea.m

```
% This program calculates & prints the area of a rectangle

% Call a fn to prompt the user & read the length and width
[length, width] = readlenwid;
% Call a fn to calculate and print the area
printrectarea(length, width)
```

readlenwid.m

```
function [ l, w] = readlenwid
% readlenwid reads & returns the length and width
% Format: readlenwid or readlenwid()

l = input ('Please enter the length: ');
w = input ('Please enter the width: ');
end
```

printrectarea.m

```
function printrectarea(len, wid)
% printrectarea prints the rectangle area
% Format: printrectarea(length, width)

% It calls a subfunction to calculate the area
area = calcrectarea(len,wid);
fprintf('For a rectangle with a length of %.2f\n',len)
fprintf('and a width of %.2f, the area is %.2f\n', ...
    wid, area);
end

function area = calcrectarea(len, wid)
% calcrectarea returns the rectangle area
% Format: calcrectarea(length, width)
area = len * wid;
end
```

An example of running this program follows:

```
>> rectarea
Please enter the length: 6
Please enter the width: 3
For a rectangle with a length of 6.00
and a width of 3.00, the area is 18.00
```

Note how the function calls and function headers match up. For example:

readlenwid function:

```
function call:[ length, width] = readlenwid;
function header: function[ l,w] = readlenwid
```

In the function call, no arguments are passed so there are no input arguments in the function header. The function returns two output arguments so there is a vector with two variables on the left side of the assignment statement in which the function is called.

printrectarea function:

```
function call: printrectarea(length, width)
function header: function printrectarea(len, wid)
```

In the function call, there are two arguments passed, so there are two input arguments in the function header. The function does not return anything, so the call to the function is a statement by itself; it is not in an assignment or output statement.

calcrectarea subfunction:

```
function call: area = calcrectarea(len,wid);
function header: function area = calcrectarea(len, wid)
```

In the function call, two arguments are passed in parentheses so there are two input arguments in the function header. The function returns one output argument so that is stored in one variable.

The **help** command can be used with the script *rectarea*, the function *readlenwid*, and with the primary function, *printrectarea*. To view the first comment in the subfunction, since it is contained within the *printrectarea.m* file, the operator > is used to specify both the primary and subfunctions:

```
>> help rectarea
  This program calculates & prints the area of a rectangle

>> help printrectarea
  printrectarea prints the rectangle area
  Format: printrectarea(length, width)
```

```
>> help printrectarea>calcrectarea
  calcrectarea returns the rectangle area
  Format: calcrectarea(length, width)
```

PRACTICE 6.5

For a right triangle with sides a, b, and c, where c is the hypotenuse and θ is the angle between sides a and c, the lengths of sides a and b are given by:

$$a = c * \cos(\theta)$$
$$b = c * \sin(\theta)$$

Write a script *righttri* that calls a function to prompt the user and read in values for the hypotenuse and the angle (in radians), and then calls a function to calculate and return the lengths of sides a and b, and a function to print out all values in a sentence format. For simplicity, ignore units. Here is an example of running the script; the output format should be exactly as shown here:

```
>> righttri
Enter the hypotenuse: 5
Enter the angle: .7854
For a right triangle with hypotenuse 5.0
  and an angle 0.79 between side a & the hypotenuse,
  side a is 3.54 and side b is 3.54
```

For extra practice, do this using two different program organizations:

- One script that calls three separate functions
- One script that calls two functions; the function that calculates the lengths of the sides will be a subfunction to the function that prints

6.3 APPLICATION: MENU-DRIVEN MODULAR PROGRAM

Many longer, more involved programs that have interaction with the user are *menu-driven*, which means that the program prints a menu of choices and then continues to loop to print the menu of choices until the user chooses to end the program. A modular menu-driven program would typically have a function that presents the menu and gets the user's choice, as well as functions to implement the action for each choice. These functions may have subfunctions. Also, the functions would error-check all user input.

As an example of such a menu-driven program, we will write a program to explore the constant e.

The constant e, called the natural exponential base, is used extensively in mathematics and engineering. There are many diverse applications of this constant. The value of the constant e is approximately 2.1718. . . . Raising e to the power of x, or e^x, is so common that this is called the exponential function. In MATLAB, as we have seen, there is a function for this, **exp**.

One way to determine the value of e is by finding a limit:

$$e = \lim_{n \to \infty} \left(1 + \frac{1}{n}\right)^n$$

As the value of n increases toward infinity, the result of this expression approaches the value of e.

An approximation for the exponential function can be found using what is called a Maclaurin series:

$$e^x \approx 1 + \frac{x^1}{1!} + \frac{x^2}{2!} + \frac{x^3}{3!} + \cdots$$

We will write a progam to investigate the value of e and the exponential function. It will be menu-driven. The menu options will be:

- Print an explanation of e.
- Prompt the user for a value of n, and then find an approximate value for e using the expression $(1 + 1/n)^n$.
- Prompt the user for a value for x. Print the value of **exp(x)** using the built-in function. Find an approximate value for e^x using the Maclaurin series just given.
- Exit the program.

The algorithm for the script main program follows:

- Call a function *eoption* to display the menu and return the user's choice.
- Loop until the user chooses to exit the program. If the user has not chosen to exit, the action of the loop is to:
    - Depending on the user's choice, do one of the following:
        - Call a function *explaine* to print an explanation of e.
        - Call a function *limite* that will prompt the user for n and calculate an approximate value for e.
        - Prompt the user for x and call a function *expfn* that will print both an approximate value for e^x and the value of the built-in **exp(x)**. Note that because any value for x is acceptable, the program does not need to error-check this value.
    - Call the function *eoption* to display the menu and return the user's choice again.

The algorithm for the *eoption* function follows:

- Use the **menu** function to display the four choices.
- Error-check (an error would occur if the user clicks on the "X" on the menu box rather than pushing one of the four buttons) by looping to display the menu until the user pushes one of the buttons.
- Return the integer value corresponding to the button push.

The algorithm for the *explaine* function is:

- Print an explanation of *e*, the **exp** function, and how to find approximate values.

The algorithm for the *limite* function is:

- Call a subfunction *askforn* to prompt the user for an integer *n*.
- Calculate and print the approximate value of *e* using *n*.

The algorithm for the subfunction *askforn* is:

- Prompt the user for a positive integer for *n*.
- Loop to print an error message and reprompt until the user enters a positive integer.
- Return the positive integer *n*.

The algorithm for the *expfn* function is:

- Receive the value of *x* as an input argument.
- Print the value of **exp(x)**.
- Assign an arbitrary value for the number of terms *n* (an alternative method would be to prompt the user for this).
- Call a subfunction *appex* to find an approximate value of **exp(x)** using a series with *n* terms.
- Print this approximate value.

The algorithm for the *appex* subfunction is:

- Receive *x* and *n* as input arguments.
- Initialize a variable for the running sum of the terms in the series (to 1 for the first term) and for a running product that will be the factorials in the denominators.
- Loops to add the *n* terms to the running sum.
- Returns the resulting sum.

The entire program consists of the following M-file script and four M-file functions:

```
eapplication.m
% This script explores e and the exponential function

% Call a function to display a menu and get a choice
choice = eoption;

% Choice 4 is to exit the program
while choice ~= 4
    switch choice
        case 1
            % Explain e
            explaine;
```

Continued

```
        case 2
            % Approximate e using a limit
            limite;
        case 3
            % Approximate exp(x) and compare to exp
            x = input('Please enter a value for x: ');
            expfn(x);
    end
    % Display menu again and get user's choice
    choice = eoption;
end
```

eoption.m

```
function choice = eoption
% eoption prints the menu of options and error-checks
% until the user pushes one of the buttons
% Format: eoption or eoption()

choice = menu('Choose an e option', 'Explanation', ...
    'Limit', 'Exponential function', 'Exit Program');
% If the user closes the menu box rather than
% pushing one of the buttons, choice will be 0
while choice == 0
    disp('Error - please choose one of the options.')
    choice = menu('Choose an e option', 'Explanation', ...
        'Limit', 'Exponential function', 'Exit Program');
end
end
```

explaine.m

```
function explaine
% explaine explains a little bit about e
% Format: explaine or explaine()

fprintf('The constant e is called the natural')
fprintf(' exponential base.\n')
fprintf('It is used extensively in mathematics and')
fprintf(' engineering.\n')
fprintf('The value of the constant e is ~ 2.1718\n')
fprintf('Raising e to the power of x is so common that\n')
fprintf('this is called the exponential function.\n')
fprintf('An approximation for e is found using a limit.\n')
fprintf('An approximation for the exponential function\n')
fprintf('can be found using a series.\n')
end
```

limite.m

```
function limite
% limite returns an approximate of e using a limit
% Format: limite or limite()
```

```
% Call a subfunction to prompt user for n
n = askforn;
fprintf('An approximation of e with n = %d is %.2f\n', ...
  n, (1 + 1/n) ^ n)
end

function outn = askforn
% askforn prompts the user for n
% Format: askforn or askforn()
% It error-checks to make sure n is a positive integer

inputnum = input('Enter a positive integer for n: ');
num2 = int32(inputnum);
while num2 ~= inputnum || num2 < 0
  inputnum = input('Invalid! Enter a positive integer: ');
  num2 = int32(inputnum);
end
outn = inputnum;
end
```

expfn.m

```
function expfn(x)
% expfn compares the built-in function exp(x)
% and a series approximation and prints
% Format: expfn(x)

fprintf('Value of built-in exp(x) is %.2f\n',exp(x))

% n is arbitrary number of terms
n = 10;
fprintf('Approximate exp(x) is %.2f\n', appex(x,n))
end

function outval = appex(x,n)
% appex approximates e to the x power using terms up to
%   x to the nth power
% Format: appex(x,n)

% Initialize the running sum in the output argument
% outval to 1 (for the first term)
outval = 1;
% runprod is the factorial in the denominator
runprod = 1;

for i = 1:n
    runprod = runprod * i;
    outval = outval + (x^i)/runprod;
end
end
```

FIGURE 6.1 Menu Figure Window for *eapplication* program

Running the script will bring up the menu seen in Figure 6.1.

Then, what happens will depend on which button(s) the user pushes. Every time the user pushes a button, the appropriate function will be called and then this menu will appear again. This will continue until the user pushes the button Exit Program. Examples will be given of running the script, with different sequences of button pushes.

In the following example, the user:

■ Closed the menu window that caused the error message and brought up a new menu
■ Chose Explanation
■ Chose Exit Program

```
>> eapplication
Error - please choose one of the options.
The constant e is called the natural exponential base.
It is used extensively in mathematics and engineering.
The value of the constant e is ~ 2.1718
Raising e to the power of x is so common that
this is called the exponential function.
An approximation for e is found using a limit.
An approximation for the exponential function
can be found using a series.
```

In the following example, the user

■ Chose Limit
  ■ When prompted for *n*, entered two invalid values before finally entering a valid positive integer.
■ Chose Exit Program

```
>> eapplication
Enter a positive integer for n: −4
Invalid! Enter a positive integer: 5.5
Invalid! Enter a positive integer: 10
An approximation of e with n = 10 is 2.59
```

To see the difference in the approximate value for *e* as *n* increases, the user kept choosing Limit and entering larger and larger values each time in the following example:

```
>> eapplication
Enter a positive integer for n: 4
An approximation of e with n = 4 is 2.44
Enter a positive integer for n: 10
An approximation of e with n = 10 is 2.59
Enter a positive integer for n: 30
An approximation of e with n = 30 is 2.67
```

```
Enter a positive integer for n: 100
An approximation of e with n = 100 is 2.70
```

In the following example, the user:

- Chose Exponential function
  - When prompted, entered 4.6 for *x*
- Chose Exponential function again
  - When prompted, entered –2.3 for *x*
- Chose Exit Program

```
>> eapplication
Please enter a value for x: 4.6
Value of built-in exp(x) is 99.48
Approximate exp(x) is 98.71
Please enter a value for x: −2.3
Value of built-in exp(x) is 0.10
Approximate exp(x) is 0.10
```

6.4 VARIABLE SCOPE

The *scope* of any variable is the workspace in which it is valid. The workspace created in the Command Window is called the *base workspace*.

As we have seen before if a variable is defined in any function it is a *local variable* to that function, which means that it is only known and used within that function. Local variables only exist while the function is executing; they cease to exist when the function stops executing. For example, in the following function that calculates the sum of the elements in a vector, there is a local loop variable *i*.

```
mysum.m
```
```
function runsum = mysum(vec)
% mysum returns the sum of a vector
% Format: mysum(vector)

runsum = 0;
for i=1:length(vec)
    runsum = runsum + vec(i);
end
end
```

Running this function does not add any variables to the base workspace, as demonstrated in the following:

```
>> clear
>> who
>> disp(mysum([5 9 1]))
    15
>> who
>>
```

In addition, variables that are defined in the Command Window cannot be used in a function.

However, scripts (as opposed to functions) *do* interact with the variables that are defined in the Command Window. For example, the function is changed to be a script *mysumscript*.

mysumscript.m

```
% This script sums a vector
vec = 1:5;
runsum = 0;
for i=1:length(vec)
    runsum = runsum + vec(i);
end
disp(runsum)
```

The variables defined in the script do become part of the base workspace:

```
>> clear
>> who
>> mysumscript
    15
>> who
Your variables are:
i runsum vec
```

Variables that are defined in the Command Window can be used in a script, but cannot be used in a function. For example, the vector *vec* could be defined in the Command Window (instead of in the script), but then used in the script:

mysumscriptii.m

```
% This script sums a vector from the Command Window

runsum = 0;
for i=1:length(vec)
    runsum = runsum + vec(i);
end
```

```
>> clear
>> vec = 1:7;
>> who
Your variables are:
vec

>> mysumscriptii
>> who
Your variables are:
i        runsum vec
>> runsum
runsum =
    28
```

Note

This, however, is very poor programming style. It is much better to pass the vector *vec* to a function.

Because the variables created in scripts and in the Command Window both use the base workspace, many programmers begin scripts with a **clear** command to eliminate variables that may have already been created elsewhere (either in the Command Window or in another script).

Instead of a *program* consisting of a script that calls other functions to do the work, in some cases programmers will write a "main function" to call the other functions. So, the program consists of all functions rather than one script and the rest functions. The reason for this is again because both scripts and the Command Window use the base workspace.

It is possible, in MATLAB as well as in other languages, to have *global variables* that can be shared by functions without passing them. Although there are some cases in which using **global** variables is efficient, it is generally regarded as poor programming style and therefore will not be explained here.

6.4.1 Persistent variables

Normally, when a function stops executing, the local variables from it are cleared. That means that every time a function is called, memory is allocated and used while the function is executing, but released when it ends. With variables that are declared as *persistent variables*, however, the value is not cleared so the next time the function is called, the variable still exists and retains its former value.

The following program demonstrates this. The script calls a function *func1*, which initializes a variable *count* to 0, then increments it, and then prints the value. Every time this function is called, the variable is created, initialized to 0, changed to 1, and then cleared when the function exits. The script then calls a function *func2*, which first declares a **persistent** variable *count*. If the variable has not yet been initialized, which will be the case the first time the function is called, it is initialized to 0. Then, like the first function, the variable is incremented and the value is printed. With the second function, however, the variable remains with its value when the function exits, so the next time the function is called the variable is incremented again.

```
persistex.m
```

```
% This script demonstrates persistent variables

% The first function has a variable "count"
fprintf('This is what happens with a "normal" variable:\n')
func1
func1

% The second fn has a persistent variable "count"
fprintf('\nThis is what happens with a persistent variable:\n')
```

Continued

```
func2
func2
```

func1.m

```
function func1
% func1 increments a normal variable "count"
%Format func1 or func1()

count = 0;
count = count + 1;
fprintf('The value of count is %d\n',count)
end
```

func2.m

```
function func2
% func2 increments a persistent variable "count"
% Format func2 or func2()

persistent count
if isempty(count)
    count = 0;
end
count = count + 1;
fprintf('The value of count is %d\n',count)
end
```

The functions can be called from the script or from the Command Window, as shown. For example, the functions are called first from the script. With the **persistent** variable, the value of *count* is incremented. Then, *func1* is called from the Command Window, and *func2* is also called from the Command Window. Since the value of the **persistent** variable was 2, this time it is incremented to 3.

```
>> persistex
This is what happens with a "normal" variable:
The value of count is 1
The value of count is 1

This is what happens with a persistent variable:
The value of count is 1
The value of count is 2

>> func1
The value of count is 1

>> func2
The value of count is 3
```

As can be seen from this, every time the function *func1* is called, whether from *persistex* or from the Command Window, the value of 1 is printed. However, with *func2* the variable count is incremented every time it is called. It is first called in this example from *persistex* twice, so the count is 1 and then 2. Then, when called from the Command Window, it is incremented to 3.

The way to restart a **persistent** variable is to use the **clear** function. The command

```
>> clear functions
```

will reinitialize all **persistent** variables (see **help clear** for more options).

PRACTICE 6.6

The following function *posnum* prompts the user to enter a positive number and loops to error-check. It returns the positive number entered by the user. It calls a subfunction in the loop to print an error message. The subfunction has a **persistent** variable to count the number of times an error has occurred. Here is an example of calling the function:

```
>> enteredvalue = posnum
Enter a positive number: −5
Error # 1 ... Follow instructions!
Does −5.00 look like a positive number to you?
Enter a positive number: −33
Error # 2 ... Follow instructions!
Does −33.00 look like a positive number to you?
Enter a positive number: 6
enteredvalue =
     6
```

Fill in the subfunction below to accomplish this.

```
posnum.m
function num = posnum
% Prompt user and error-check until the
% user enters a positive number

num = input('Enter a positive number: ');
while num < 0
    errorsubfn(num)
    num = input('Enter a positive number: ');
end

function errorsubfn(num)
```

Of course, the numbering of the error messages will continue if the function is executed again without clearing it first.

6.5 DEBUGGING TECHNIQUES

Any error in a computer program is called a *bug*. This term is thought to date back to the 1940s, when a problem with an early computer was found to have been caused by a moth in the computer's circuitry! The process of finding errors in a program, and correcting them, is still called *debugging*.

6.5.1 Types of errors

There are several different kinds of errors that can occur in a program, which fall into the categories of *syntax errors*, *runtime errors*, and *logical errors*.

Syntax errors are mistakes in using the language. Examples of syntax errors are missing a comma or a quotation mark, or misspelling a word. MATLAB itself will flag syntax errors and give an error message. For example, the following string is missing the end quote:

```
>> mystr = 'how are you;
??? mystr = 'how are you;
              |
Error: A MATLAB string constant is not terminated properly.
```

If this type of error is typed in a script or function using the Editor, the Editor will flag it.

Another common mistake is to spell a variable name incorrectly; MATLAB will also catch this error.

```
>> value = 5;
>> newvalue = valu + 3;
??? Undefined function or variable 'valu'.
```

Runtime, or execution-time, errors are found when a script or function is executing. With most languages, an example of a runtime error would be attempting to divide by zero. However, in MATLAB, this will generate a warning message. Another example would be attempting to refer to an element in an array that does not exist.

```
runtimeEx.m

% This script shows an execution-time error

vec = 3:5;

for i = 1:4
    disp(vec(i))
end
```

The previous script initializes a vector with three elements, but then attempts to refer to a fourth. Running it prints the three elements in the vector, and then an error message is generated when it attempts to refer to the fourth element. Note that the script gives an explanation of the error, and it gives the line number in the script in which the error occurred.

```
>> runtimeEx
   3
   4
   5
??? Attempted to access vec(4); index out of bounds because
numel(vec) = 3.

Error in ==> runtimeEx at 6
    disp(vec(i))
```

Logical errors are more difficult to locate, because they do not result in any error message. A logical error is a mistake in reasoning by the programmer, but it is not a mistake in the programming language. An example of a logical error would be dividing by 2.54 instead of multiplying to convert inches to centimeters. The results printed or returned would be incorrect, but this might not be obvious.

All programs should be robust and should wherever possible anticipate potential errors, and guard against them. For example, whenever there is input into a program, the program should error-check and make sure that the input is in the correct range of values. Also, before dividing, any denominator should be checked to make sure that it is not zero.

Despite the best precautions, there are bound to be errors in programs.

6.5.2 Tracing

Many times, when a program has loops and/or selection statements and is not running properly, it is useful in the debugging process to know exactly which statements have been executed. For example, following is a function that attempts to display "In middle of range" if the argument passed to it is in the range 3 to 6, and "Out of range" otherwise.

testifelse.m

```
function testifelse(x)
% testifelse will test the debugger
% Format: testifelse(Number)

if 3 < x < 6
    disp('In middle of range')
else
    disp('Out of range')
end
end
```

However, it seems to print "In middle of range" for all values of *x*:

```
>> testifelse(4)
In middle of range

>> testifelse(7)
In middle of range

>> testifelse(-2)
In middle of range
```

One way of following the flow of the function, or *tracing* it, is to use the **echo** function. The **echo** function, which is a toggle, will display every statement as it is executed as well as results from the code. For scripts, just **echo** can be typed, but for functions, the name of the function must be specified. For example, the general form is

```
echo functionname on/off

>> echo testifelse on
>> testifelse(-2)
% This function will test the debugger
if 3 < x < 6
    disp('In middle of range')
In middle of range
end
```

We can see from this result that the action of the <u>if</u> clause was executed.

6.5.3 Editor/Debugger
MATLAB has many useful functions for debugging, and debugging can also be done through its Editor, which is more properly called the Editor/Debugger.

Typing **help debug** at the prompt in the Command Window will show some of the debugging functions. Also, in the Help Browser, clicking on the Search tab and then typing "debugging" will display basic information about the debugging processes.

It can be seen in the previous example that the action of the <u>if</u> clause was executed and it printed "In middle of range", but just from that it cannot be determined why this happened. There are several ways to set *breakpoints* in a file (script or function) so that the variables or expressions can be examined. These can be done from the Editor/Debugger, or commands can be typed from the Command Window. For example, the following **dbstop** command will set a breakpoint in the fifth line of this function (which is the action of the <u>if</u> clause), which allows the values of variables and/or expressions to be examined at that point in the execution. The function **dbcont** can be used to continue the execution, and **dbquit** can be used to quit the debug mode. Note that the prompt becomes K>> in debug mode.

```
>> dbstop testifelse 5
>> testifelse(-2)
5       disp('In middle of range')
K>> x
x =
    -2

K>> 3 < x
ans =
    0

K>> 3 < x < 6
ans =
    1

K>> dbcont
In middle of range
end
>>
```

By typing the expressions $3 < x$ and then $3 < x < 6$, we can determine that the expression $3 < x$ will return either 0 or 1. Both 0 and 1 are less than 6, so the expression will always be **true**, regardless of the value of x!

6.5.4 Function stubs

Another common debugging technique that is used when there is a script main program that calls many functions is to use *function stubs*. A function stub is a place holder, used so that the script will work even though that particular function hasn't been written yet. For example, a programmer might start with a script main program that consists of calls to three functions that accomplish all of the tasks.

```
mainscript.m
% This program gets values for x and y, and
%   calculates and prints z

[x, y] = getvals;
z = calcz(x,y);
printall(x,y,z)
```

The three functions have not yet been written, however, so function stubs are put in place so that the script can be executed and tested. The function stubs consist of the proper function headers, followed by a simulation of what the function will eventually do. For example, the first two functions put arbitrary values in for the output arguments, and the last function prints.

getvals.m

```
function [ x, y] = getvals
x = 33;
y = 11;
```

calcz.m

```
function z = calcz (x, y)
z = 2.2;
```

printall.m

```
function printall (x, y, z)
disp (x, y, z)
```

Then, the functions can be written and debugged one at a time. It is much easier to write a working program using this method than to attempt to write everything at once—then, when errors occur, it is not always easy to determine where the problem is!

SUMMARY

Common Pitfalls

- Not matching up arguments in a function call with the input arguments in a function header.
- Not having enough variables in an assignment statement to store all of the values returned by a function through the output arguments.
- Attempting to call a function that does not return a value from an assignment statement, or from an output statement.
- Not using the same name for the function and the file in which it is stored.
- Not thoroughly testing functions for all possible inputs and outputs.
- Forgetting that **persistent** variables are updated every time the function in which they are declared is called—whether from a script or from the Command Window.

Programming Style Guidelines

- If a function is calculating one or more values, return these value(s) from the function by assigning them to output variable(s).
- Give the function and the file in which it is stored the same name.
- Function headers and function calls must correspond. The number of arguments passed to a function must be the same as the number of input arguments in the function header. If the function returns values, the number

of variables in the left side of an assignment statement should match the number of output arguments returned by the function.

■ If arguments are passed to a function in the function call, do not replace these values by using **input** in the function itself.

■ Functions that calculate and return value(s) will not normally also print them.

■ Functions should not normally be longer than one page in length.

■ Do not declare variables in the Command Window and then use them in a script, or vice versa.

■ Pass all values to be used in functions to input arguments in the functions.

■ When writing large programs with many functions, start with the main program script and use function stubs, filling in one function at a time while debugging.

MATLAB Reserved Words	
global	persistent

MATLAB Functions and Commands	
echo	dbcont
dbstop	dbquit

Exercises

1. Write a function that will receive as an input argument a temperature in degrees Fahrenheit, and will return the temperature in both degrees Celsius and Kelvin. The conversion factors are $C = (F - 32) * 5/9$ and $K = C + 273.15$.

2. Write a function that will receive as an input argument a number of kilometers (K). The function will convert the kilometers to miles and to U.S. nautical miles, and return both results. The conversions are $1\ K = 0.621$ miles and 1 U.S. nautical mile $= 1.852\ K$.

3. A vector can be represented by its rectangular coordinates x and y or by its polar coordinates r and θ. For positive values of x and y, the conversions from rectangular to polar coordinates in the range from 0 to 2π are $r = \sqrt{x^2 + y^2}$ and $\theta = \arctan(y/x)$. The function for arctan is **atan**. Write a function *recpol* to receive as input arguments the rectangular coordinates and return the corresponding polar coordinates.

4. A vector can be represented by its rectangular coordinates x and y or by its polar coordinates r and θ. For positive values of *x* and *y*, the conversions from rectangular to polar coordinates in the range from 0 to 2π are $r = \sqrt{x^2 + y^2}$ and $\theta = \arctan(y/x)$.

Write a function to receive as input arguments the rectangular coordinates and return the corresponding polar coordinates.

5. Write a function to calculate the volume and surface area of a hollow cylinder. It receives as input arguments the radius of the cylinder base and the height of the cylinder. The volume is given by $\pi r^2 h$, and the surface area is $2 \pi r h$.

6. Write a function that will receive the radius of a circle and will print both the radius and diameter of the circle in a sentence format. This function will not return any value; it simply prints.

7. Write a function that will receive as an input argument a length in inches, and will print in a sentence format the length in both inches and centimeters (1 in. $= 2.54$ cm). Note that this function will not return any value.

8. Write a function that will receive an integer n and a character as input arguments, and will print the character n times.

9. Convert the *printstars* script from Chapter 4 to a function that receives as inputs the number of rows and columns, and prints a box of asterisks with the specified number of rows and columns.

10. Convert the *multtable* function from Chapter 4 to a function that receives as input arguments the number of rows and columns and prints a multiplication table (rather than returning the matrix).

11. Write a function that will receive a matrix as an input argument, and prints it in a table format.

12. Write a function that receives a matrix as an input argument, and prints a random row from the matrix.

13. Write a function that receives a count as an input argument, and prints the value of the count in a sentence that would read "It happened 1 time." if the value of the count is 1, or "It happened xx times." if the value of the count (xx) is greater than 1.

14. Write a function that will print an explanation of temperature conversions. The function does not receive any input arguments; it simply prints.

15. Write a function that receives an x vector, a minimum value, and a maximum value, and plots **sin(x)** from the specified minimum to the specified maximum.

16. Write a function that prompts the user for a value of an integer n, and returns the value of n. No input arguments are passed to this function.

17. Write a function that prompts the user for a value of an integer n, and returns a vector of values from 1 to n. The function should error-check to make sure that the user enters an integer. No input arguments are passed to this function.

18. Write a script that will:
   ▪ Call a function to prompt the user for an angle in degrees.
   ▪ Call a function to calculate and return the angle in radians. (*Note:* π radians $= 180°$.)
   ▪ Call a function to print the result.
   Write all of the functions as well. Note that the solution to this problem involves four M-files: one which acts as a main program (the script), and three for the functions.

19. Modify the program in Exercise 18 so that the function to calculate the angle is a subfunction to the function that prints.

20. Write a program to calculate and print the area and circumference of a circle. There should be one script and three functions to accomplish this (one that prompts for the radius, one that calculates the area and circumference, and one that prints).

21. The lump sum S to be paid when interest on a loan is compounded annually is given by $S = P(1 + i)^n$, where P is the principal invested, i is the interest rate, and n is the number of years. Write a program that will plot the amount S as it increases through the years from 1 to n. The main script will call a function to prompt the user for the number of years (and error-check to make sure that the user enters a positive integer). The script will then call a function that will plot S for years 1 through n. It will use 0.05 for the interest rate and $10,000 for P.

22. The following script *prtftlens* loops to:
    ■ Call a function to prompt the user for a length in feet.
    ■ Call a function to convert the length to inches.
    ■ Call a function to print both.

prtftlens.m

```
for i = 1:3
    lenf = lenprompt();
    leni = convert_f_to_i(lenf);
    printlens(lenf, leni)
end
```

Write all of the functions.

23. Write a program to write a length conversion chart to a file. It will print lengths in feet, from 1 to an integer specified by the user, in one column and the corresponding length in meters (1 foot = 0.3048 m) in a second column. The main script will call one function that prompts the user for the maximum length in feet; this function must error-check to make sure that the user enters a valid positive integer. The script then calls a function to write the lengths to a file.

24. For a prism that has as its base an n-sided polygon and height h, the volume V and surface area A are given by:

$$V = \frac{n}{4}hS^2 \cot\frac{\pi}{n}$$
$$A = \frac{n}{2}S^2 \cot\frac{\pi}{n} + nSh$$

where S is the length of the sides of the polygons. Write a script that calls a function *getprism* that prompts the user for the number of sides n, the height h, and the length of the sides S, and returns these three values. It then calls a function *calc_v_a* that calculates and returns the volume and surface area, and then finally a function *printv_a* that prints the results. The built-in function in MATLAB for cotangent is cot.

25. The resistance R in ohms of a conductor is given by $R = \frac{E}{I}$ where E is the potential in volts and I is the current in amperes. Write a script that will:

- Call a function to prompt the user for the potential and the current.
- Call a function that will print the resistance; this will call a subfunction to calculate and return the resistance.

Write the functions as well.

26. The power in watts is given by $P = EI$. Modify the program in Exercise 25 to calculate and print both the resistance and the power. Modify the subfunction so that it calculates and returns both values.

27. The distance between any two points (x_1,y_1) and (x_2,y_2) is given by:

$$distance = \sqrt{(x_1 - x_2)^2 + (y_1 - y_2)^2}$$

The area of a triangle is:

$$area = \sqrt{s^*(s - a)^*(s - b)^*(s - c)}$$

where a, b, and c are the lengths of the sides of the triangle, and s is equal to half the sum of the lengths of the three sides of the triangle. Write a script that will prompt the user to enter the coordinates of three points that determine a triangle (e.g., the x and y coordinates of each point). The script will then calculate and print the area of the triangle. It will call one function to calculate the area of the triangle. This function will call a subfunction that calculates the length of the side formed by any two points (the distance between them).

Satellite navigation systems have become ubiquitous. Navigation systems based in space such as the global positioning system (GPS) can send data to handheld personal devices. The coordinate systems that are used to represent locations present these data in several formats.

28. The geographic coordinate system is used to represent any location on Earth as a combination of altitude and longitude values. These values are angles that can be written in the decimal degrees (DD) form or the degrees, minutes, seconds (DMS) form just like time. For example, 24.5° is equivalent to 24°30'0". Write a script that will prompt the user for an angle in DD form and will print in sentence format the same angle in DMS form. The script should error-check for invalid user input. The angle conversion is to be done by calling a separate function in the script. An example of running the script is:

```
>> DMSscript
Enter an angle in decimal degrees form: 24.5588
24.56 degrees is equivalent to 24 degrees, 33 minutes, 31.68
seconds
```

29. Write a program to write a temperature conversion chart to a file. The main script will:
- Call a function that explains what the program will do.
- Call a function to prompt the user for the minimum and maximum temperatures in degrees Fahrenheit, and return both values. This function checks to make sure that the minimum is less than the maximum, and calls a subfunction to swap the values if not.

- Call a function to write temperatures to a file: the temperature in degrees F from the minimum to the maximum in one column, and the corresponding temperature in degrees Celsius in another column. The conversion is C = (F − 32) * 5/9.

30. A bar is a unit of pressure. Polyethylene water pipes are manufactured in pressure grades, which indicate the amount of pressure in bars that the pipe can support for water at a standard temperature. The following script *printpressures* prints some common pressure grades, as well as the equivalent pressure in atm (atmospheres) and psi (pounds per square inch). The conversions are:

    ```
    1 bar = 0.9869 atm = 14.504 psi
    ```

    The script calls a function to convert from bars to atm and psi, and calls another function to print the results. You may assume that the bar values are integers.

    printpressures.m
    ```
    % prints common water pipe pressure grades
    commonbar =[ 4 6 10] ;
    for bar = commonbar
        [ atm, psi] = convertbar(bar);
        print_press(bar,atm,psi)
    end
    ```

31. The following script (called *circscript*) loops *n* times to prompt the user for the circumference of a circle (where *n* is a random integer). Error-checking is ignored to focus on functions in this program. For each, it calls one function to calculate the radius and area of that circle, and then calls another function to print these values. The formulas are $r = c/(2\pi)$ and $a = \pi r^2$ where *r* is the radius, c is the circumference, and *a* is the area. Write the two functions.

    circscript.m
    ```
    n = round(rand* 4)+1;
    for i = 1:n
        circ = input('Enter the circumference of the circle: ');
        [ rad area] = radarea(circ);
        dispra(rad,area)
    end
    ```

32. Write a script that will ask the user to choose his or her favorite science class, and print a message regarding that course. It will call a function to display a menu of choices (using the **menu** function); this function will error-check to make sure that the user pushes one of the buttons. The function will return the number corresponding to the user's choice. The script will then print a message.

33. Write a menu-driven program to convert a time in seconds to other units (minutes, hours, and so on). The main script will loop to continue until the user chooses to exit. Each time in the loop, the script will generate a random time in seconds, call a function to present a menu of options, and print the converted time. The conversions

must be made by individual functions (e.g., one to convert from seconds to minutes). All user entries must be error-checked.

34. Write a menu-driven program to investigate the constant π. Model it after the program that explores the constant e. **Pi** (π) is the ratio of a circle's circumference to its diameter. Many mathematicians have found ways to approximate π. For example, Machin's formula is

$$\frac{\pi}{4} = 4 \arctan\left(\frac{1}{5}\right) - \arctan\left(\frac{1}{239}\right)$$

Leibniz found that π can be approximated by:

$$\pi = \frac{4}{1} - \frac{4}{3} + \frac{4}{5} - \frac{4}{7} + \frac{4}{9} - \frac{4}{11} + \cdots$$

This is called a sum of a series. There are six terms shown in this series. The first term is 4, the second term is $-4/3$, the third term is 4/5, and so forth. For example, the menu-driven program might have the following options:
- Print the result from Machin's formula.
- Print the approximation using Leibniz's formula, allowing the user to specify how many terms to use.
- Print the approximation using Leibniz's formula, looping until a "good" approximation is found.
- Exit the program.

35. Modify the function *func2* from Section 6.4.1 that has a **persistent** variable *count*. Instead of having the function print the value of *count*, the value should be returned.

36. Write a function *per2* that receives one number as an input argument. The function has a **persistent** variable that sums the values passed to it. Here are the first two times the function is called:

```
>> per2(4)
ans =
    4

>> per2(6)
ans =
    10
```

37. What would be the output from the following program? Think about it, write down your answer, and then type it in to verify.

```
testscope.m
answer = 5;
fprintf('Answer is %d\n',answer)
pracfn
pracfn
fprintf('Answer is %d\n',answer)
printstuff
fprintf('Answer is %d\n',answer)
```

pracfn.m

```
function pracfn
persistent count
if isempty(count)
    count = 0;
end
count = count + 1;
fprintf('This function has been called %d times.\n',count)
```

printstuff.m

```
function printstuff
answer = 33;
fprintf('Answer is %d\n',answer)
pracfn
fprintf('Answer is %d\n',answer)
```

38. Assume a matrix variable *mat*, as in the following example:

```
mat =
    4   2   4   3   2
    1   3   1   0   5
    2   4   4   0   2
```

The following **for** loop:

```
[r c] = size(mat);
for i = 1:r
  sumprint(mat(i,:))
end
```

prints this result:

```
The sum is now 15
The sum is now 25
The sum is now 37
```

Write the function *sumprint*.

String Manipulation

KEY TERMS

string	leading blanks	string concatenation
substring	trailing blanks	delimiter
control characters	vectors of characters	token
white space characters	empty string	

A *string* in the MATLAB® software consists of any number of characters and is contained in single quotes. Actually, strings are vectors in which every element is a single character, which means that many of the vector operations and functions that we have already seen work with strings.

MATLAB also has many built-in functions that are specifically written to manipulate strings. In some cases, strings contain numbers, and it is useful to convert from strings to numbers and vice versa; MATLAB has functions to do this as well.

There are many applications for using strings, even in fields that are predominantly numerical. For example, when data files consist of combinations of numbers and characters, it is often necessary to read each line from the file as a string, break the string into pieces, and convert the parts that contain numbers to number variables that can be used in computations. In this chapter the string manipulation techniques necessary for this will be introduced, and in Chapter 9 applications in file input/output will be demonstrated.

7.1 CREATING STRING VARIABLES

A string consists of any number of characters (including, possibly, none). The following are examples of strings:

```
' '
'x'
'cat'
'Hello there'
'123'
```

A *substring* is a subset or part of a string. For example, 'there' is a substring within the string 'Hello there'.

Characters include letters of the alphabet, digits, punctuation marks, white space, and control characters. *Control characters* are characters that cannot be printed, but accomplish a task (e.g., a backspace or tab). *White space characters* include the space, tab, newline (which moves the cursor down to the next line), and carriage return (which moves the cursor to the beginning of the current line). *Leading blanks* are blank spaces at the beginning of a string, for example, ' hello' and *trailing blanks* are blank spaces at the end of a string.

There are several ways that string variables can be created. One is using assignment statements:

```
>> word = 'cat';
```

Another method is to read into a string variable. Recall that to read into a string variable using the **input** function, the second argument 's' must be included:

```
>> strvar = input('Enter a string: ', 's')
Enter a string: xyzabc
strvar =
xyzabc
```

If leading or trailing blanks are typed by the user, these will be stored in the string. For example, in the following the user entered four blanks and then 'xyz':

```
>> s = input('Enter a string: ','s')
Enter a string:     xyz
s =
    xyz
```

7.1.1 Strings as vectors

Strings are treated as *vectors of characters*, or in other words, a vector in which every element is a single character, so many vector operations can be performed. For example, the number of characters in a string can be found using the **length** function:

Note
There is a difference between an *empty string*, which has a length of 0, and a string consisting of a blank space, which has a length of 1.

```
>> length('cat')
ans =
    3

>> length(' ')
ans =
    1
```

```
>> length('')
ans =
     0
```

Expressions can refer to an individual element (a character within the string), or a subset of a string or a transpose of a string:

```
>> mystr = 'Hi';
>> mystr(1)
ans =
H

>> mystr'
ans =
H
i

>> sent = 'Hello there';
>> length(sent)
ans =
    11
>> sent(4:8)
ans =
lo th
```

A character matrix can be created that consists of strings in every row. This is created as a column vector of strings, but the end result is that it is a matrix in which every element is a character:

Note
A blank space in a string is a valid character within the string.

```
>> wordmat = ['Hello';'Howdy']
wordmat =
Hello
Howdy

>> size(wordmat)
ans =
     2     5
```

This created a 2 × 5 matrix of characters.

With a character matrix, we can refer to an individual element, which is a character, or an individual row, that is one of the strings:

```
>> wordmat(2,4)
ans =
d

>> wordmat(1,:)
ans =
Hello
```

Since rows within a matrix must always be the same length, the shorter strings must be padded with blanks so that all strings have the same length; otherwise an error will occur.

```
>> greetmat = ['Hello '; 'Goodbye']
??? Error using ==> vertcat
CAT arguments dimensions are not consistent.

>> greetmat = ['Hello  '; 'Goodbye']
greetmat =
Hello
Goodbye
>> size(greetmat)
ans =
         2        7
```

PRACTICE 7.1

Prompt the user for a string. Print the length of the string and also the last character in the string. Make sure that this works regardless of what the user enters.

7.2 OPERATIONS ON STRINGS

MATLAB has many built-in functions that work with strings. Some of the string manipulation functions that perform the most common operations will be described here.

7.2.1 Concatenation

String concatenation means to join strings together. Of course, since strings are just vectors of characters, the method of concatenating vectors works for strings, also. For example, to create one long string from two strings, it is possible to join them by putting them in square brackets:

```
>> first = 'Bird';
>> last = 'house';
>> [first last]
ans =
Birdhouse
```

The function **strcat** concatenates horizontally, meaning that it creates one longer string from the inputs.

```
>> first = 'Bird';
>> last = 'house';
>> strcat(first,last)
ans =
Birdhouse
```

There is a difference between these two methods of concatenating, however, if there are leading or trailing blanks in the strings. The method of using the square brackets will concatenate the strings, including all leading and trailing blanks.

```
>> str1 = 'xxx    ';
>> str2 = '   yyy';
>> [str1 str 2]
ans =
xxx       yyy

>> length(ans)
ans =
    12
```

The **strcat** function, however, will remove trailing blanks (but not leading blanks) from strings before concatenating. Note that in these examples, the trailing blanks from *str1* are removed, but the leading blanks from *str2* are not.

```
>> strcat(str1,str2)
ans =
xxx    yyy

>> length(ans)
ans =
    9

>> strcat(str2,str1)
ans =
   yyyxxx

>> length(ans)
ans =
    9       .
```

The function **strvcat** will concatenate vertically, meaning that it will create a column vector of strings (or, in other words, a matrix of characters).

```
>> strvcat(first,last)
ans =
Bird
house

>> size(ans)
ans =
    2    5
```

Note that **strvcat** will pad with extra trailing blanks automatically, in this case to make both strings have a length of 5.

We have seen already that the **char** function can be used to convert from an ASCII code to a character, such as the following example:

```
>> char(97)
ans =
a
```

The **char** function can also be used to create a matrix of characters. When using the **char** function to create a matrix, it will automatically pad the strings within the rows with trailing blanks as necessary so that they are all the same length, just like **strvcat**.

```
>> clear greetmat
>> greetmat = char('Hello','Goodbye')
greetmat =
Hello
Goodbye

>> size(greetmat)
ans =
     2       7
```

PRACTICE 7.2

Create the following string variables:

```
v1 = 'Mechanical';
v2 = 'Engineering';
```

Then, get the length of each string. Create a new variable *v3*, which is a substring of *v2* that stores just 'Engineer'. Create a matrix consisting of the values of *v1* and *v2* in separate rows.

7.2.2 Creating customized strings

There are several built-in functions that create customized strings, including **blanks** and **sprintf**.

The **blanks** function will create a string consisting of *n* blank characters (which are kind of hard to see!!). However, in MATLAB, if the mouse is moved to highlight the result in *ans*, the blanks can be seen.

```
>> blanks(4)
ans =

>> length(ans)
ans =
     4
```

It is usually most useful to use this function when concatenating strings, and a number of blank spaces is desired in between. For example, this will insert five blank spaces in between the words:

```
>> [first blanks(5) last]
ans =
Bird     house
```

Displaying the transpose of the **blanks** function can also be used to move the cursor down. In the Command Window, it would look like this:

```
>> disp(blanks(4)')



>>
```

This is useful in a script or function to create space in the output, and is essentially equivalent to printing the newline character four times.

The **sprintf** function works exactly like the **fprintf** function, *but instead of printing it creates a string*. Here are several examples in which the output is not suppressed so the value of the string variable is shown:

```
>> sent1 = sprintf('The value of pi is %.2f', pi)
sent1 =
The value of pi is 3.14

>> sent2 = sprintf('Some numbers: %5d, %2d', 33, 6)
sent2 =
Some numbers:    33,  6

>> length(sent2)
ans =
    23
```

In the following example, on the other hand, the output of the assignment is suppressed so the string is created including a random integer and stored in the string variable. Then, some exclamation points are concatenated to that string.

```
>> phrase = sprintf('A random integer is %d', ...
                round(rand*5+5));
>> strcat(phrase, '!!!')
ans =
A random integer is 7!!!
```

All of the formatting options that can be used in the **fprintf** function can also be used in the **sprintf** function.

Applications of customized strings: prompts, labels, arguments to functions

One very useful application of the **sprintf** function is to include numbers in strings that are used for plot titles and axis labels. For example, assume that a file *expnoanddata.dat* stores an experiment number, followed by the experiment data. In this case the experiment number is "123," and then the rest of the file consists of the actual data.

```
123  4.4  5.6  2.5  7.2  4.6  5.3
```

The following script would load these data and plot them with a title that includes the experiment number.

plotexpno.m

```
% This script loads a file that stores an experiment number
% followed by the actual data. It plots the data and puts
% the experiment # in the plot title

load expnoanddata.dat
experNo = expnoanddata(1);
data = expnoanddata(2:end);
plot(data,'ko')
xlabel('Sample #')
ylabel('Weight')
title(sprintf('Data from experiment %d', experNo))
```

The script loads all numbers from the file into a row vector. It then separates the vector; it stores the first element, which is the experiment number, in a variable *experNo*, and the rest of the vector in a variable *data* (the rest being from the second element to the end). It then plots the data, using **sprintf** to create the title, which includes the experiment number as seen in Figure 7.1.

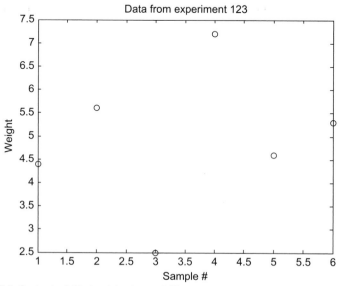

FIGURE 7.1 Customized title in plot using **sprintf**

PRACTICE 7.3

In a loop, create and print strings with file names *file1.dat*, *file2.dat*, and so on for file numbers 1 through 5.

QUICK QUESTION!

How could we use the **sprintf** function in customizing prompts for the **input** function?

Answer: For example, if it is desired to have the contents of a string variable printed in a prompt, **sprintf** can be used:

```
>> username = input('Please enter your name: ', 's');
Please enter your name: Bart

>> prompt = sprintf('%s, Enter your id #: ',username);
>> id_no = input(prompt)
Bart, Enter your id #: 177
id_no =
   177
```

Another way of accomplishing (in a script or function) would be:

```
fprintf('%s, Enter your id #: ',username);
id_no = input(' ');
```

Note: The calls to the **sprintf** and **fprintf** functions are identical except that the **fprintf** prints (so there is no need for a prompt in the **input** function), whereas the **sprintf** creates a string that can then be displayed by the **input** function. In this case using **sprintf** seems cleaner than using **fprintf** and then having an empty string for the prompt in **input**.

As another example, the following program prompts the user for endpoints (x_1, y_1) and (x_2, y_2) of a line segment, and calculates the midpoint of the line segment, which is the point (x_m, y_m). The coordinates of the midpoint are found by:

$$x_m = \frac{1}{2}(x_1 + x_2) \quad y_m = \frac{1}{2}(y_1 + y_2)$$

The script *midpoint* calls a function *entercoords* to separately prompt the user for the x and y coordinates of the two endpoints, calls a function *findmid* twice to calculate separately the x and y coordinates of the midpoint, and then prints this midpoint. When the program is executed, the output looks like this:

```
>> midpoint
Enter the x coord of the first endpoint: 2
Enter the y coord of the first endpoint: 4
Enter the x coord of the second endpoint: 3
Enter the y coord of the second endpoint: 8
The midpoint is (2.5, 6.0)
```

In this example, the word 'first' or 'second' is passed to the *entercoords* function so that it can use whichever word is passed in the prompt. Two methods are shown to customize this prompt, using **sprintf** for the prompt and using **fprintf** and then an empty prompt (for no reason other than to demonstrate the differences).

```
midpoint.m
```

```
% This program finds the midpoint of a line segment

[x1, y1] = entercoords('first');
[x2, y2] = entercoords('second');

midx = findmid(x1,x2);
midy = findmid(y1,y2);

fprintf('The midpoint is (%.1f, %.1f)\n',midx,midy)
```

```
entercoords.m
```

```
function[ xpt, ypt] = entercoords(word)
% entercoords reads in & returns the coordinates of
%  the specified endpoint of a line segment
% Format: entercoords(word) where word is 'first'
%          or 'second'

% Two different methods are used to customize the
%  prompt to show the difference
fprintf('Enter the x coord of the %s endpoint: ', word)
xpt = input(' ');

prompt = sprintf('Enter the y coord of the %s endpoint: ', ...
   word);
ypt = input(prompt);
end
```

```
findmid.m
```

```
function mid = findmid(pt1,pt2)
% findmid calculates a coordinate (x or y) of the
%  midpoint of a line segment
% Format: findmid(coord1, coord2)

mid = 0.5 * (pt1 + pt2);
end
```

7.2.3 Removing white space characters

MATLAB has functions that will remove trailing blanks from the end of a string and/or leading blanks from the beginning of a string.

The **deblank** function will remove blank spaces from the end of a string. For example, if some strings are padded in a character matrix so that all are the same length, it is frequently desired to then remove those extra blank spaces to use the string in its original form.

```
>> names = char('Sue', 'Cathy','Xavier')
names =
Sue
Cathy
Xavier

>> name1 = names(1,:)
name1 =
Sue

>> length(name1)
ans =
     6

>> name1 = deblank(name1);
>> length(name1)
ans =
     3
```

Note
The **deblank** function only removes trailing blanks from a string, not leading blanks.

The **strtrim** function will remove both leading and trailing blanks from a string, but not blanks in the middle of the string. In the following example, the three blanks in the beginning and four blanks in the end are removed, but not the two blanks in the middle. Highlighting the result in the Command Window with the mouse would show the blank spaces.

```
>> strvar = [blanks(3) 'xx' blanks(2) 'yy' blanks(4)]
strvar =
   xx  yy
>> length(strvar)
ans =
    13

>> strtrim(strvar)
ans =
xx  yy
>> length(ans)
ans =
     6
```

7.2.4 Changing case
MATLAB has two functions that convert strings to all uppercase letters, or all lowercase, called **upper** and **lower**.

```
>> mystring = 'AbCDEfgh';
>> lower(mystring)
ans =
abcdefgh
```

```
>> upper(ans)
ans =
ABCDEFGH
```

PRACTICE 7.4

Assume that these expressions are typed sequentially in the Command Window. Think about it, write down what you think the results will be, and then verify your answers by actually typing them.

```
wxyzstring = '123456789012345';
longstring = '   abc   de f   '

length(longstring)

shortstring = strtrim(longstring)

length(shortstring)

upper(shortstring)

news = sprintf('The first part is %s', ...
     shortstring(1:3))
```

7.2.5 Comparing strings

There are several functions that compare strings and return **logical true** if they are equivalent, or **logical false** if not. The function **strcmp** compares strings, character by character. It returns **logical true** if the strings are completely identical (which infers that they must be of the same length, also) or **logical false** if the strings are not the same length or any corresponding characters are not identical. Note that for strings, these functions are used to determine whether strings are equal to each other or not, not the equality operator ==. Here are some examples of these comparisons:

```
>> word1 = 'cat';
>> word2 = 'car';
>> word3 = 'cathedral';
>> word4 = 'CAR';
>> strcmp(word1,word2)
ans =
     0

>> strcmp(word1,word3)
ans =
     0
```

```
>> strcmp(word1,word1)
ans =
     1

>> strcmp(word2,word4)
ans =
     0
```

The function **strncmp** compares only the first n characters in strings and ignores the rest. The first two arguments are the strings to compare, and the third argument is the number of characters to compare (the value of n).

```
>> strncmp(word1,word3,3)
ans =
     1

>> strncmp(word1,word3,4)
ans =
     0
```

QUICK QUESTION!

How can we compare strings, ignoring whether the characters in the string are uppercase or lowercase?

Answer: See the following Programming Concept and Efficient Method.

THE PROGRAMMING CONCEPT

Ignoring the case when comparing strings can be done by changing all characters in the strings to either upper- or lowercase; for example, in MATLAB using the **upper** or **lower** function:

```
>> strcmp(upper(word2),upper(word4))
ans =
     1
```

Note
These functions are used in MATLAB to test strings for equality, not the == operator.

THE EFFICIENT METHOD

The function **strcmpi** compares the strings but ignores the case of the characters.

```
>> strcmpi(word2,word4)
ans =
     1
```

There is also a function **strncmpi** that compares n characters, ignoring the case.

7.2.6 Finding, replacing, and separating strings

There are functions that find and replace strings, or parts of strings, within other strings and functions that separate strings into substrings.

The function **findstr** receives two strings as input arguments. It finds all occurrences of the shorter string within the longer, and returns the subscripts of the beginning of the occurrences. The order of the strings does not matter with **findstr**; it will always find the shorter string within the longer, whichever that is. The shorter string can consist of one character, or any number of characters. If there is more than one occurrence of the shorter string within the longer one, **findstr** returns a vector with all indices. Note that what is returned is the index of the beginning of the shorter string.

```
>> findstr('abcde', 'd')
ans =
    4

>> findstr('d','abcde')
ans =
    4

>> findstr('abcde', 'bc')
ans =
    2

>> findstr('abcdeabcdedd', 'd')
ans =
    4    9    11    12
```

The function **strfind** does essentially the same thing, except that the order of the arguments does make a difference. The general form is **strfind(string, substring)**; it finds all occurrences of the substring within the string, and returns the subscripts.

```
>> strfind('abcdeabcde','e')
ans =
    5    10
```

For both **strfind** and **findstr**, if there are no occurrences, the empty vector is returned.

```
>> strfind('abcdeabcde','ef')
ans =
    [ ]
```

QUICK QUESTION!

How can you find how many blanks there are in a string (e.g., 'how are you')?

Answer: The **strfind** function will return an index for every occurrence of a substring within a string, so the result is a vector of indices. The **length** of this vector of indices would be the number of occurrences. For example, the following finds the number of blank spaces in the variable *phrase*:

```
>> phrase = 'Hello, and how are you doing?';
>> length(strfind(phrase,' '))
ans =
     5
```

If it is desired to get rid of any leading and trailing blanks first (in case there are any), the **strtrim** function would be used first.

```
>> phrase = '  Well, hello there!  ';
>> length(strfind(strtrim(phrase),' '))
ans =
     2
```

Let's expand this, and write a script that creates a vector of strings that are phrases. The output is not suppressed so that the strings can be seen when the script is executed. It loops through this vector and passes each string to a function *countblanks*. This function counts the number of blank spaces in the string, not including any leading or trailing blanks.

phraseblanks.m

```
% This script creates a column vector of phrases
% It loops to call a function to count the number
%   of blanks in each one and prints that

phrasemat = char('Hello and how are you? ', ...
   'Hi there everyone!', 'How is it going? ', 'Whazzup? ')
[r c] = size(phrasemat);

for i = 1:r
    % Pass each row (each string) to countblanks function
    howmany = countblanks(phrasemat(i,:));
    fprintf('Phrase %d had %d blanks\n',i,howmany)
end
```

countblanks.m

```
function num = countblanks(phrase)
% countblanks returns the # of blanks in a trimmed string
% Format: countblanks(string)

num = length(strfind(strtrim(phrase), ' '));
end
```

For example, running this script would result in:

```
>> phraseblanks
phrasemat =
Hello and how are you?
Hi there everyone!
How is it going?
Whazzup?

Phrase 1 had 4 blanks
Phrase 2 had 2 blanks
Phrase 3 had 3 blanks
Phrase 4 had 0 blanks
```

The function **strrep** finds all occurrences of a substring within a string, and replaces them with a new substring. The order of the arguments matters. The format is

```
strrep(string, oldsubstring, newsubstring)
```

The following example replaces all occurrences of the substring 'e' with the substring 'x':

```
>> strrep('abcdeabcde','e','x')
ans =
abcdxabcdx
```

All strings can be any length, and the lengths of the old and new substrings do not have to be the same.

In addition to the string functions that find and replace, there is a function that separates a string into two substrings. The **strtok** function breaks a string into two pieces; it can be called several ways. The function receives one string as an input argument. It looks for the first *delimiter*, which is a character or set of characters that act as a separator within the string.

By default, the delimiter is any white space character. The function returns a *token* that is the beginning of the string, up to (but not including) the first delimiter. It also returns the rest of the string, which includes the delimiter. Assigning the returned values to a vector of two variables will capture both of these. The format is

```
[token rest] = strtok(string)
```

where *token* and *rest* are variable names. For example,

```
>> sentence1 = 'Hello there';
>> [word rest] = strtok(sentence1)
word =
Hello
rest =
 there

>> length(word)
ans =
   5
```

```
>> length(rest)
ans =
      6
```

Note that the rest of the string includes the blank space delimiter.

By default, the delimiter for the token is a white space character (meaning that the token is defined as everything up to the blank space), but alternate delimiters can be defined. The format

```
[ token rest] = strtok(string, delimeters)
```

returns a token that is the beginning of the string, up to the first character contained within the delimiters string, and also the rest of the string. In the following example, the delimiter is the character 'l'.

```
>> [word rest] = strtok(sentence1,'l')
word =
He
rest =
llo there
```

Leading delimiter characters are ignored, whether it is the default white space or a specified delimiter. For example, the leading blanks are ignored here:

```
>> [firstpart lastpart] = strtok('  materials science')
firstpart =
materials

lastpart =
  science
```

QUICK QUESTION!

What do you think **strtok** returns if the delimiter is not in the string?

Answer: The first result returned will be the entire string, and the second will be the empty string.

```
>> [first rest] = strtok('ABCDE')
```

```
first =
ABCDE

rest =
  Empty string: 1-by-0
```

PRACTICE 7.5

Think about what would be returned by the following sequence of expressions and statements, and then type them into MATLAB to verify your results.

```
strcmp('hello', 'height')

strncmp('hello', 'height',2)
```

```
strncmpi('yes', 'YES', 1)

name = 'Smith, Carly';
ind = findstr(name,',')
first = name(1:ind-1)

last = name(ind+2:end)

[ f rest] = strtok(name, ',')

l = rest(3:end)
```

QUICK QUESTION!

The function **date** returns the current date as a string (e.g., 07-Feb-2011). How could we write a function to return the day, month, and year as separate output arguments?

Answer: We could use **strrep** to replace the '-' characters with blanks, and then use **strtok** with the blank as the default delimiter to break up the string (twice), or more simply we could just use **strtok** and specify the '-' character as the delimiter.

```
separatedate.m
function[ todayday, todaymo, todayyr] = separatedate
% separatedate separates the current date into day,
% month, and year
% Format: separatedate or separatedate()

[todayday rest] = strtok(date,'-');
[todaymo todayyr] = strtok(rest,'-');
todayyr = todayyr(2:end);
end
```

Since we need to separate the string into three parts, we need to use the **strtok** function twice. The first time the string is separated into '07' and '-Feb-2011' using **strtok**. Then, the second string is separated into 'Feb' and '-2011' using **strtok**. (Since leading delimiters are ignored the second '-' is found as the delimiter in '-Feb-2011'.) Finally, we need to remove the '-' from the string '-2011'; this can be done by just indexing from the second character to the end of the string.

Note: No input arguments are passed to the *separatedate* function; instead, the **date** function returns the current date as a string. Also, note that all three output arguments are strings.

An example of calling this function follows:

```
>> [d m y] = separatedate
d =
07
m =
Feb
y =
2011
```

7.2.7 Evaluating a string

The function **eval** is used to evaluate a string. If the string contains a call to a function, then that function call is executed. For example, in the following, the string 'plot(x)' is evaluated to be a call to the **plot** function, and it produces the plot shown in Figure 7.2.

```
>> x = [2 6 8 3];
>> eval('plot(x)')
```

This would be useful if the user entered the name of the type of plot to use. In this example, the user chooses the type of plot to use for some quiz grades. The string that the user enters (in this case, 'bar') is concatenated with the string '(x)' to create the string 'bar(x)'; this is then evaluated as a call to the **bar** function as seen in Figure 7.3. The name of the plot type is also used in the title.

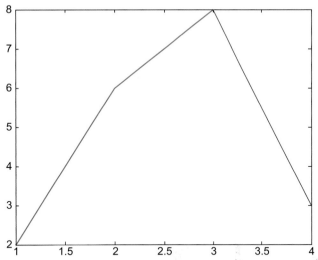

FIGURE 7.2 Plot type passed to the **eval** function

```
>> x = [9 7 10 9];
>> whatplot = input('What type of plot?: ', 's');
What type of plot? : bar
>> eval([whatplot '(x)'])
>> title(whatplot)
>> xlabel('Student #')
>> ylabel('Quiz Grade')
```

PRACTICE 7.6

Create an *x* vector. Prompt the user for 'sin', 'cos', or 'tan' and create a string with that function of *x* (e.g., 'sin(x)' or 'cos(x)'). Use **eval** to create a *y* vector using the specified function.

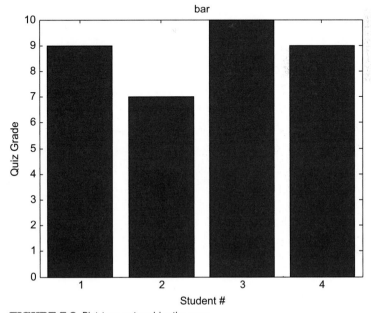

FIGURE 7.3 Plot type entered by the user

7.3 THE "IS" FUNCTIONS FOR STRINGS

There are several "is" functions for strings, which return **logical true** or **false**. The function **isletter** returns **logical true** for every character in a string if the character is a letter of the alphabet. The function **isspace** returns **logical true** if the character is a white space character. If strings are passed to these functions, they will return **logical true** or **false** for every element, or in other words, every character.

```
>> isletter('EK127')
ans =
     1    1    0    0    0

>> isspace('a b')
ans =
     0    1    0
```

The **ischar** function will return **logical true** if the vector argument is a character vector (in other words, a string), or **logical false** if not.

```
>> vec = 'EK127';
>> ischar(vec)
ans =
     1

>> vec = 3:5;
>> ischar(vec)
ans =
     0
```

7.4 CONVERTING BETWEEN STRING AND NUMBER TYPES

MATLAB has several functions that convert numbers to strings in which each character element is a separate digit and vice versa.

To convert numbers to strings, MATLAB has the functions **int2str** for integers and **num2str** for real numbers (which also works with integers). The function **int2str** would convert, for example, the integer 4 to the string 4.

```
>> num = 38;
num =
     38

>> s1 = int2str(num)
s1 =
38
>> length(num)
ans =
     1
```

```
>> length(s1)
ans =
     2
```

The variable *num* is a scalar that stores one number, whereas *s1* is a string that stores two characters, 3 and 8. Even though the result of the first two assignments is "38," note that the indentation in the Command Window is different for the number and the string.

The **num2str** function, which converts real numbers, can be called in several ways. If only the real number is passed to the **num2str** function, it will create a string that has four decimal places, which is the default in MATLAB for displaying real numbers. The precision can also be specified (which is the number of digits), and format strings can also be passed, as shown in the following:

```
>> str2 = num2str(3.456789)
str2 =
3.4568

>> length(str2)
ans =
     6

>> str3 = num2str(3.456789,3)
str3 =
3.46

>> str = num2str(3.456789,'%6.2f')
str =
3.46
```

Note that in the last example, MATLAB removed the leading blanks from the string.

The function **str2num** does the reverse; it takes a string in which a number is stored and converts it to the type **double**:

```
>> num = str2num('123.456')
num =
   123.4560
```

If there is a string in which there are numbers separated by blanks, the **str2num** function will convert this to a vector of numbers (of the default type double). For example,

```
>> mystr = '66 2 111';
>> numvec = str2num(mystr)
numvec =
    66     2    111
```

Note

These are different from the functions such as **char** and **double** that convert characters to ASCII equivalents and vice versa.

```
>> sum(numvec)
ans =
    179
```

PRACTICE 7.7

Think about what would be returned by the following sequence of expressions and statements, and then type them into MATLAB to verify your results.

```
isletter('?')

isspace('Oh no!')

str = '12 33';
ischar(str)

v = str2num(str)

ischar(v)

sum(v)

num = 234;
size(num)

snum = int2str(num);
size(snum)
```

QUICK QUESTION!

Let's say that we have a string that consists of an angle followed by either *d* for degrees or *r* for radians. For example, it may be a string entered by the user:

```
degrad = input('Enter angle and d/r: ', 's');
Enter angle and d/r: 54r
```

How could we separate the string into the angle and the character, and then get the sine of that angle using either **sin** or **sind**, as appropriate (**sin** for radians or **sind** for degrees)?
Answer: First, we could separate this string into its two parts:

```
>> angle = degrad(1:end-1)
angle =
```

```
54
>> dorr = degrad(end)
dorr =
r
```

Then, using an **if-else** statement, we would decide whether to use the **sin** or **sind** function, based on the value of the variable *dorr*. Let's assume that the value is 'r' so we want to use **sin**.

The variable *angle* is a string so the following would not work:

```
>> sin(angle)
??? Function 'sin' is not defined for values
of class 'char'.
>>
```

QUICK QUESTION!—CONT'D

Instead, we could either use **str2num** to convert the string to a number, or use concatenation to create a string 'sin(54)' (or whatever the value of the variable *angle* is) and pass that to the **eval** function:

```
>> eval(['sin(' angle ')'])
ans =
```

```
    -0.5588

>> sin(str2num(angle))
ans =
    -0.5588
```

A complete script to accomplish this is shown in the box of code.

angleDorR.m

```
% Prompt the user for angle and 'd' for degrees
% or 'r' for radians; print the sine of the angle

% Read in the response as a string and then
% separate the angle and character
degrad = input('Enter angle and d/r: ', 's');
angle = degrad(1:end-1);
dorr = degrad(end);

% Error-check to make sure that user enters 'd' or 'r'
while dorr ~= 'd' && dorr ~= 'r'
    disp('Error! Enter d or r with the angle.')
    degrad = input('Enter angle and d/r: ', 's');
    angle = degrad(1:end-1);
    dorr = degrad(end);
end
% Convert angle to number
anglenum = str2num(angle);
fprintf('The sine of %.1f ', anglenum)
% Choose sin or sind function
if dorr == 'd'
    fprintf('degrees is %.3f.\n', sind(anglenum))
else
    fprintf('radians is %.3f.\n', sin(anglenum))
end
```

```
>> angleDorR
Enter angle and d/r: 3.1r
The sine of 3.1 radians is 0.042.
```

```
>> angleDorR
Enter angle and d/r: 53t
Error! Enter d or r with the angle.
Enter angle and d/r: 53d
The sine of 53.0 degrees is 0.799.
```

SUMMARY

Common Pitfalls

- Putting arguments to **strfind** in incorrect order (the order matters for **strfind** but not for **findstr**).
- Confusing **sprintf** and **fprintf**. The syntáx is the same, but **sprintf** creates a string whereas **fprintf** prints.
- Trying to create a vector of strings with varying lengths (the easiest way is to use **strvcat** or **char**, which will pad with extra blanks automatically).
- Forgetting that when using **strtok**, the second argument returned (the "rest" of the string) contains the delimiter.
- When breaking a string into pieces, forgetting to convert the numbers in the strings to actual numbers that can then be used in calculations.

Programming Style Guidelines

- Trim trailing blanks from strings stored in matrices before using.
- Make sure that the correct string comparison function is used; for example, use **strcmpi** if ignoring case is desired.

MATLAB Functions and Commands		
strcat	strcmp	date
strvcat	strncmp	eval
blanks	strcmpi	isspace
sprintf	strncmpi	ischar
deblank	findstr	int2str
strtrim	strfind	num2str
upper	strtok	str2num
lower	strrep	

Exercises

1. Write a function that will receive a name and department as separate strings and will create and return a code consisting of the first two letters of the name and the last two letters of the department. The code should be uppercase letters. For example,

   ```
   >> namedept('Robert','Mechanical')
   ans =
   ROAL
   ```

2. Write a function that will generate two random integers, each in the range from 10 to 30. It will then return a string consisting of the two integers joined together, for example, if the random integers are 11 and 29, the string that is returned will be '1129'.

3. Write a function *ranlowlet* that will return a random lowercase letter of the alphabet. Do *not* build in the ASCII equivalents of any characters; rather, use built-in functions to determine them (e.g., you may know that the ASCII equivalent of 'a' is 97, but do

not use 97 in your function; use a built-in function that would return that value instead).

```
>> let = ranlowlet
let =
a

>> fprintf('The random letter is %c\n', ranlowlet)
The random letter is y
```

4. Write a function that will prompt the user separately for a first and last name and will create and return a string with the form 'last, first'.

5. Write a function that will prompt the user separately for a filename and extension and will create and return a string with the form 'filename.ext'.

6. Write a script that will, in a loop, prompt the user for four course numbers. Each will be a string of length 5 of the form 'CS101'. These strings are to be stored in a character matrix.

7. The following script calls a function *getstr* that prompts the user for a string, error-checking until the user enters something (the error would occur if the user just hits the Enter key without any other characters first). The script then prints the length of the string. Write the *getstr* function.

```
thestring = getstr();
fprintf('Thank you, your string is %d characters long\n', ...
    length(thestring))
```

8. Write a function that will receive one input argument, which is an integer *n*. The function will prompt the user for a number in the range from 1 to *n* (the actual value of *n* should be printed in the prompt) and return the user's input. The function should error-check to make sure that the user's input is in the correct range.

9. Write a script that will create *x* and *y* vectors. Then, it will ask the user for a color (red, blue, or green) and for a plot style (circle or star). It will then create a string *pstr* that contains the color and plot style, so that the call to the **plot** function would be **plot(x,y,pstr)**. For example, if the user enters "blue" and star (*), the variable *pstr* would contain 'b*'.

10. Write a script that will generate a random integer, ask the user for a field width, and print the random integer with the specified field width. The script will use **sprintf** to create a string such as 'The # is %4d\n' (if, for example, the user entered 4 for the field width), which is then passed to the **fprintf** function. To print (or create a string using **sprintf**) either the % or \ character, there must be two of them in a row.

11. What does the **blanks** function return when a 0 is passed to it? A negative number? Write a function *myblanks* that does exactly the same thing as the **blanks** function. Here are some examples of calling it:

```
>> fprintf('Here is the result:%s!\n', myblanks(0))
```

```
Here is the result:!

>> fprintf('Here is the result:%s!\n', myblanks(7))
Here is the result: !
```

12. Write a function that will receive two strings as input arguments, and will return a character matrix with the two strings in separate rows. Rather than using the **char** function to accomplish this, the function should pad with extra blanks as necessary and create the matrix using square brackets.

13. The functions that label the x and y axes and title on a plot expect string arguments. These arguments can be string variables. Write a script that will prompt the user for an integer n, create an x vector with integer values from 1 to n, a y vector that is x^2, and then plot with a title that says "x^2 with n values" where the number is actually in the title.

14. Load files named *file1.dat*, *file2.dat*, and so on in a loop. To test this, create just two files with these names in your Current Folder first.

15. Write a function that will receive two input arguments: a character matrix that is a column vector of strings, and a string. It will loop to look for the string within the character matrix. The function will return the row number in which the string is found if it is in the character matrix, or the empty vector if not.

16. If the strings passed to **strfind** or **findstr** are the same length, what are the only two possible results that could be returned?

17. Either in a script or in the Command Window, create a string variable that stores a string in which numbers are separated by the character 'x', such as '12x3x45x2'. Create a vector of the numbers, and then get the sum (e.g., for the example given it would be 62 but the solution should be general).

18. Assembly language instructions frequently are in the form of a word that represents the operator and then the operands separated by a comma. For example, the string 'ADD n,m' is an instruction to add n + m. Write a function called *assembly_add* that will receive a string in this form and will return the sum of n + m. For example,

```
>> assembly_add('ADD 10,11')
ans =
    21
```

Cryptography, or encryption, is the process of converting plaintext (e.g., a sentence or paragraph) into something that should be unintelligible, called the ciphertext. *The reverse process is code-breaking, or cryptanalysis, which relies on searching the encrypted message for weaknesses and deciphering it from that point. Modern security systems are heavily reliant on these processes.*

19. In cryptography, the intended message sometimes consists of the first letter of every word in a string. Write a function called *crypt* that will receive a string with the encrypted message and return the message.

```
>> estring = 'The early songbird tweets';
>> m = crypt(estring)
m =
Test
```

20. Using the functions **char** and **double**, one can shift words. For example, one can convert from lowercase to uppercase by subtracting 32 from the character codes:

```
>> orig = 'ape';
>> new = char(double(orig)-32)
new =
APE

>> char(double(new)+32)
ans =
ape
```

We've "encrypted" a string by altering the character codes. Figure out the original string. Try adding and subtracting different values (do this in a loop) until you decipher it:

```
Jmkyvih$mx$syx$} ixC
```

21. Write a function *rid_multiple_blanks* that will receive a string as an input argument. The string contains a sentence that has multiple blank spaces in between some of the words. The function will return the string with only one blank in between words. For example,

```
>> mystr = 'Hello  and how   are   you?';
>> rid_multiple_blanks(mystr)
ans =
Hello and how are you?
```

22. Words in a sentence variable (just a string variable) called *mysent* are separated by right slashes (/) instead of blank spaces. For example, *mysent* might have this value:

```
'This/is/not/quite/right'
```

Write a function *slashtoblank* that will receive a string in this form and will return a string in which the words are separated by blank spaces. This should be general and work regardless of the value of the argument. No loops are allowed in this function; the built-in string function(s) must be used.

```
>> mysent = 'This/is/not/quite/right';
>> newsent = slashtoblank(mysent)
newsent =
This is not quite right
```

23. Create the following two variables:

```
>> var1 = 123;
>> var2 = '123';
```

Then, add 1 to each of the variables. What is the difference?

24. A file name is supposed to be in the form *filename.ext*. Write a function that will determine whether a string is in the form of a name followed by a dot followed by a three-character extension, or not. The function should return 1 for **logical true** if it is in that form, or 0 for **false** if not.

25. The built-in **clock** function returns a vector with six elements representing the year, month, day, hours, minutes, and seconds. The first five elements are integers whereas the last is a **double** value, but calling it with **fix** will convert all to integers. The built-in **date** function returns the day, month, and year as a string. For example,

```
>> fix(clock)
ans =
       2011     4     25     14     25     49

>> date
ans =
25-Apr-2011
```

Write a script that will call both of these built-in functions, and then compare results to make sure that the year is the same. The script will have to convert one from a string to a number, or the other from a number to a string to compare.

26. Find out how to pass a vector of integers to **int2str** or real numbers to **num2str**.

27. Write a function *wordscramble* that will receive a word in a string as an input argument. It will then randomly scramble the letters and return the result. The following are examples of calling the function:

```
>> wordscramble('fantastic')
ans =
safntcait

>> sc = wordscramble('hello')
sc =
hleol
```

28. Two variables store strings that consist of a letter of the alphabet, a blank space, and a number (in the form 'R 14.3'). Write a script that would initialize two such variables. Then, use string manipulating functions to extract the numbers from the strings and add them together.

29. Write a script that will first initialize a string variable that will store x and y coordinates of a point in the form 'x 3.1 y 6.4'. Then, use string manipulating functions to extract the coordinates and plot them.

30. Modify the script in Exercise 29 to be more general: the string could store the coordinates in any order; for example, it could store 'y 6.4 x 3.1'.

31. Write a script that will be a temperature converter. The script prompts the user for a temperature in degrees Fahrenheit, and then uses the **menu** function to let the user choose whether to convert that temperature to degrees Celsius or degrees Kelvin. The user's temperature should be in the title of the menu. The script will then print the converted temperature. The conversions are C = (F − 32) * 5/9 and K = C + 273.15.

32. Write a function called *readthem* that prompts the user for a string consisting of a number immediately followed by a letter of the alphabet. The function error-checks to make sure that the first part of the string is actually a number, and to make sure that the last character is actually a letter of the alphabet. The function returns the number and letter as separate output arguments. Note: If a string 'S' is not a number, **str2num(S)** returns the empty vector. An example of calling the function follows:

```
>> [num let] = readthem
Please enter a number immediately followed
by a letter of the alphabet
Enter a # and a letter: 3.3&
Error! Enter a # and a letter: xyz4.5t
Error! Enter a # and a letter: 3.21f

num =
    3.2100

let =
f
```

33. Assume that you have the following function and that it has not yet been called.

strfunc.m

```
function strfunc(instr)
persistent mystr
if isempty(mystr)
    mystr = ' ';
end
mystr = strcat(instr,mystr);
fprintf('The string is %s\n',mystr)
end
```

What would be the result of the following sequential expressions?

```
strfunc('hi')
```

```
strfunc('hello')
```

34. Explain in words what the following function accomplishes (not step by step, but what the end result is).

```
function out = dostr(inp)
persistent str
[w r] = strtok(inp);
str = strcat(str,w);
out = str;
end
```

35. Write the *beautyofmath* script described in Chapter 4, Exercise 19, as a string problem.

Data Structures: Cell Arrays and Structures

CONTENTS

Data structures are variables that store more than one value. For it to make sense to store more than one value in a variable, the values should somehow be logically related. There are many different kinds of data structures. We have already been working with one kind, arrays (e.g., vectors and matrices). An array is a data structure in which all of the values are logically related in that they are of the same type, and represent in some sense "the same thing." So far, that has been true for the vectors and matrices that we have used. We use vectors and matrices when we want to be able to loop through them (or essentially have this done for us using vectorized code).

A *cell array* is a kind of data structure that stores values of different types. Cell arrays can be vectors or matrices; the different values are referred to as the elements of the array. One very common use of a cell array is to store strings of different lengths. Cell arrays actually store pointers to the stored data.

Structures are data structures that group together values that are logically related, but are not the same thing and not necessarily the same type. The different values are stored in separate *fields* of the structure.

One use of structures is to set up a *database* of information. For example, a professor might want to store for every student in a class the student's name, university ID number, grades on all assignments and quizzes, and so forth. In many programming languages and database programs, the terminology is that within a database file, there would be one *record* of information for each student; each separate piece of information (name, quiz 1 score, and so on) would be called a *field* of the record. In the MATLAB® software these records are called **structs**.

Both cell arrays and structures can be used to store values that are different types in a single variable. The main difference between them is that cell arrays are indexed, and can therefore be used with loops or vectorized code. Structures, on the other hand, are not indexed; the values are referenced using the names of the fields, which can be more mnemonic than indexing.

8.1 CELL ARRAYS

One type of data structure that MATLAB has but is not found in many programming languages is a *cell array*. A cell array in MATLAB is an array, but unlike the vectors and matrices we have used so far, elements in cell arrays are cells that can store different types of values.

8.1.1 Creating cell arrays

There are several ways to create cell arrays. For example, we will create a cell array in which one element will store an integer, one element will store a character, one element will store a vector, and one element will store a string. Just like with the arrays we have seen so far, this could be a *1 × 4* row vector, a *4 × 1* column vector, or a *2 × 2* matrix. The syntax for creating vectors and matrices is the same as before. Values within rows are separated by spaces or commas, and rows are separated by semicolons. However, for cell arrays, curly braces are used rather than square brackets. For example, the following creates a row vector cell array with the four types of values:

```
>> cellrowvec = {23, 'a', 1:2:9, 'hello'}
cellrowvec =
    [ 23]      'a'      [ 1 x 5 double]      'hello'
```

To create a column vector cell array, the values are instead separated by semicolons:

```
>> cellcolvec = {23; 'a'; 1:2:9; 'hello'}
cellcolvec =
    [        23]
    'a'
    [1 x 5 double]
    'hello'
```

This method creates a *2 × 2* cell array matrix:

```
>> cellmat = {23 'a'; 1:2:9 'hello'}
cellmat =
      [      23]      'a'
      [1 x 5 double]    'hello'
```

Another method of creating a cell array is to simply assign values to specific array elements and build it up element by element. However, as explained before, extending an array element by element is a very inefficient and time-consuming method.

It is much more efficient, if the size is known ahead of time, to preallocate the array. For cell arrays, this is done with the **cell** function. For example, to preallocate a variable *mycellmat* to be a *2 × 2* cell array, the **cell** function would be called as follows:

```
>> mycellmat = cell(2,2)
mycellmat =
      []    []
      []    []
```

Note that this is a function call, so the arguments to the function are in parentheses; a matrix is created in which all of the elements are empty vectors. Then, each element can be replaced by the desired value. How to refer to each element to accomplish this will be explained next.

8.1.2 Referring to and displaying cell array elements and attributes

Just like with the other vectors we have seen so far, we can refer to individual elements of cell arrays. However, with cell arrays, there are two different ways to do this. The elements in cell arrays are cells. These cells can contain different types of values. With cell arrays, you can refer to the cells, or to the contents of the cells.

Using curly braces for the subscripts will reference the contents of a cell; this is called *content indexing*. For example, this refers to the contents of the second element of the cell array *cellrowvec*; *ans* will have the type **char**:

```
> cellrowvec{2}
ans =
a
```

Row and column subscripts are used to refer to the contents of an element in a matrix (again using curly braces):

```
>> cellmat{1,1}
ans =
   23
```

Values can be assigned to cell array elements. For example, after preallocating the variable *mycellmat* in the previous section, the elements can be initialized:

```
>> mycellmat{1,1} = 23
mycellmat =
    [23]      []
    []        []
```

Using parentheses for the subscripts references the cells; this is called as **cell indexing**. For example, this refers to the second cell in the cell array *cellrowvec*; *ans* will be a *1 × 1* **cell** array:

```
>> cellcolvec(2)
ans =
    'a'
```

When an element of a cell array is itself a data structure, only the type of the element is displayed when the cells are shown. For example, in the previous cell arrays, the vector is shown just as "*1 × 5* double" (this is a high-level view of the cell array). This is what will be displayed with cell indexing; content indexing would display its contents:

```
>> cellmat(2,1)
ans =
    [1 x 5 double]

>> cellmat{2,1}
ans =
    1    3    5    7    9
```

Since this element is a vector, parentheses are used to refer to its elements. For example, the fourth element of the previous vector is:

```
>> cellmat{2,1}(4)
ans =
    7
```

Note that the index into the cell array is given in curly braces, and then parentheses are used to refer to an element of the vector.

One can also refer to subsets of cell arrays, such as in the following:

```
>> cellcolvec{2:3}
ans =
a

ans =
    1    3    5    7    9
```

Note, however, that MATLAB stored *cellcolvec{2}* in the default variable *ans*, and then replaced that with the value of *cellcolvec{3}*. Using content indexing returns them as a *comma-separated list*. However, they could be stored in two separate variables by having a vector of variables on the left side of an assignment:

```
>> [c1 c2] = cellcolvec{2:3}
c1 =
a

c2 =
     1     3     5     7     9
```

Using cell indexing, the two cells would be put in a new cell array (in this case, in *ans*):

```
>> cellcolvec(2:3)
ans =
    'a'
    [1 x 5 double]
```

There are several methods for displaying cell arrays. The **celldisp** function displays the contents of all elements of the cell array:

```
>> celldisp(cellrowvec)
cellrowvec{1} =
     23

cellrowvec{2} =
a

cellrowvec{3} =
     1     3     5     7     9

cellrowvec{4} =
hello
```

The function **cellplot** puts a graphical display of the cell array into a Figure Window; however, it is a high-level view and basically just displays the same information as typing the name of the variable (so, for instance, it would not show the contents of the vector in the previous example). In other words, it shows the cells, not their contents.

Many of the functions and operations on arrays that we have already seen also work with cell arrays. For example, here are some related to dimensioning:

```
>> length(cellrowvec)
ans =
     4

>> size(cellcolvec)
ans =
     4     1
```

```
>> cellrowvec{end}
ans =
hello
```

To delete an element from a vector cell array, use cell indexing:

```
>> cellrowvec
mycell =

    [23]        'a'        [1 x 5 double]        'hello'

>> length(cellrowvec)
ans =
     4

>> cellrowvec(2) = []
cellrowvec =

    [23]        [1 x 5 double]        'hello'

>> length(cellrowvec)
ans =
     3
```

For a matrix, an entire row or column can be deleted using cell indexing:

```
>> cellmat
mycellmat =
    [           23]        'a'
    [1 x 5 double]        'hello'

>> cellmat(1,:) = []
mycellmat =
    [1 x 5 double]        'hello'
```

8.1.3 Storing strings in cell arrays

One good application of a cell array is to store strings of different lengths. Since cell arrays can store different types of values in the elements, that means strings of different lengths can be stored in the elements.

```
>> names = {'Sue', 'Cathy', 'Xavier'}
names =

    'Sue'        'Cathy'        'Xavier'
```

This is extremely useful, because unlike vectors of strings created using **char** or **strvcat**, these strings do not have extra trailing blanks. The length of each string can be displayed using a **for** loop to loop through the elements of the cell array:

```
>> for i = 1:length(names)
     disp(length(names{i}))
end
  3

  5

  6
```

It is possible to convert from a cell array of strings to a character array, and vice versa. MATLAB has several functions that facilitate this. For example, the function **cellstr** converts from a character array padded with blanks to a cell array in which the trailing blanks have been removed.

```
>> greetmat = char('Hello','Goodbye');

>> cellgreets = cellstr(greetmat)
cellgreets =
     'Hello'
     'Goodbye'
```

The **char** function can convert from a cell array to a character matrix:

```
>> names = {'Sue', 'Cathy', 'Xavier'};
>> cnames = char(names)
cnames =

Sue
Cathy
Xavier

>> size(cnames)
ans =
   3     6
```

The function **iscellstr** will return **logical true** if a cell array is a cell array of all strings, or **logical false** if not.

```
>> iscellstr(names)
ans =
   1

>> iscellstr(cellcolvec)
ans =
   0
```

We will see several examples of cell arrays containing strings of varying lengths in later chapters, including advanced file input functions and customizing plots.

PRACTICE 8.1

Write an expression that would display a random element from a cell array (without assuming that the number of elements in the cell array is known). Create two different cell arrays and try the expression on them to make sure that it is correct.

For more practice, write a function that will receive one cell array as an input argument and will display a random element from it.

8.2 STRUCTURES

Structures are data structures that group values together that are logically related in *fields* of the structure. An advantage of structures is that the fields are named, which helps to make it clear what values are stored in the structure. However, structure variables are not arrays. They do not have elements that are indexed, so it is not possible to loop through the values in a structure.

8.2.1 Creating and modifying structure variables

Creating structure variables can be accomplished by simply storing values in fields using assignment statements, or by using the **struct** function.

In our first example, assume that the local Computer Super Mart wants to store information on the software packages that it sells. For each one, they will store the following:

- item number
- cost to the store
- price to the customer
- character code indicating the type of software

An individual structure variable for a given software package might look like this:

package

item_no	cost	price	code
123	19.99	39.95	g

The name of the structure variable is *package*; it has four fields: *item_no*, *cost*, *price*, and *code*.

One way to initialize a structure variable is to use the **struct** function, which pre-allocates the structure. The names of the fields are passed as strings; each is followed by the value for that field (so, pairs of field names and values are passed to **struct**).

```
>> package = struct('item_no',123,'cost',19.99,...
      'price',39.95,'code','g')

package =
   item_no: 123
```

Note

Some programmers use names that begin with an uppercase letter for structure variables (e.g., *Package*) to make them easily distinguishable.

```
     cost: 19.9900
    price: 39.9500
     code: 'g'
```

Note that in the Workspace Window, the variable *package* is listed as a *1×1* struct. MATLAB, since it is written to work with arrays, assumes the array format. Just like a single number is treated as a *1 × 1* double, a single structure is treated as a *1 × 1* struct. Later in this chapter we will see how to work more generally with vectors of structs.

An alternative method of creating this structure, which is not as efficient, involves using the ***dot operator*** to refer to fields within the structure. The name of the structure variable is followed by a dot, or period, and then the name of the field within that structure. Assignment statements can be used to assign values to the fields.

```
>> package.item_no = 123;
>> package.cost = 19.99;
>> package.price = 39.95;
>> package.code = 'g';
```

By using the dot operator in the first assignment statement, a structure variable is created with the field *item_no*. The next three assignment statements add more fields to the structure variable. Again, extending the structure in this manner is not as efficient as using **struct**.

Adding a field to a structure later is accomplished as shown here, by using an assignment statement.

An entire structure variable can be assigned to another. This would make sense, for example, if the two structures had some values in common. Here, for example, the values from one structure are copied into another and then two fields are selectively changed.

```
>> newpack = package;
>> newpack.item_no = 111;
>> newpack.price = 34.95
newpack =
  item_no: 111
     cost: 19.9900
    price: 34.9500
     code: 'g'
```

To print from a structure, the **disp** function will display either the entire structure or an individual field.

```
>> disp (package)
  item_no: 123
     cost: 19.9900
    price: 39.9500
     code: 'g'
```

```
>> disp(package.cost)
   19.9900
```

However, when using **fprintf** only individual fields can be printed; the entire structure cannot be printed.

```
>> fprintf('%d %c\n', package.item_no, package.code)
123 g
```

The function **rmfield** removes a field from a structure. It returns a new structure with the field removed, but does not modify the original structure (unless the returned structure is assigned to that variable). For example, the following would remove the *code* field from the *newpack* structure, but store the resulting structure in the default variable *ans*. The value of *newpack* remains unchanged.

```
>> rmfield(newpack, 'code')
ans =
   item_no: 111
      cost: 19.9900
     price: 34.9500

>> newpack
newpack =
   item_no: 111
      cost: 19.9000
     price: 34.9500
      code: 'g'
```

To change the value of *newpack*, the structure that results from calling **rmfield** must be assigned to *newpack*.

```
>> newpack = rmfield(newpack, 'code')
newpack =
   item_no: 111
      cost: 19.9000
     price: 34.9500
```

PRACTICE 8.2

A silicon wafer manufacturer stores, for every part in its inventory, a part number, quantity in the factory, and the cost for each.

onepart

part_no	quantity	costper
123	4	33

Create this structure variable using **struct**. Print the cost in the form $xx.xx.

8.2.2 Passing structures to functions

An entire structure can be passed to a function, or individual fields can be passed. For example, here are two different versions of a function that calculates the profit on a software package. The profit is defined as the price minus the cost.

In the first version, the entire structure variable is passed to the function, so the function must use the dot operator to refer to the *price* and *cost* fields of the input argument.

calcprof.m

```
function profit = calcprof(packstruct)
% calcprofit calculates the profit for a
% software package
% Format: calcprof(structure w/ price & cost fields)

profit = packstruct.price - packstruct.cost;
end
```

```
>> calcprof(package)
ans =
   19.9600
```

In the second version, just the *price* and *cost* fields are passed to the function using the dot operator in the function call. These are passed to two scalar input arguments in the function header, so there is no reference to a structure variable in the function itself, and the dot operator is not needed in the function.

calcprof2.m

```
function profit = calcprof2(oneprice, onecost)
% Calculates the profit for a software package
% Format: calcprof2(price, cost)

profit = oneprice - onecost;
end
```

```
>> calcprof2(package.price, package.cost)
ans =
   19.9600
```

It is important, as always with functions, to make sure that the arguments in the function call correspond one-to-one with the input arguments in the function header. In the case of *calcprof*, a structure variable is passed to an input argument, which is a structure. For the second function *calcprof2*, two individual fields, which are **double** values, are passed to two **double** input arguments.

8.2.3 Related structure functions

There are several functions that can be used with structures in MATLAB. The function **isstruct** will return **logical** 1 for **true** if the variable argument is a structure variable, or 0 if not. The **isfield** function returns **logical true** if a field-name (as a string) is a field in the structure argument, or **logical false** if not.

```
>> isstruct(package)
ans =
     1

>> isfield(package,'cost')
ans =
     1
```

The **fieldnames** function will return the names of the fields that are contained in a structure variable.

```
>> pack_fields = fieldnames(package)
pack_fields =
    'item_no'
    'cost'
    'price'
    'code'
```

Since the names of the fields are of varying lengths, the **fieldnames** function returns a cell array with the names of the fields.

Curly braces are used to refer to the elements, since *pack_fields* is a cell array. For example, we can refer to the length of one of the field names:

```
>> length(pack_fields{2})
ans =
     4
```

QUICK QUESTION!

How can we ask the user for a field in a structure and either print its value or an error if it is not actually a field?
Answer: The isfield function can be used to determine whether it is a field of the structure. Then, by concatenating

that field name to the structure variable and dot, and then passing the entire string to **eval**, the expression would be evaluated as the actual field in the structure. The following is the code

```
inputfield = input('Which field would you like to see: ','s');
if isfield(package, inputfield)
     fprintf('The value of the %s field is: %c\n', ...
          inputfield, eval([ 'package.' inputfield] ))
else
   fprintf('Error: %s is not a valid field\n', inputfield)
end
```

that would produce this output (assuming the *package* variable was initialized as shown previously):

```
Which field would you like to see: code
The value of the code field is: g
```

PRACTICE 8.3

Modify the code from the preceding Quick Question to use **sprintf** rather than **eval**.

8.2.4 Vectors of structures

In many applications, including database applications, information would normally be stored in a *vector of structures*, rather than in individual structure variables. For example, if the Computer Super Mart is storing information on all of the software packages that it sells. It would likely be in a vector of structures such as the following:

packages

	item_no	cost	price	code
1	123	19.99	39.95	g
2	456	5.99	49.99	l
3	587	11.11	33.33	w

In this example, *packages* is a vector that has three elements. It is shown as a column vector. Each element is a structure consisting of four fields: *item_no, cost, price,* and *code*. It may look like a matrix, which has rows and columns, but it is instead a vector of structures.

This vector of structures can be created several ways. One method is to create a structure variable, as shown earlier, to store information on one software package.

This can then be expanded to be a vector of structures.

```
>> packages = struct('item_no',123,'cost',19.99,...
      'price',39.95,'code','g');
>> packages(2) = struct('item_no',456,'cost', 5.99,...
      'price',49.99,'code','l');
>> packages(3) = struct('item_no',587,'cost',11.11,...
      'price',33.33,'code','w');
```

The first assignment statement shown here creates the first structure in the structure vector, the next one creates the second structure, and so on. This actually creates a *1 × 3* row vector.

Alternatively, the first structure could be treated as a vector to begin with; for example,

```
>> packages(1) = struct('item_no',123,'cost',19.99,...
        'price',39.95,'code','g');
>> packages(2) = struct('item_no',456,'cost', 5.99,...
        'price',49.99,'code','l');
>> packages(3) = struct('item_no',587,'cost',11.11,...
        'price',33.33,'code','w');
```

Both of these methods, however, involve extending the vector. As we have already seen, preallocating any vector in MATLAB is more efficient than extending it. There are several methods of preallocating the vector. By starting with the last element, MATLAB would create a vector with that many elements. Then, the elements from 1 through end-1 could be initialized. For example, for a vector of structures that has three elements, start with the third element.

```
>> packages(3) = struct('item_no',587,'cost',11.11,...
        'price',33.33,'code','w');
>> packages(1) = struct('item_no',123,'cost',19.99,...
        'price',39.95,'code','g');
>> packages(2) = struct('item_no',456,'cost', 5.99,...
        'price',49.99,'code','l');
```

Another method is to create one element with the values from one structure, and use **repmat** to replicate it to the desired size. The remaining elements can then be modified. The following creates one structure and then replicates this into a *1* × *3* matrix.

```
>> packages = repmat(struct('item_no',123,'cost',19.99,...
        'price',39.95,'code','g'),1,3);
>> packages(2) = struct('item_no',456,'cost', 5.99,...
        'price',49.99,'code','l');
>> packages(3) = struct('item_no',587,'cost',11.11,...
        'price',33.33,'code','w');
```

Typing the name of the variable will display only the size of the structure vector and the names of the fields:

```
>> packages
packages =
1 x 3 struct array with fields:
    item_no
    cost
    price
    code
```

The variable *packages* is now a vector of structures, so each element in the vector is a structure. To display one element in the vector (one structure), an index into the vector would be specified. For example, to refer to the second element:

```
>> packages(2)
ans =
  item_no: 456
     cost: 5.9900
    price: 49.9900
     code: '1'
```

To refer to a field, it is necessary to refer to the particular structure, and then the field within it. This means using an index into the vector to refer to the structure, and then the dot operator to refer to a field. For example:

```
>> packages(1).code
ans =
g
```

Thus, there are essentially three levels to this data structure. The variable *packages* is the highest level, which is a vector of structures. Each of its elements is an individual structure. The fields within these individual structures are the lowest level. The following loop displays each element in the *packages* vector.

```
>> for i = 1:length(packages)
       disp(packages(i))
   end

   item_no:123
      cost: 19.9900
     price: 39.9500
      code: 'g'

   item_no:456
      cost: 5.9900
     price: 49.9900
      code: '1'

   item_no:587
      cost: 11.1100
     price: 33.3300
      code: 'w'
```

To refer to a particular field for all structures, in most programming languages it would be necessary to loop through all elements in the vector and use the dot operator to refer to the field for each element. However, this is not the case in MATLAB.

THE PROGRAMMING CONCEPT

For example, to print all of the costs, a <u>for</u> loop could be used:

```
>> for i=1:3
      fprintf('%f\n',packages(i).cost)
   end
19.990000
5.990000
11.110000
```

THE EFFICIENT METHOD

However, **fprintf** would do this automatically in MATLAB:

```
>> fprintf('%f\n',packages.cost)
19.990000
5.990000
11.110000
```

Using the dot operator in this manner to refer to all values of a field would result in the values being stored successively in the default variable *ans* since this method results in a comma-separated list:

```
>> packages.cost
ans =
    19.9900

ans =
    5.9900

ans =
    11.1100
```

However, the values can all be stored in a vector:

```
>> pc = [packages.cost]
pc =
    19.9900   5.9900   11.1100
```

Using this method, MATLAB allows the use of functions on all of the same fields within a vector of structures. For example, to sum all three cost fields, the vector of cost fields is passed to the **sum** function:

```
>> sum([packages.cost])
ans =
    37.0900
```

For vectors of structures, the entire vector (e.g., *packages*) could be passed to a function, or just one element (e.g., *packages(1)*), which would be a structure, or a field within one of the structures (e.g., *packages(2).price*).

The following is an example of a function that receives the entire vector of structures as an input argument, and prints all of it in a nice table format.

printpackages.m

```
function printpackages(packstruct)
% printpackages prints a table showing all
% values from a vector of 'packages' structures
% Format: printpackages(package structure)

fprintf('\nItem # Cost Price Code\n\n')
no_packs = length(packstruct);
for i = 1:no_packs
    fprintf('%6d %6.2f %6.2f %3c\n', ...
        packstruct(i).item_no, ...
        packstruct(i).cost, ...
        packstruct(i).price, ...
        packstruct(i).code)
end
end
```

The function loops through all of the elements of the vector, each of which is a structure, and uses the dot operator to refer to and print each field. An example of calling the function follows:

```
>> printpackages(packages)

Item #   Cost  Price  Code

  123   19.99  39.95     g
  456    5.99  49.99     l
  587   11.11  33.33     w
```

PRACTICE 8.4

A silicon wafer manufacturer stores, for every part in their inventory, a part number, how many are in the factory, and the cost for each. First, create a vector of structs called *parts* so that when displayed it has the following values:

```
>> parts
parts =
1 x 3 struct array with fields:
    partno
    quantity
    costper

>> parts(1)
ans =
```

```
        partno: 123
      quantity: 4
       costper: 33

>> parts(2)
ans =
        partno: 142
      quantity: 1
       costper: 150

>> parts(3)
ans =
        partno: 106
      quantity: 20
       costper: 7.5000
```

Next, write general code that will, for any values and any number of structures in the variable *parts*, print the part number and the total cost (quantity of the parts multiplied by the cost of each) in a column format.

For example, if the variable *parts* stores the previous values, the result would be:

```
123 132.00
142 150.00
106 150.00
```

The previous example involved a vector of structs. In the next example, a somewhat more complicated data structure will be introduced: a vector of structs in which some fields are vectors themselves. The example is a database of information that a professor might store for her class. This will be implemented as a vector of structures. The vector will store all of the class information.

Every element in the vector will be a structure, representing all information about one particular student. For every student, the professor wants to store (for now, this would be expanded later):

- name (a string)
- university ID number
- quiz scores (a vector of four quiz scores)

The vector variable, called *student*, might look like the following:

student

	name	id_no	quiz 1	quiz 2	quiz 3	quiz 4
1	C, Joe	999	10.0	9.5	0.0	10.0
2	Hernandez, Pete	784	10.0	10.0	9.0	10.0
3	Brownnose, Violet	332	7.5	6.0	8.5	7.5

Each element in the vector is a struct with three fields (*name, id_no, quiz*). The *quiz* field is a vector of quiz grades. The *name* field is a string.

This data structure could be defined as follows:

```
>> student(3) = struct('name','Brownnose, Violet',...
    'id_no',332,'quiz', [7.5 6 8.5 7.5]);
>> student(1) = struct('name','C, Joe',...
    'id_no',999,'quiz', [10 9.5 0 10]);
>> student(2) = struct('name','Hernandez, Pete',...
    'id_no',784,'quiz', [10 10 9 10]);
```

Once the data structure has been initialized, in MATLAB we could refer to different parts of it. The variable *student* is the entire array; MATLAB just shows the names of the fields.

```
>> student
student =
1 x 3 struct array with fields:
    name
    id_no
    quiz
```

To see the actual values, one would have to refer to individual structures and/or fields.

```
>> student(1)
ans =
      name: 'C, Joe'
     id_no: 999
      quiz: [10 9.5000 0 10]

>> student(1).quiz
ans =
    10.0000   9.5000      0  10.0000

>> student(1).quiz(2)
ans =
    9.5000

>> student(3).name(1)
ans =
B
```

With a more complicated data structure like this, it is important to be able to understand different parts of the variable. The following are examples of expressions that refer to different parts of this data structure:

- *student* is the entire data structure, which is a vector of structs
- *student(1)* is an element from the vector, which is an individual struct
- *student(1).quiz* is the *quiz* field from the structure, which is a vector of doubles

- *student(1).quiz(2)* is an individual **double** quiz grade
- *student(3).name(1)* is the first letter of the third student's name

One example of using this data structure would be to calculate and print the quiz average for each student. The following function accomplishes this. The *student* structure, as defined before, is passed to this function. The algorithm in the function follows:

- Print column headings
- Loop through the individual students. For each,
  - Sum the quiz grades
  - Calculate the average
  - Print the student's name and quiz average

With the programming method, a second (nested) loop would be required to find the running sum of the quiz grades. However, as we have seen, the **sum** function can be used to sum the vector of all quiz grades for each student. The function is defined as follows:

```
printAves.m

function printAves(student)
% This function prints the average quiz grade
% for each student in the vector of structs

fprintf('%-20s %-10s\n', 'Name', 'Average')
for i = 1:length(student)
    qsum = sum([ student(i).quiz] );
    no_quizzes = length(student(i).quiz);
    ave = qsum / no_quizzes;
    fprintf('%-20s %.1f\n', student(i).name, ave);
end
```

Here is an example of calling the function:

```
>> printAves(student)
Name                Average
C, Joe              7.4
Hernandez, Pete     9.8
Brownnose, Violet   7.4
```

8.2.5 Nested structures

A *nested structure* is a structure in which at least one member is itself a structure. For example, a structure for a line segment might consist of fields representing the two points at the ends of the line segment. Each of these points would be represented as a structure consisting of the x and y coordinates.

```
                    lineseg

            endpoint1  endpoint2
              x    y    x    y
            ┌────┬────┬────┬────┐
            │ 2  │ 4  │ 1  │ 6  │
            └────┴────┴────┴────┘
```

This shows a structure variable called *lineseg* that has two fields, *endpoint1* and *endpoint2*. Each of these is a structure consisting of two fields, *x* and *y*.

One method of defining this is to nest calls to the **struct** function:

```
>> lineseg = struct('endpoint1',struct('x',2,'y',4), ...
                    'endpoint2',struct('x',1,'y',6))
```

This method is the most efficient.

Another method would be to create structure variables first for the points, and then use these for the fields in the **struct** function (instead of using another **struct** function).

```
>> pointone = struct('x', 5, 'y', 11);
>> pointtwo = struct('x', 7, 'y', 9);
>> lineseg = struct('endpoint1', pointone,...
                    'endpoint2', pointtwo);
```

A third method, the least efficient, would be to build the nested structure one field at a time. Since this is a nested structure with one structure inside of another, the dot operator must be used twice here to get to the actual *x* and *y* coordinates.

```
>> lineseg.endpoint1.x = 2;
>> lineseg.endpoint1.y = 4;
>> lineseg.endpoint2.x = 1;
>> lineseg.endpoint2.y = 6;
```

Once the nested structure has been created, we can refer to different parts of the variable *lineseg*. Just typing the name of the variable shows only that it is a structure consisting of two fields, *endpoint1* and *endpoint2*, each of which is a structure.

```
>> lineseg
lineseg =
     endpoint1: [1 x 1 struct]
     endpoint2: [1 x 1 struct]
```

Typing the name of one of the nested structures will display the field names and values within that structure:

```
>> lineseg.endpoint1
ans =
     x: 2
     y: 4
```

Using the dot operator twice will refer to an individual coordinate, such as in the following example:

```
>> lineseg.endpoint1.x
ans =
     2
```

QUICK QUESTION!

How could we write a function *strpoint* that returns a string '(x,y)' containing the *x* and *y* coordinates? For example, it might be called separately to create strings for the two endpoints and then printed as shown here:

```
>> fprintf('The line segment consists of %s and %s \n',...
   strpoint(lineseg.endpoint1),...
   strpoint(lineseg.endpoint2))
The line segment consists of (2, 4) and (1, 6)
```

Answer: Since an *endpoint* structure is passed to an input argument in the function, the dot operator is used within the function to refer to the *x* and *y* coordinates. The **sprintf** function is used to create the string that is returned.

strpoint.m

```
function ptstr = strpoint(ptstruct)
% strpoint receives a struct containing x and y
% coordinates and returns a string '(x,y)'
% Format: strpoint(structure with x and y fields)
ptstr = sprintf('(%d, %d)', ptstruct.x, ptstruct.y);
end
```

8.2.6 Vectors of nested structures

Combining vectors and nested structures, it is possible to have a vector of structures in which some fields are structures themselves. Here is an example in which a company manufactures cylinders from different materials for industrial use. Information on them is stored in a data structure in a program.

The variable *cyls* is a vector of structures, each of which has fields *code*, *dimensions*, and *weight*. The *dimensions* field is a structure itself consisting of fields *rad* and *height* for the radius and height of each cylinder.

cyls

	code	dimensions		weight
		rad	height	
1	x	3	6	7
2	a	4	2	5
3	c	3	6	9

The following is an example of initializing this data structure by preallocating:

```
>> cyls(3) = struct('code', 'c', 'dimensions',...
      struct('rad', 3, 'height', 6), 'weight', 9);
>> cyls(1) = struct('code', 'x', 'dimensions',...
      struct('rad', 3, 'height', 6), 'weight', 7);
>> cyls(2) = struct('code', 'a', 'dimensions',...
      struct('rad', 4, 'height', 2), 'weight', 5);
```

Alternatively, it could be initialized by using the dot operator (which is not as efficient):

```
>> cyls(3).code = 'c';
>> cyls(3).dimensions.rad = 3;
>> cyls(3).dimensions.height = 6;
>> cyls(3).weight = 9;
>> cyls(1).code = 'x';
>> cyls(1).dimensions.rad = 3;
>> cyls(1).dimensions.height = 6;
>> cyls(1).weight = 7;
>> cyls(2).code = 'a';
>> cyls(2).dimensions.rad = 4;
>> cyls(2).dimensions.height = 2;
>> cyls(2).weight = 5;
```

There are several layers in this variable. For example,

- `cyls` is the entire data structure, which is a vector of structs
- `cyls(1)` is an individual element from the vector, which is a struct
- `cyls(2).code` is the *code* field from the struct *cyls(2)*; it is a character
- `cyls(3).dimensions` is the *dimensions* field from the struct *cyls(3)*; it is a struct itself
- `cyls(1).dimensions.rad` is the *rad* field that is from the struct *cyls(1). dimensions*; it is a **double** number

For these cylinders, one desired calculation may be the volume of each cylinder, which is defined as $\pi * r^2 * h$, where r is the radius and h is the height. The following function *printcylvols* prints the volume of each cylinder, along with its code for identification purposes. It calls a subfunction to calculate each volume.

```
printcylvols.m
```

```
function printcylvols(cyls)
% printcylvols prints the volumes of each cylinder
% in a specialized structure
% Format: printcylvols(cylinder structure)
```

Continued

```
% It calls a subfunction to calculate each volume

for i = 1:length(cyls)
  vol = cylvol(cyls(i).dimensions);
   fprintf('Cylinder %c has a volume of %.1f in^3\n', ...
     cyls(i).code, vol);
end
end

function cvol = cylvol(dims)
% cylvol calculates the volume of a cylinder
%  Format:  cylvol(dimensions  struct  w/  fields  'rad',
% 'height')

cvol = pi * dims.rad ^ 2 * dims.height;
end
```

The following is an example of calling this function:

```
>> printcylvols(cyls)
Cylinder x has a volume of 169.6 in^3
Cylinder a has a volume of 100.5 in^3
Cylinder c has a volume of 169.6 in^3
```

Note that the entire data structure, *cyls*, is passed to the function. The function loops through every element, each of which is a structure. It prints the *code* field for each, which is given by *cyls(i).code*. To calculate the volume of each cylinder, only the radius and height are needed, so rather than passing the entire structure to the subfunction *cylvol* (which would be *cyls(i)*), only the *dimensions* field is passed (*cyls(i).dimensions*). The function then receives the *dimensions* structure as an input argument, and uses the dot operator to refer to the *rad* and *height* fields within it.

PRACTICE 8.5

Modify the function *cylvol* to calculate the surface area of the cylinder in addition to the volume.

SUMMARY

Common Pitfalls

- Confusing the use of parentheses (cell indexing) versus curly braces (content indexing) for a cell array
- Forgetting to index into a vector using parentheses or referring to a field of a structure using the dot operator

Programming Style Guidelines

■ Use arrays when values are the same type and represent in some sense the same thing.
■ Use cell arrays or structures when the values are logically related but not the same type nor the same thing.
■ Use cell arrays rather than character matrices when storing strings of different lengths.
■ Use cell arrays rather than structures when it is desired to loop through the values or to vectorize the code.
■ Use structures rather than cell arrays when it is desired to use names for the different values rather than indices.

MATLAB Functions and Commands		
cell	iscellstr	isfield
celldisp	struct	fieldnames
cellplot	rmfield	
cellstr	isstruct	

MATLAB Operators
cell arrays { }
dot operator for structs .

Exercises

1. Create a cell array that stores phrases, such as:

    ```
    exclaimcell = {'Bravo', 'Fantastic job'};
    ```

    Pick a random phrase to print.
2. Create the following cell array:

    ```
    >> ca = {'abc', 11, 3:2:9, zeros(2)}
    ```

    Use the **reshape** function to make it a 2 × 2 matrix. Then, write an expression that would refer to just the last column of this cell array.
3. Create a 2 × 2 cell array by using the **cell** function to preallocate and then put values in the individual elements. Then, insert a row in the middle so that the cell array is now 3 × 2.
4. Create three cell array variables that store people's names, verbs, and nouns. For example:

    ```
    names = {'Harry', 'Xavier', 'Sue'};
    verbs = {'loves', 'eats'};
    nouns = {'baseballs', 'rocks', 'sushi'};
    ```

Write a script that will initialize these cell arrays, and then print sentences using one random element from each cell array (e.g., 'Xavier eats sushi').

5. Write a script that will prompt the user for strings and read them in, store them in a cell array (in a loop), and then print them out.

6. Create a row vector cell array to store the string 'xyz', the number 33.3, the vector 2:6, and the **logical** expression 'a' < 'c'. Use the transpose operator to make this a column vector, and use **reshape** to make it a 2×2 matrix. Use **celldisp** to display all elements.

7. Write a function *convstrs* that will receive a cell array of strings and a character 'u' or 'l'. If the character is 'u', it will return a new cell array with all of the strings in uppercase. If the character is 'l', it will return a new cell array with all of the strings in lowercase. If the character is neither 'u' nor 'l', *or* if the cell array does not contain all strings, the cell array that is returned will be identical to the input cell array.

8. Write a function *buildstr* that will receive a character and a positive integer *n*. It will create and return a cell array with strings of increasing lengths, from 1 to the integer *n*. It will build the strings with successive characters in the ASCII encoding.

```
>> buildstr('a',4)
ans =
    'a'      'ab'      'abc'      'abcd'

>> buildstr('F', 5)
ans =
    'F'      'FG'      'FGH'      'FGHI'      'FGHIJ'
```

9. Write a script that will create and display a cell array that will loop to store strings of lengths 1, 2, 3, and 4. The script will prompt the user for the strings. It will error-check, and print an error message and repeat the prompt if the user enters a string with an incorrect length.

10. Write a script that will loop three times, each time prompting the user for a vector, and will store the vectors in elements in a cell array. It will then loop to print the lengths of the vectors in the cell array.

11. Create a cell array variable that would store for a student his or her name, university ID number, and GPA. Print this information.

12. Create a structure variable that would store for a student his or her name, university ID number, and GPA. Print this information.

Note

This is just a structure exercise; MATLAB can handle complex numbers automatically as will be seen in Chapter 15.

13. A complex number is a number of the form $a + ib$, where a is called the real part, b is called the imaginary part, and $i = \sqrt{-1}$. Write a script that prompts the user separately to enter values for the real and imaginary parts, and stores them in a structure variable. It then prints the complex number in the form $a + ib$. The script should just print the value of a, then the string '$+ i$', and then the value of b. For example, if the script is named *compnumstruct*, running it would result in:

```
>> compnumstruct
Enter the real part: 2.1
Enter the imaginary part: 3.3
The complex number is 2.1 + i3.3
```

14. Modify the preceding script to call a function to prompt the user for the real and imaginary parts of the complex number, and also call a function to print the complex number.

15. Given a vector of structures defined by the following statements:

```
kit(2).sub.id = 123;
kit(2).sub.wt = 4.4;
kit(2).sub.code = 'a';
kit(2).name = 'xyz';
kit(2).lens =[ 4 7] ;
kit(1).name = 'rst';
kit(1).lens = 5:6;
kit(1).sub.id = 33;
kit(1).sub.wt = 11.11;
kit(1).sub.code = 'q';
```

Which of the following expressions are valid? If the expression is valid, give its value. If it is not valid, explain why.

```
kit(1).sub
```

```
kit(2).lens(1)
```

```
kit(1).code
```

```
kit(2).sub.id == kit(1).sub.id
```

```
strfind(kit(1).name, 's')
```

16. Create a vector of structures *experiments* that stores information on subjects used in an experiment. Each struct has four fields: *num*, *name*, *weights*, and *height*. The field *num* is an integer, *name* is a string, *weights* is a vector with two values (both of which are double values), and *height* is a struct with the fields *feet* and *inches* (both of which are integers). The following is an example of what the format might look like.

experiments

	num	name	weights 1	weights 2	height feet	height inches
1	33	Joe	200.34	202.45	5	6
2	11	Sally	111.45	111.11	7	2

Write a function *printhts* that will receive a vector in this format and will print the height of each subject in inches. This function calls another function *howhigh* that receives a height struct and returns the total height in inches. This function could also be called separately. Here is an example of calling the *printhts* function (assuming the preceding data), which calls the *howhigh* function:

```
>> printhts(experiments)
Joe is 66 inches tall
Sally is 86 inches tall
```

Here is an example of calling just the *howhigh* function:

```
>> howhigh(experiments(2).height)
ans =
    86
```

17. Create a data structure to store information about the elements in the periodic table of elements. For every element, store the name, atomic number, chemical symbol, class, atomic weight, and a seven-element vector for the number of electrons in each shell. Create a structure variable to store the information. An example for lithium follows:

```
Lithium 3 Li alkali_metal 6.94 2 1 0 0 0 0 0
```

18. In chemistry, the pH of an aqueous solution is a measure of its acidity. A solution with a pH of 7 is said to be *neutral*, a solution with a pH greater than 7 is *basic*, and a solution with a pH less than 7 is *acidic*. Create a vector of structures with various solutions and their pH values. Write a function that will determine acidity. Add another field to every structure for this.

19. A team of engineers is designing a bridge to span the Podunk River. As part of the design process, the local flooding data must be analyzed. The following information on each storm that has been recorded in the last 40 years is stored in a file: a code for the location of the source of the data, the amount of rainfall (in inches), and the duration of the storm (in hours), in that order. For example, the file might look like this:

```
321    2.4    1.5
111    3.3    12.1
    etc.
```

- Create a data file. Write the first part of the program: design a data structure to store the storm data from the file, and also the intensity of each storm. The intensity is the rainfall amount divided by the duration. Write a function to read the data from the file (use **load**), copy from the matrix into a vector of structs, and then calculate the intensities. Write another function to print all of the information in a neatly organized table.
- Add a function to the program to calculate the average intensity of the storms.

- Add a function to the program to print all of the information given on the most intense storm. Use a subfunction for this function that will return the index of the most intense storm.

20. A script stores information on potential subjects for an experiment in a vector of structures called *subjects*. The following shows an example of what the contents might be:

```
>> subjects
subjects =
1 x 3 struct array with fields:
  name
  sub_id
  height
  weight
>> subjects(1)
ans =
    name: 'Joey'
  sub_id: 111
  height: 6.7000
  weight: 222.2000
```

For this particular experiment, the only subjects who are eligible are those whose height or weight is lower than the average height or weight of all subjects. The script will print the names of those who are eligible. Create a vector with sample data in a script, and then write the code to accomplish this. Don't assume that the length of the vector is known; the code should be general.

21. Quiz data for a class is stored in a file. Each line in the file has the student ID number (which is an integer) followed by the quiz scores for that student. For example, if there are four students and three quizzes for each, the file might look like this:

```
44    7     7.5    8
33    5.5   6      6.5
37    8     8      8
24    6     7      8
```

First create the data file, and then store the data in a script in a vector of structures. Each element in the vector will be a structure that has two members: the integer student ID number and a vector of quiz scores. The structure will look like this:

students

	id_no	quiz 1	2	3
1	44	7	7.5	8
2	33	5.5	6	6.5
3	37	8	8	8
4	24	6	7	8

To accomplish this, first use the **load** function to read all the information from the file into a matrix. Then, using nested loops, copy the data into a vector of structures as specified. Then, the script will calculate and print the quiz average for each student. For example, for the previous file:

```
Student        Quiz Ave
44               7.50
33               6.00
37               8.00
24               7.00
```

22. Create a nested struct to store a person's name, address, and phone numbers. The struct should have three fields for the name, address, and phone. The address fields and phone fields will be structs.

23. Design a nested structure to store information on constellations for a rocket design company. Each structure should store the constellation's name and information on the stars in the constellation. The structure for the star information should include the star's name, core temperature, distance from the sun, and whether it is a binary star or not. Create variables and sample data for your data structure.

   To remain competitive, every manufacturing enterprise must maintain strict quality control measures. Extensive testing of new machines and products must be incorporated into the design cycle. Once manufactured, rigorous testing for imperfections and documentation is an important part of the feedback loop to the next design cycle.

24. Quality control involves keeping statistics on the quality of products. A company tracks its products and any failures that occur. For every imperfect part, a record is kept that includes the part number, a character code, a string that describes the failure, and the cost of both labor and material to fix the part. Create a vector of structures and create sample data for this company. Write a script that will print the information from the data structure in an easy-to-read format.

25. A manufacturer is testing a new machine that mills parts. Several trial runs are made for each part, and the resulting parts that are created are weighed. A file stores, for every part, the part identification number, the ideal weight for the part, and also the weights from five trial runs of milling this part. Create a file in this format. Write a script that will read this information and store it in a vector of structures. For every part print whether the average of the trial weights was less than, greater than, or equal to the ideal weight.

26. Create a data structure to store information on the planets in our solar system. For every planet, store its name, distance from the sun, and whether it is an inner planet or an outer planet.

27. Write a script that creates a vector of line segments (where each is a nested structure as shown in this chapter). Initialize the vector using any method. Print a table showing the values, such as shown in the following:

```
Line      From          To
====      =======       =======
1         ( 3, 5)       ( 4, 7)
2         ( 5, 6)       ( 2, 10)
          etc.
```

28. Investigate the built-in function **cell2struct** that converts a cell array into a vector of structs.

29. Investigate the built-in function **struct2cell** that converts a struct to a cell array.

Advanced File Input and Output

KEY TERMS

file input and output

file types

lower-level file I/O functions

open the file

close the file

file identifier

permission strings

end of the file

This chapter extends the input and output concepts that were introduced in Chapter 2. In that chapter, we saw how to read values entered by the user using the **input** function, and also the output functions **disp** and **fprintf** that display information in windows on the screen. For *file input and output* (file I/O), we used the **load** and **save** functions that can read from a data file into a matrix, and write from a matrix to a data file. We also saw that there are three different operations that can be performed on files: reading from files, writing to files (implying writing to the beginning of a file), and appending to a file (writing to the end of a file).

There are many different *file types*, which use different filename extensions. Thus far, using **load** and **save**, we have worked with files in the ASCII format that typically use either the extension *.dat* or *.txt*. The **load** command works only if there are the same number of values in each line and the values are the same type, so that the data can be stored in a matrix, and the **save** command only writes from a matrix to a file. If the data to be written or file to be read are in a different format, *lower-level file I/O functions* must be used.

The MATLAB® software has functions that can read and write data from different file types such as spreadsheets; typically, Excel spreadsheets have the filename extension *.xls*. Excel also has its own binary file type that uses the extension *.mat*. These are usually called MAT-files, and can be used to store variables that have been created in MATLAB.

In this chapter, we will introduce lower-level file input and output functions, as well as functions that work with different file types.

9.1 LOWER-LEVEL FILE I/O FUNCTIONS

When reading from a data file, the **load** function works as long as the data in the file are "regular"—in other words, the same kind of data on every line and in the same format on every line—so that they can be read into a matrix. However, data files are not always set up in this manner. When it is not possible to use **load**, MATLAB has what are called lower-level file input functions that can be used. The file must be opened first, which involves finding or creating the file and positioning an indicator at the beginning of the file. This indicator then moves through the file as it is being read from. When the reading has been completed, the file must be closed.

Similarly, the **save** function can write or append matrices to a file, but if the output is not a simple matrix, there are lower-level functions that write to files. Again, the file must be opened first and closed when the writing has been completed.

In general, the steps involved are:

■ *open the file*
■ read from the file, write to the file, or append to the file
■ *close the file*

First, the steps involved in opening and closing the file will be described. Several functions that perform the middle step of reading from or writing to the file will be described subsequently.

9.1.1 Opening and closing a file

Files are opened with the **fopen** function. By default, the **fopen** function opens a file for reading. If another mode is desired, a "permission string" is used to specify which, for example, for writing or appending. The **fopen** function returns –1 if it is not successful in opening the file, or an integer value that becomes the *file identifier* if it is successful. This file identifier is then used to refer to the file when calling other file I/O functions. The general form is

```
fid = fopen('filename', 'permission string');
```

where *fid* is a variable (it can be named anything) and the *permission strings* include:

r	reading (this is the default)
w	writing
a	appending

See **help fopen** for others.

After the **fopen** is attempted, the value returned should be tested to make sure that the file was successfully opened. For example, if attempting to open to read and the file does not exist, the **fopen** will not be successful. Since the **fopen** function returns –1 if the file was not found, this can be tested to decide whether to print an error message or to carry on and use the file. For example, if it is desired to read from a file *samp.dat*:

```
fid = fopen ('samp.dat');
if fid == -1
   disp('File open not successful')
else
   % Carry on and use the file!
end
```

Files should be closed when the program has finished reading from or writing or appending to them. The function that accomplishes this is the **fclose** function, which returns 0 if the file close was successful, or –1 if not. Individual files can be closed by specifying the file identifier, or if more than one file is open, all open files can be closed by passing the string 'all' to the **fclose** function. The general forms are:

```
closeresult = fclose(fid);

closeresult = fclose('all');
```

The **fclose** function should also be checked with an **if-else** statement to make sure that it was successful, so the outline of the code will be:

```
fid = fopen('filename', 'permission string' );
if fid == -1
   disp('File open not successful')
else

   % do something with the file!

   closeresult = fclose(fid);
   if closeresult == 0
      disp('File close successful')
   else
      disp('File close not successful')
   end
end
```

9.1.2 Reading from files

There are several lower-level functions that read from files. The function **fscanf** reads formatted data into a matrix, using conversion formats such as %d for integers, %s for strings, and %f for floats (**double** values). The **textscan** function reads text data from a file and stores the data in a cell array; it also uses conversion formats. The **fgetl** and **fgets** functions both read strings from a file one line at a time; the difference is that the **fgets** keeps the newline character if there is one at the end of the line, whereas the **fgetl** function gets rid of it. All of these functions require first opening the file, and then closing it when finished.

Since the **fgetl** and **fgets** functions read one line at a time, these functions are typically inside some form of a loop. The **fscanf** and **textscan** functions can read the entire data file into one data structure. In terms of level, these two functions are somewhat in between the **load** function and the lower-level functions such as **fgetl**. The file must be opened using **fopen** first, and should be closed using **fclose** after the data have been read. However, no loop is required; they will read in the entire file automatically into a data structure.

We will concentrate first on the **fgetl** function, which reads strings from a file one line at a time. The **fgetl** function affords more control over how the data are read than other input functions. The **fgetl** function reads one line of data from a file into a string; string functions can then be used to manipulate the data. Since **fgetl** only reads one line, it is normally placed in a loop that keeps going until the *end of the file* is reached. The function **feof** returns **logical true** if the end of the file has been reached. The function call **feof(fid)** would return **logical true** if the end of the file has been reached for the file identified by *fid*, or **logical false** if not. A general algorithm for reading from a file into strings would be:

- Attempt to open the file.
  - Check to ensure that the file open was successful.
- If opened, loop until the end of the file is reached.
  - For each line in the file,
    - read it into a string
    - manipulate the data
- Attempt to close the file.
  - Check to make sure that the file close was successful.

The following is the generic code to accomplish these tasks:

```
fid = fopen('filename');
if fid == -1
    disp('File open not successful')
else
    while feof(fid) == 0
        % Read one line into a string variable
        aline = fgetl(fid);
        % Use string functions to extract numbers, strings,
        % etc. from the line
        % Do something with the data!
    end
    closeresult = fclose(fid);
```

```
      if closeresult == 0
          disp('File close successful')
      else
          disp('File close not successful')
      end
   end
```

The permission string could be included in the call to the **fopen** function. For example:

```
fid = fopen('filename', 'r');
```

but the 'r' is not necessary since reading is the default. The condition on the **while** loop can be interpreted as saying "while the file end-of-file is false." Another way to write this is

```
while ~feof(fid)
```

which can be interpreted similarly as "while we're not at the end of the file."

For example, assume that there is a data file *subjexp.dat* that has on each line a number followed by a space followed by a character code. The **type** function can be used to display the contents of this file (since the file does not have the default extension *.m*, the extension on the filename must be included).

```
>> type subjexp.dat
5.3 a
2.2 b
3.3 a
4.4 a
1.1 b
```

The **load** function would not be able to read this into a matrix since it contains both numbers and text. Instead, the **fgetl** function can be used to read each line as a string and then string functions are used to separate the numbers and characters. For example, the following just reads each line and prints the number with two decimal places and then the rest of the string:

```
fileex.m
```

```
% Reads from a file one line at a time using fgetl
% Each line has a number and a character
% The script separates and prints them

% Open the file and check for success
fid = fopen('subjexp.dat');
if fid == -1
    disp('File open not successful')
else
    while feof(fid) == 0
        aline = fgetl(fid);
```

Continued

```
        % Separate each line into the number and character
        % code and convert to a number before printing
        [ num charcode] = strtok(aline);
        fprintf('%.2f %s\n', str2num(num), charcode)
    end

    % Check the file close for success
    closeresult = fclose(fid);
    if closeresult == 0
        disp('File close successful')
    else
        disp('File close not successful')
    end
end
```

The following is an example of executing the previous script:

```
>> fileex
5.30 a
2.20 b
3.30 a
4.40 a
1.10 b
File close successful
```

In this example, in the loop each time the **fgetl** function reads one line into a string variable. The string function **strtok** is then used to store the number and the character in separate variables, both of which are string variables (the second variable actually stores the blank space and the letter). If it is desired to perform calculations using the number, the function **str2num** would be used to convert the number stored in the string variable into a **double** value.

PRACTICE 9.1

Modify the script *fileex* to sum the numbers from the file. Create your own file in this format first.

Instead of using the **fgetl** function to read one line at a time, once a file has been opened the **fscanf** function can be used to read from this file directly into a matrix. However, the matrix must be manipulated somewhat to get it back into the original form from the file. The format of using the function is:

```
mat = fscanf(fid, 'format', [dimensions] )
```

The **fscanf** reads into the matrix variable *mat* columnwise from the file identified by *fid*. The 'format' includes conversion characters much like those used in the **fprintf** function. The dimensions specify the desired dimensions of *mat*; if the number of values in the file is not known, **inf** can be used for the second dimension. For example, the following would read in the same file just specified; each line contains a number, followed by a space and then a character.

```
>> fid = fopen('subjexp.dat');

>> mat = fscanf(fid,'%f %c',[2 inf])
mat =
     5.3000    2.2000    3.3000    4.4000    1.1000
    97.0000   98.0000   97.0000   97.0000   98.0000

>> fclose(fid);
```

The **fopen** opens the file for reading. The **fscanf** then reads from each line one double and one character, and places each pair in separate columns in the matrix (in other words, every line in the file becomes a column in the matrix). Note that the space in the format string is important: '%f %c' specifies that there is a float, a space, and a character. The dimensions specify that the matrix is to have two rows, by however many columns are necessary (equal to the number of lines in the file). Since matrices store values that are all the same type, the characters are stored as their ASCII equivalents in the character encoding (e.g., 'a' is 97).

Once this matrix has been created, it may be more useful to separate the rows into vector variables and to convert the second back to characters, which can be accomplished as follows:

```
>> nums = mat(1,:);

>> charcodes = char(mat(2,:))
charcodes =
abaab
```

Of course, the results from **fopen** and **fclose** should be checked but were omitted here for simplicity.

PRACTICE 9.2

Write a script to read in this file using **fscanf**, and sum the numbers.

QUICK QUESTION!

Instead of using the dimensions [2 inf] in the **fscanf** function, could we use [inf 2]?

Answer: No, [inf 2] would not work. Because **fscanf** reads each row from the file into a column in the matrix, the number of rows in the resulting matrix is known but the number of columns is not.

QUICK QUESTION!

Why is the space in the conversion string '%f %c' important? Would the following also work?

```
>> mat = fscanf(fid,'%f%c',[2 inf])
```

Answer: No, it would not work. The conversion string '%f %c' specifies that there is a real number, then a space, and then a character. Without the space in the conversion string, it would specify a real number immediately followed by a character (which would be the space in the file). Then, the next time it would be attempting to read the next real number but the file position indicator would be pointing to the character on the first line; the error would cause the **fscanf** function to halt. The end result follows:

```
>> fid = fopen('subjexp.dat');
>> mat = fscanf(fid,'%f%c',[2 inf])
mat =
    5.3000
   32.0000
```

The 32 is the numerical equivalent of the space character ' ', as seen here.

```
>> double(' ')
ans =
   32
```

Another option for reading from a file is to use the **textscan** function. The **textscan** function reads text data from a file and stores the data in a cell array. The **textscan** function is called, in its simplest form, as

```
cellarray = textscan(fid, 'format');
```

where the 'format' includes conversion characters much like those used in the **fprintf** function. For example, to read the file *subjexp.dat* we could do the following (again, for simplicity, omitting the error-check of **fopen** and **fclose**):

```
>> fid = fopen('subjexp.dat');
>> subjdata = textscan(fid,'%f %c');
>> fclose(fid)
```

The format string '%f %c' specifies that on each line there is a **double** value followed by a space followed by a character. This creates a *1 × 2* cell array variable called *subjdata*. The first element in this cell array is a column vector of doubles (the first column from the file); the second element is a column vector of characters (the second column from the file), as shown here:

```
>> subjdata
subjdata =
    [5 x 1 double]    [5 x 1 char]
>> subjdata{1}
ans =
    5.3000
    2.2000
    3.3000
    4.4000
    1.1000
```

```
>> subjdata{2}
ans =
a
b
a
a
b
```

To refer to individual values from the vector, it is necessary to index into the cell array using curly braces and then index into the vector using parentheses. For example, to refer to the third number in the first element of the cell array:

```
>> subjdata{1}(3)
ans =
    3.3000
```

A script that reads in these data and echo prints is shown here:

textscanex.m

```
% Reads data from a file using textscan
fid = fopen('subjexp.dat');
if fid == -1
    disp('File open not successful')
else
    % Reads numbers and characters into separate elements
    % in a cell array
    subjdata = textscan(fid,'%f %c');
    len = length(subjdata{1});
    for i= 1:len
        fprintf('%.1f %c\n',subjdata{1}(i),subjdata{2}(i))
    end

    closeresult = fclose(fid);
    if closeresult == 0
        disp('File close successful')
    else
        disp('File close not successful')
    end
end
```

Executing this script produces the following results:

```
>> textscanex
5.3 a
2.2 b
3.3 a
4.4 a
1.1 b
File close successful
```

PRACTICE 9.3

Modify the script *textscanex* to calculate the average of the column of numbers.

Comparison of input file functions

To compare the use of these input file functions, consider the example of a file called *xypoints.dat* that stores the x and y coordinates of some data points in the following format:

```
>> type xypoints.dat
x2.3y4.56
x7.7y11.11
x12.5y5.5
```

What we want is to be able to store the x and y coordinates in vectors so that we can plot the points. The lines in this file store combinations of characters and numbers, so the **load** function cannot be used. It is necessary to separate the characters from the numbers so that we can create the vectors. The following is the outline of the script to accomplish this:

```
fileInpCompare.m
```

```
fid = fopen('xypoints.dat');

if fid == −1
    disp('File open not successful')
else
    % Create x and y vectors for the data points
    % This part will be filled in using different methods

    % Plot the points
    plot(x,y,'k*')
    xlabel('x')
    ylabel('y')

    % Close the file
    closeresult = fclose(fid);
    if closeresult == 0
        disp('File close successful')
    else
        disp('File close not successful')
    end
end
```

We will now complete the middle part of this script using four different methods: **fgetl**, **fscanf** (two ways), and **textscan**.

To use the **fgetl** function, it is necessary to loop until the end-of-file is reached, reading each line as a string, and parsing the string into the various components and converting the strings containing the actual x and y coordinates to numbers. This would be accomplished as follows:

```
% using fgetl
x = [ ] ;
y = [ ] ;
while feof(fid) == 0
    aline = fgetl(fid);
    aline = aline(2:end);
    [xstr rest] = strtok(aline,'y');
    x = [x str2num(xstr)] ;
    ystr = rest(2:end);
    y = [y str2num(ystr)] ;
end
```

To instead use the **fscanf** function, we need to specify the format of every line in the file as a character, a number, a character, a number, and the newline character. Since the matrix that will be created will store every line from the file in a separate column, the dimensions will be $4 \times n$, where *n* is the number of lines in the file (and since we do not know that, **inf** is specified instead). The *x* characters will be in the first row of the matrix (the ASCII equivalent of 'x' in each element), the x coordinates will be in the second row, the ASCII equivalent of 'y' will be in the third row, and the fourth row will store the y coordinates. The code would be:

```
% using fscanf

mat = fscanf(fid, '%c%f%c%f\n', [4 inf] );
x = mat(2,:);
y = mat(4,:);
```

Note that the newline character in the format string is necessary. The data file itself was created by typing in the MATLAB Editor/Debugger, and to move down to the next line the Enter key was used, which is equivalent to the newline character. It is an actual character that is at the end of every line in the file. It is important to note that if the **fscanf** function is looking for a number, it will skip over whitespace characters including blank spaces and newline characters. However, if it is looking for a character, it would read a whitespace character.

In this case, after reading in 'x2.3y4.56' from the first line of the file, if we had as the format string '%c%f%c%f' (without the '\n'), it would then attempt to read again using '%c%f%c%f', but the next character it would read for the first '%c' would be the newline character—and then it would find the 'x' on the second line for the '%f'—not what is intended! (The difference between this and the previous example is that before we read a number followed by a character on each line. Thus, when looking for the next number it would skip over the newline character.)

Since we know that every line in the file contains both 'x' and 'y', not just any random characters, we can build that into the format string:

```
% using fscanf method 2

mat = fscanf(fid, 'x%fy%f\n', [2 inf] );
x = mat(1,:);
y = mat(2,:);
```

In this case the characters 'x' and 'y' are not read into the matrix, so the matrix only has the x coordinates (in the first row) and the y coordinates (in the second row).

Finally, to use the **textscan** function, we could put '%c' in the format string for the 'x' and 'y' characters, or build those in as with **fscanf**. If we build those in, the format string essentially specifies that there are four columns in the file, but it will only read the columns with the numbers into column vectors in the cell array *xydat*. The reason that the newline character is not necessary is that with **textscan**, the format string specifies what the columns look like in the file, whereas with **fscanf**, it specifies the format of every line in the file. Thus, it is a slightly different way of viewing the file format.

```
% using textscan

xydat = textscan(fid,'x%fy%f');
x = xydat{1} ;
y = xydat{2} ;
```

To summarize, we have now seen four methods of reading from a file. The function **load** will work only if the values in the file are all the same type and the same number of values are on every line in the file, so that they can be read into a matrix. If this is not the case, lower-level functions must be used. To use these, the file must be opened first and then closed when the reading has been completed.

The **fscanf** function will read into a matrix, converting the characters to their ASCII equivalents. The **textscan** function will instead read into a cell array that stores each column from the file into separate column vectors of the cell array. Finally, the **fgetl** function can be used in a loop to read each line from the file as a separate string; string manipulating functions must then be used to break the string into pieces and convert to numbers.

QUICK QUESTION!

If a data file is in the following format, which file input function(s) could be used to read it in?

```
48    25    23    23
12    45     1    31
31    39    42    40
```

Answer: Any of the file input functions could be used, but since the file consists of only numbers, the **load** function would be the easiest.

9.1.3 Writing to files

There are several lower-level functions that can write to files. We will concentrate on the **fprintf** function, which can be used to write to a file and also to append to a file.

To write one line at a time to a file, the **fprintf** function can be used. Like the other low-level functions, the file must be opened first for writing (or appending) and should be closed once the writing has been completed. We have, of course, been using **fprintf** to write to the screen. The screen is the default output device, so if a file identifier is not specified, the output goes to the screen; otherwise, it goes to the specified file. The default file identifier number is 1 for the screen. The general form is

```
fprintf(fid, 'format', variable(s));
```

The **fprintf** function actually returns the number of bytes that was written to the file, so if it is not desired to see that number, the output should be suppressed with a semicolon as shown here.

Note
When writing to the screen, the value returned by **fprintf** is not seen, but could be stored in a variable.

The following is an example of writing to a file named *tryit.txt*:

```
>> fid = fopen('tryit.txt', 'w');
>> for i = 1:3
       fprintf(fid,'The loop variable is %d\n', i);
   end
>> fclose(fid);
```

The permission string in the call to the **fopen** function specifies that the file is opened for writing to it. Just like when reading from a file, the results from **fopen** and **fclose** should really be checked to make sure they were successful. The **fopen** function attempts to open the file for writing. If the file already exists, the contents are erased so it is as if the file had not existed. If the file does not currently exist (which would be the norm), a new file is created. The **fopen** could fail, for example, if there isn't space to create this new file.

To see what was written to the file, we could then open it (for reading) and loop to read each line using **fgetl**:

```
>> fid = fopen('tryit.txt');

>> while ~feof(fid)
        aline = fgetl(fid)
    end

aline =
The loop variable is 1

aline =
The loop variable is 2

aline =
The loop variable is 3

>> fclose(fid);
```

Of course, we could also just display the contents using **type**.

Here is another example in which a matrix is written to a file. First, a 2×4 matrix is created, and then it is written to a file using the format string `'%d %d\n'`, which means that each column from the matrix will be written as a separate line in the file.

```
>> mat = [20 14 19 12; 8 12 17 5]
mat =

    20    14    19    12
     8    12    17     5

>> fid = fopen('randmat.dat','w');
>> fprintf(fid,'%d %d\n',mat);
>> fclose(fid);
```

Since this is a matrix, the **load** function can be used to read it in.

```
>> load randmat.dat
>> randmat
randmat =
    20     8
    14    12
    19    17
    12     5

>> randmat'
ans =
    20    14    19    12
     8    12    17     5
```

Transposing the matrix will display in the form of the original matrix. If this is desired to begin with, the matrix variable *mat* can be transposed before using **fprintf** to write to the file. (Of course, it would be much simpler in this case to just use **save** instead!)

PRACTICE 9.4

Create a *3 × 5* matrix of random integers, each in the range from 1 to 100. Write this to a file called *myrandmat.dat* in a *3 × 5* format using fprintf, so that the file appears identical to the original matrix. Load the file to confirm that it was created correctly.

9.1.4 Appending to files

The **fprintf** function can also be used to append to an existing file. The permission string is 'a', so the general form of the **fopen** would be:

```
fid = fopen('filename, 'a');
```

Then, using **fprintf** (typically in a loop), we would write to the file starting at the end of the file. The file would then be closed using **fclose**. What is written to the end of the file doesn't have to be in the same format as what is already in the file when appending.

9.2 WRITING AND READING SPREADSHEET FILES

MATLAB has functions **xlswrite** and **xlsread** that will write to and read from Excel spreadsheet files that have the extension *.xls*. (Note: This works under Windows environments provided that Excel is loaded. Under other environments, problems may be encountered if Excel cannot be loaded as a COM server.) For example, the following will create a *5 × 3* matrix of random integers, and then write it to a spreadsheet file *ranexcel.xls* that has five rows and three columns:

```
>> ranmat = round(rand(5,3)*100)
ranmat =

    96    77    62
    24    46    80
    61     2    93
    49    83    74
    90    45    18

>> xlswrite('ranexcel',ranmat)
```

The **xlsread** function will read from a spreadsheet file. For example, use the following to read from the file *ranexcel.xls*:

```
>> ssnums = xlsread('ranexcel')
ssnums =
    96    77    62
    24    46    80
    61     2    93
    49    83    74
    90    45    18
```

In both cases the *.xls* extension on the filename is the default, so it can be omitted.

These are shown in their most basic forms, when the matrix and/or spreadsheet contains just numbers and the entire spreadsheet is read or matrix is written. There are many qualifiers that can be used for these functions, however. For example, the following would read from the spreadsheet file *texttest.xls* that contains:

a	123	Cindy
b	333	Suzanne
c	432	David
d	987	Burt

```
>> [nums, txt] = xlsread('texttest.xls')
nums =
   123
   333
   432
   987
txt =
    'a'    ' '    'Cindy'
    'b'    ' '    'Suzanne'
    'c'    ' '    'David'
    'd'    ' '    'Burt'
```

This reads the numbers into a **double** vector variable *nums* and the text into a cell array *txt* (the **xlsread** function always returns the numbers first and then the text). The cell array is 4×3. It has three columns since the file had three columns, but since the middle column had numbers (which were extracted and stored in the vector *nums*), the middle column in the cell array *txt* consists of empty strings.

A loop could then be used to echo print the values from the spreadsheet in the original format:

```
>> for i = 1:length(nums)
       fprintf('%c %d %s\n', txt{i,1}, ...
           nums(i), txt{i,3})
   end
```

```
a 123 Cindy
b 333 Suzanne
c 432 David
d 987 Burt
```

9.3 USING MAT-FILES FOR VARIABLES

In addition to the functions that manipulate data files, MATLAB has functions that allow reading variables from and saving variables to files. These files are called MAT-files (because the extension on the filename is *.mat*), and they store the names and contents of variables. Variables can be written to MAT-files, appended to them, and read from them. (*Note:* MAT-files are very different from the data files that we have worked with so far. Rather than just storing data, MAT-files store the variable names in addition to their values. These files are typically used only within MATLAB; they are not used to share data with other programs.)

9.3.1 Writing variables to a file

The **save** command can be used to write variables to a MAT-file, or to append variables to a MAT-file. By default, the **save** function writes to a MAT-file. It can either save the entire current workspace (all variables that have been created), or a subset of the workspace (including, for example, just one variable). The **save** function will save the MAT-file in the Current Folder so it is important to set that correctly first.

To save all workspace variables in a file, the command is

```
save filename
```

The *.mat* extension is added automatically to the filename. The contents of the file can be displayed using **who** with the -file qualifier:

```
who -file filename
```

For example, in the following session in the Command Window, three variables are created; these are then displayed using **who**. Then, the variables are saved to a file named *sess1.mat*. The **who** function is then used to display the variables stored in that file.

```
>> mymat = rand(3,5);
>> x = 1:6;
>> y = x.^2;
>> who
Your variables are:
mymat    x    y
```

```
>> save sess1

>> who -file sess1
Your variables are:
mymat    x    y
```

To save just one variable to a file, the format is

```
save filename variablename
```

For example, just the matrix variable *mymat* is saved to a file called *sess2*:

```
>> save sess2 mymat
>> who -file sess2
Your variables are:
mymat
```

9.3.2 Appending variables to a MAT-file

Appending to a file adds to what has already been saved in a file, and is accomplished using the `-append` option. For example, assuming that the variable *mymat* has already been stored in the file *sess2.mat* as just shown, this would append the variable *x* to the file:

```
>> save -append sess2 x
>> who -file sess2
Your variables are:
mymat    x
```

Without specifying variable(s), just **save -append** would add all variables from the Command Window to the file. When this happens, if the variable is not in the file, it is appended. If there is a variable with the same name in the file, it is replaced by the current value from the Command Window.

9.3.3 Reading from a MAT-file

The **load** function can be used to read from different types of files. As with the **save** function, by default the file will be assumed to be a MAT-file, and **load** can load all variables from the file or only a subset. For example, in a new Command Window session in which no variables have yet been created, the **load** function could load from the files created in the previous section:

```
>> who
>> load sess2
>> who
Your variables are:
mymat    x
```

A subset of the variables in a file can be loaded by specifying them in the form:

```
load filename variable list
```

SUMMARY

Common Pitfalls
- Misspelling a file name, which causes a file open to be unsuccessful
- Using a lower-level file I/O function, when **load** or **save** could be used
- Forgetting that **fscanf** reads columnwise into a matrix, so every line in the file is read into a column in the resulting matrix
- Forgetting that **fscanf** converts characters to their ASCII equivalents
- Forgetting that **textscan** reads into a cell array (so curly braces are necessary to index)
- Forgetting to use the permission string 'a' for appending to a file (which means the data already in the file would be lost if 'w' was used!)

Programming Style Guidelines
- Use **load** when the file contains the same kind of data on every line and in the same format on every line.
- Always close files that were opened.
- Always check to make sure that files were opened and closed successfully.
- Make sure that all data are read from a file; for example, use a conditional loop to loop until the end of the file is reached rather than using a <u>for</u> loop.
- Be careful to use the correct formatting string when using **fscanf** or **textscan**.
- Store groups of related variables in separate MAT-files.

MATLAB Functions and Commands		
fopen	fgetl	xlswrite
fclose	fgets	xlsread
fscanf	feof	
textscan	fprintf	

Exercises

1. Write a script that will read from a file x and y data points in the following format:

```
x  0   y  1
x  1.3 y  2.2
x  2.2 y  6
x  3.4 y  7.4
```

The format of every line in the file is the letter 'x,' a space, the x value, space, the letter 'y,' space, and the y value. First, create the data file with 10 lines in this format. Do this by using the Editor/Debugger, and then File Save As *xypts.dat*. The script will attempt to open the data file and error-check to make sure that it was opened. If so, it uses a <u>for</u> loop and **fgetl** to read each line as a string. In the loop, it creates x and y vectors for the data points. After the loop, it plots these points and attempts to close the file. The script should print whether the file was successfully closed.

2. Modify the script from the previous problem. Assume that the data file is in exactly that format, but do not assume that the number of lines in the file is known. Instead of using a <u>for</u> loop, loop until the end of the file is reached. The number of points, however, should be in the plot title.

Medical organizations store a lot of very personal information on their patients. There is an acute need for improved methods of storing, sharing, and encrypting these medical records. Being able to read from and write to the data files is just the first step.

3. For a biomedical experiment, the names and weights of some patients have been stored in a file named *patwts.dat*. For example, the file might look like this:

```
Darby George 166.2
Helen Dee 143.5
Giovanni Lupa 192.4
Cat Donovan 215.1
```

Create this data file first. Then, write a script *readpatwts* that will first attempt to open the file. If the file open is not successful, an error message should be printed. If it is successful, the script will read the data into strings, one line at a time. Print for each person the name in the form 'last,first' followed by the weight. Also, calculate and print the average weight. Finally, print whether the file close was successful. For example, the result of running the script would look like this:

```
>> readpatwts
George,Darby 166.2
Dee,Helen 143.5
Lupa,Giovanni 192.4
Donovan,Cat 215.1
The ave weight is 179.30
File close successful
```

4. Create a data file to store blood donor information for a biomedical research company. For every donor, store the person's name, blood type, Rh factor, and blood pressure information. The blood type is A, B, AB, or O. The Rh factor is + or −. The blood pressure consists of two readings: systolic and diastolic (both are **double** numbers). Write a script to read from your file into a data structure and print the information from the file.

5. Create a file called *parts_inv.dat* that stores on each line a part number, cost, and quantity in inventory, in the following format:

```
123  5.99  52
456  3.97  100
333  2.22  567
```

Use **fscanf** to read this information, and print the total dollar amount of inventory (the sum of the cost multiplied by the quantity for each part).

6. A data file called *mathfile.dat* stores three characters on each line: an operand (a single-digit number), an operator (a one-character operator, such as +, −, /, \, *, ^), and then another operand (a single-digit number). For example, it might look like this:

```
>> type mathfile.dat
5+2
4-3
8-1
3+3
```

You are to write a script that will use **fgetl** to read from the file, one line at a time, perform the specified operation, and print the result.

7. Create a file that stores on each line a letter, a space, and a real number. For example, it might look like this:

```
e 5.4
f 3.3
c 2.2
f 1.1
c 2.2
```

Write a script that uses **textscan** to read from this file. It will print the sum of the numbers in the file. The script should error-check the file open and close, and print error messages as necessary.

8. Create a file named *phonenos.dat* that contains phone numbers in the following form:

```
6012425932
6178987654
8034562468
```

Read the phone numbers from the file and print them in the following form:

```
601-242-5932
```

Use **load** to read the phone numbers.

9. Create the file *phonenos.dat* as in Exercise 8. Use **textscan** to read the phone numbers, and then print them in the previous format.

10. Create the file *phonenos.dat* as in Exercise 8. Use **fgetl** to read the phone numbers in a loop, and then print them in the previous format.

11. Modify any of the previous scripts to write the phone numbers in the new format to a new file.

12. Write a script that will prompt the user for the name of a file from which to read. Loop to error-check until the user enters a valid filename that can be opened. (Note: This would be part of a longer program that would actually do something with the file, but for this problem all you have to do is error-check until the user enters a valid filename that can be read from.)

13. Write a script to read in division codes and sales for a company from a file that has the following format:

```
A 4.2
B 3.9
```

Print the division with the highest sales.

14. Assume that a file named *testread.dat* stores the following:

```
110x0.123y5.67z8.45
120x0.543y6.77z11.56
```

Assume that the following are typed *sequentially*. What would the values be?

```
tstid = fopen('testread.dat')

fileline = fgetl(tstid)

[beg endline] = strtok(fileline,'y')

length(beg)

feof(tstid)
```

15. Create a data file to store information on hurricanes. Each line in the file should have the name of the hurricane, its speed in miles per hour, and the diameter of its eye in miles. Then, write a script to read this information from the file and create a vector of structures to store it. Print the name and area of the eye for each hurricane.

16. Write a script that will loop to prompt the user for n circle radii. The script will call a function to calculate the area of each circle, and will write the results in sentence form to a file.

17. The wind chill factor (WCF) measures how cold it feels with a given air temperature (T, in degrees Fahrenheit) and wind speed (V, in miles per hour). One formula for the WCF follows:

$$WCF = 35.7 + 0.6\,T - 35.7\,(V^{0.16}) + 0.43\,T\,(V^{0.16})$$

Create a table showing WCFs for temperatures ranging from –20 to 55 in steps of 5, and wind speeds ranging from 0 to 55 in steps of 5. Write this to a file named *wcftable.dat*.

18. Create a data file that has points in a three-dimensional space stored in the following format:

```
x  2.2  y  5.3  z  1.8
```

Do this by creating x, y, and z vectors and then use **fprintf** to create the file in the specified format.

19. Create a file that contains college department names and enrollments. For example, it might look like this:

```
Aerospace 201
Civil 45
Mechanical 66
```

Write a script that will read the information from this file and create a new file that has just the first four characters from the department names, followed by the enrollments. The new file will be in this form:

```
Aero 201
Civi 45
Mech 66
```

20. A software package writes data to a file in a format that includes curly braces around each line and commas separating the values. For example, a data file *mm.dat* might look like this:

```
{33, 2, 11}
{45, 9, 3}
```

Use the **fgetl** function in a loop to read these data in. Create a matrix that stores just the numbers, and write the matrix to a new file. Assume that each line in the original file contains the same number of numbers.

21. A file stores sales data (in millions) by quarter. For example, the format may look like this:

```
2007Q1   4.5
2007Q2   5.2
```

Create the described file and then append the next quarter's data to it.

22. Create a spreadsheet that has on each line an integer student identification number followed by three quiz grades for that student. Read that information from the spreadsheet into a matrix, and print the average quiz score for each student.

23. The **xlswrite** function can write the contents of a cell array to a spreadsheet. A manufacturer stores information on the weights of selected parts in a cell array. Each row stores the part identifier code followed by weights of certain sample parts. To simulate this, create the following cell array:

```
>> parts = {'A22', 4.41 4.44 4.39 4.39
            'Z29', 8.88 8.95 8.84 8.92}
```

Then, write the cell array to a spreadsheet file.

24. A spreadsheet *popdata.xls* stores the population every 20 years for a small town that underwent a boom and then decline. Create this spreadsheet (include the header

row) and then read the headers into a cell array and the numbers into a matrix. Plot the data using the header strings on the axis labels.

Year	Population
1920	4021
1940	8053
1960	14994
1980	9942
2000	3385

25. Create a multiplication table and write it to a spreadsheet.
26. Read numbers from any spreadsheet file, and write the variable to a MAT-file.
27. Clear out any variables that you have in your Command Window. Create a matrix variable and two vector variables.
   - Make sure that you have your Current Folder set.
   - Store all variables to a MAT-file.
   - Store just the two vector variables in a different MAT-file.
   - Verify the contents of your files using **who**.
28. Create a set of random matrix variables with descriptive names (e.g., *ran2by2int*, *ran3by3double*, etc.) for use when testing matrix functions. Store all of these in a MAT-file.
29. A data file is created as a **char** matrix and then saved to a file. For example:

```
>> cmat = char('hello', 'ciao', 'goodbye')
cmat =
hello
ciao
goodbye
>> save stringsfile.dat cmat -ascii
```

Can the **load** function be used to read this in? What about **textscan**?
30. Create a file of strings as in Exercise 29 but create the file by opening a new M-file, type in the strings, and then save it as a data file. Can the **load** function be used to read this in? What about **textscan**?
31. Environmental engineers are trying to determine whether the underground aquifers in a region are being drained by a new spring water company in the area. Well depth data have been collected every year at several locations in the area. Create a data file that stores on each line the year, an alphanumeric code representing the location, and the measured well depth that year. Write a script that will read the data from the file and determine whether the average well depth has declined.

Advanced Functions

KEY TERMS

anonymous functions
function handle
function function
nested functions

recursive functions
variable number of
  arguments
outer function

inner function
recursion
infinite recursion

Functions were introduced in Chapter 2, and then expanded on in Chapter 6. In this chapter, several advanced features of functions and types of functions will be described. *Anonymous functions* are simple one-line functions that are called using their *function handle*. Other uses of function handles will also be demonstrated, including *function functions*. All of the functions that we have seen so far have had a well-defined number of input and output arguments, but we will see that the number of arguments can be varied. *Nested functions* are also introduced, which are functions contained within other functions. Finally, *recursive functions* are functions that call themselves. A recursive function can return a value, or may simply accomplish a task such as printing.

10.1 ANONYMOUS FUNCTIONS

An anonymous function is a very simple, one-line function. The advantage of an anonymous function is that it does not have to be stored in an M-file. This can greatly simplify programs, since often calculations are very simple, and the use of anonymous functions reduces the number of M-files necessary for a program. Anonymous functions can be created in the Command Window or in any script. The syntax for an anonymous function follows:

```
fnhandle = @ (arguments) functionbody;
```

where *fnhandle* stores the ***function handle***; it is essentially a way of referring to the function. The handle is assigned to the variable name using the @ operator. The arguments, in parentheses, correspond to the argument(s) that are passed to the function, just like any other kind of function. The `functionbody` is the body of the function, which is any valid MATLAB® expression. For example, here is an anonymous function that calculates and returns the area of a circle:

```
>> cirarea = @ (radius) pi * radius .^ 2;
```

The function handle name is *cirarea*. One argument is passed to the input argument *radius*. The body of the function is the expression `pi * radius .^ 2`. The .^ operator is used so that a vector of radii can be passed to the function.

The function is then called using the handle and passing arguments to it. The function call using the function handle looks just like a function call using a function name:

```
>> cirarea (4)
ans =
    50.2655

>> cirarea (1:4)
ans =
    3.1416   12.5664   28.2743   50.2655
```

Unlike functions stored in M-files, if no argument is passed to an anonymous function, the parentheses must still be in the function definition and in the function call. For example, following is an anonymous function that prints a random real number with two decimal places, as well as a call to this function:

```
>> prtran = @ () fprintf ('%.2f\n',rand);
>> prtran ()
0.95
```

Typing just the name of the function handle will display its contents, which is the function definition.

```
>> prtran
prtran =
    @ () fprintf ('%.2f\n',rand)
```

This is why parentheses must be used to call the function, even though no arguments are passed.

An anonymous function can be saved to a MAT-file, and then it can be loaded when needed.

```
>> cirarea = @ (radius) pi * radius .^ 2;
>> save anonfns cirarea
>> clear
>> load anonfns
```

```
>> who
Your variables are:
cirarea

>> cirarea
cirarea =
    @ (radius) pi * radius .^ 2
```

Other anonymous functions could be appended to this MAT-file. Even though an advantage of anonymous functions is that they do not have to be saved in individual M-files, it is frequently useful to save groups of related anonymous functions in a single MAT-file. Anonymous functions that are frequently used can be saved in a MAT-file and then loaded from this MAT-file in every MATLAB Command Window.

PRACTICE 10.1

Create your own anonymous functions to calculate the areas of circles, rectangles, and something else (you decide!). Store these anonymous functions in a file called *myareas.mat*.

10.2 USES OF FUNCTION HANDLES

Function handles can also be created for functions other than anonymous functions, both built-in and user-defined functions. For example, the following would create a function handle for the built-in **factorial** function:

```
>> facth = @factorial;
```

The @ operator gets the handle of the function, which is then stored in a variable *facth*.

The handle could then be used to call the function, just like the handle for the anonymous functions, such as the following:

```
>> facth(5)
ans =
    120
```

Using the function handle to call the function instead of using the name of the function does not in itself demonstrate why it is useful, so an obvious question would be why function handles are necessary for functions other than anonymous functions.

10.2.1 Function functions

One reason for using function handles is to be able to pass functions to other functions—these are called *function functions*. For example, let's say we have a function that creates an x vector. The y vector is created by evaluating a function at each of the x points, and then these points are plotted.

fnfnexamp.m

```
function fnfnexamp(funh)
% fnfnexamp receives the handle of a function
% and plots that function of x (which is 1:.25:6)
% Format: fnfnexamp(function handle)

x = 1:.25:6;
y = funh(x);
plot(x,y,'ko')
xlabel('x')
ylabel('fn(x)')
title(func2str(funh))
end
```

What we want to do is pass a function to be the value of the input argument *funh*, such as **sin**, **cos**, or **tan**. Simply passing the name of the function does not work:

```
>> fnfnexamp(sin)
??? Error using ==> sin
Not enough input arguments.
```

Instead, we have to pass the handle of the function:

```
>> fnfnexamp(@sin)
```

which creates the *y* vector as **sin(x)** and then brings up the plot as seen in Figure 10.1. The function **func2str** converts a function handle to a string; this is used for the title.

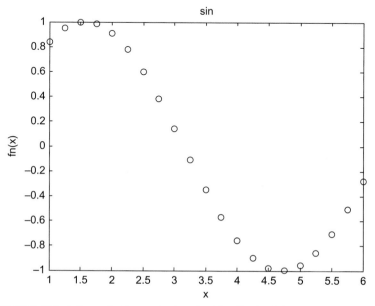

FIGURE 10.1 Plot of **sin** created by passing handle of function to plot

Passing the handle to the **cos** function instead would graph cosine instead of sine:

```
>> fnfnexamp(@cos)
```

We could also pass the handle of any user-defined or anonymous function to the *fnfnexamp* function.

There is also a built-in function **str2func** that will convert a string to a function handle. A string containing the name of a function could be passed as an input argument, and then converted to a function handle.

fnstrfn2.m

```
function fnstrfn2(funstr)
% fnstrfn2 receives the name of a function as a string
% it converts this to a function handle and
% then plots the function of x (which is 1:.25:6)
% Format: fnstrfn2(function name as string)
x = 1:.25:6;
funh = str2func(funstr);
y = funh(x);
plot(x,y,'ko')
xlabel('x')
ylabel('fn(x)')
title(funstr)
end
```

This would be called by passing a string to the function, and would create the same plot:

```
>> fnstrfn2('sin')
```

PRACTICE 10.2

Write a function that will receive as input arguments an *x* vector and a function handle, and will create a vector *y* that is the function of *x* (whichever function handle is passed) and will also plot the data from the *x* and *y* vectors with the function name in the title.

MATLAB has some built-in function functions. One built-in function function is **fplot**, which plots a function between limits that are specified. The form of the call to **fplot** is

```
fplot(fnhandle, [xmin    xmax] )
```

For example, to pass the **sin** function to **fplot** one would pass its handle (see Figure 10.2 for the result):

```
>> fplot(@sin, [-pi pi])
```

The **fplot** function is a nice shortcut—it is not necessary to create *x* and *y* vectors, and it plots a continuous curve rather than discrete points.

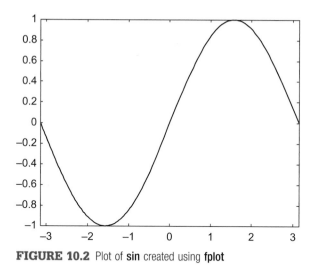

FIGURE 10.2 Plot of **sin** created using **fplot**

The function **feval** will evaluate a function handle and execute the function for the specified argument. For example, the following is equivalent to **sin(3.2)**:

```
>> feval (@sin, 3.2)
ans =
    -0.0584
```

10.3 VARIABLE NUMBERS OF ARGUMENTS

The functions that we've written thus far contain a fixed number of input and output arguments. For example, in the following function that we have defined previously, there is one input argument and two output arguments:

areacirc.m

```
function [ area, circum] = areacirc (rad)
% areacirc returns the area and
% the circumference of a circle
% Format: areacirc (radius)

area = pi * rad .* rad;
circum = 2 * pi * rad;
end
```

However, this is not always the case. It is possible to have a *variable number of arguments*, both input and output arguments. A built-in cell array **varargin** can be used to store a variable number of input arguments and a built-in cell array **varargout** can be used to store a variable number of output arguments. These are cell arrays because the arguments could be different types, and only cell arrays can store different kinds of values in the various elements. The function **nargin** returns the number of input arguments that were passed to the function, and the function **nargout** determines how many output arguments are expected to be returned from a function.

10.3.1 Variable number of input arguments

For example, the following function *areafori* has a variable number of input arguments, either one or two. The name of the function stands for "area, feet or inches." If only one argument is passed to the function, it represents the radius in feet. If two arguments are passed, the second can be a character 'i' indicating that the result should be in inches (for any other character, the default of feet is assumed).

The function uses the built-in cell array **varargin**, which stores any number of input arguments. The function **nargin** returns the number of input arguments that were passed to the function. In this case, the radius is the first argument passed so it is stored in the first element in **varargin**. If a second argument is passed (if **nargin** is 2), it is a character that specifies the units.

areafori.m

```
function area = areafori(varargin)
% areafori returns the area of a circle in feet
% The radius is passed, and potentially the unit of
% inches is also passed, in which case the result will be
% given in inches instead of feet
% Format: areafori(radius) or areafori(radius,'i')

n = nargin; % number of input arguments
radius = varargin{ 1} ; % Given in feet by default
if n == 2
    unit = varargin{ 2} ;
    % if inches is specified, convert the radius
    if unit == 'i'
        radius = radius * 12;
    end
end
area = pi * radius .^ 2;
end
```

Note
Curly braces are used to refer to the elements in the cell array **varargin**.

Some examples of calling this function follow:

```
>> areafori(3)
ans =
    28.2743

>> areafori(1,'i')
ans =
    452.3893
```

In this case, it was assumed that the radius will always be passed to the function. The function header can therefore be modified to indicate that the radius will be passed, and then a variable number of remaining input arguments (either none or one):

areafori2.m

```
function area = areafori2(radius, varargin)
% areafori2 returns the area of a circle in feet
% The radius is passed, and potentially the unit of
% inches is also passed, in which case the result will be
% given in inches instead of feet
```

Continued

```
% Format: areafori(radius) or areafori(radius,'i')

n = nargin; % number of input arguments

if n == 2
    unit = varargin{1};
    % if inches is specified, convert the radius
    if unit == 'i'
        radius = radius * 12;
    end
end
area = pi * radius .^ 2;
end
```

```
>> areafori2(1,'i')
ans =
    452.3893

>> areafori2(3)
ans =
    28.2743
```

Note that **nargin** returns the total number of input arguments, not just the number of arguments in the cell array **varargin**.

There are basically two formats for the function header with a variable number of input arguments. For a function with one output argument, the options are:

```
function outarg = fnname(varargin)
```

```
function outarg = fnname(input arguments, varargin)
```

Either some input arguments are built into the function header, and **varargin** stores anything else that is passed, or all of the input arguments go into **varargin**.

PRACTICE 10.3

The sum of a geometric series is given by

$$1 + r + r^2 + r^3 + r^4 + \ldots + r^n$$

Write a function called *geomser* that will receive a value for *r* and calculate and return the sum of the geometric series. If a second argument is passed to the function, it is the value of *n*; otherwise, the function generates a random integer for *n* (in the range from 5 to 30). Note that loops are not necessary to accomplish this. The following examples of calls to this function illustrate what the result should be:

```
>> geomser(1,5) % 1 + 1^1 + 1^2 + 1^3 + 1^4 + 1^5
ans =
    6
```

```
>> g = geomser(2,4) % 1 + 2^1 + 2^2 + 2^3 + 2^4
g =
   31

>> geomser(1) % 1 + 1^1 + 1^2 + 1^3 + ... ?
ans =
   12
```

Note that in the last example, a random integer was generated for n (which must have been 11). Use the following header for the function, and fill in the rest:

```
function sgs = geomser(r, varargin)
```

10.3.2 Variable number of output arguments

A variable number of output arguments can also be specified. For example, one input argument is passed to the following function *typesize*. The function will always return a character specifying whether the input argument was a scalar ('s'), vector ('v'), or matrix ('m'). This character is returned through the output argument *arrtype*.

Additionally, if the input argument was a vector, the function returns the length of the vector, and if the input argument was a matrix, the function returns the number of rows and the number of columns of the matrix. The output argument **varargout** is used, which is a cell array. So, for a vector the length is returned through **varargout**, and for a matrix both the number of rows and columns are returned through **varargout**.

```
typesize.m
```

```
function[ arrtype, varargout] = typesize(inputval)
% typesize returns a character 's' for scalar, 'v'
% for vector, or 'm' for matrix input argument
% also returns length of a vector or dimensions of matrix
% Format: typesize(inputArgument)

[r c] = size(inputval);

if r==1 && c==1
    arrtype = 's';
elseif r==1 || c==1
    arrtype = 'v';
    varargout{ 1} = length(inputval);
else
    arrtype = 'm';
    varargout{ 1} = r;
    varargout{ 2} = c;
end
end
```

```
>> typesize(5)
ans =
s

>> [arrtype, len] = typesize(4:6)
arrtype =
v
len =
     3

>> [arrtype, r, c] = typesize([4:6;3:5])
arrtype =
m

r =
     2
c =
     3
```

In the examples shown here, the user must actually know the type of the argument to determine how many variables to have on the left side of the assignment statement. An error will result if there are too many variables.

```
>>[arrtype, r,c] = typesize(4:6)
Error in ==> typesize at 7
[r c] = size(inputval);

??? Output argument "varargout{2}" (and maybe others) not
assigned during call to "\path\typesize.m>typesize".
```

The function **nargout** can be called to determine how many output arguments were used to call a function. For example, in the following function *mysize*, a matrix is passed to the function. The function behaves like the built-in function **size** in that it returns the number of rows and columns. However, if three variables are used to store the result of calling this function, it also returns the total number of elements:

```
mysize.m
```

```
function[ row col varargout] = mysize(mat)
% mysize returns dimensions of input argument
% and possibly also total # of elements
% Format: mysize(inputArgument)

[row col] = size(mat);

if nargout == 3
    varargout{ 1} = row* col;
end
end
```

```
>> [r c] = mysize(eye(3))
r =
     3
c =
     3
>> [r c elem] = mysize(eye(3))
r =
     3
c =
     3
elem =
     9
```

In the first call to the *mysize* function in the previous example, the value of **nargout** was 2, so the function only returned the output arguments *row* and *col*. In the second call, since there were three variables on the left of the assignment statement, the value of **nargout** was 3; thus, the function also returned the total number of elements.

There are basically two formats for the function header with a variable number of output arguments:

```
function varargout = fnname(input args)

function [output args, varargout] = fnname(input args)
```

Either some output arguments are built into the function header, and **varargout** stores anything else that is returned or all go into **varargout**. The function is called as follows:

```
[ variables] = fnname(input args);
```

Note

The function **nargout** does not return the number of output arguments in the function header, but the number of output arguments expected from the function (meaning, the number of variables in the vector in the left side of the assignment statement when calling the function).

QUICK QUESTION!

A temperature in degrees Centigrade is passed to a function called *converttemp*. How could we write this function so that it converts this temperature to degrees Fahrenheit, and possibly also to degrees Kelvin, depending on the number of output arguments? The conversions follow:

$$F = \frac{9}{5}C + 32$$

$$K = C + 273.15$$

Here are possible calls to the function:

```
>> df = converttemp(17)
df =
```

```
    62.6000
>> [df dk] = converttemp(17)
df =
    62.6000
dk =
   290.1500
```

Answer: We could write the function two different ways: one with only **varargout** in the function header, and one that has an output argument for the degrees F and also **varargout** in the function header.

Continued

converttemp.m

```
function[ degreesF, varargout] = converttemp(degreesC)
% converttemp converts temperature in degrees C
% to degrees F and maybe also K
% Format: converttemp(C temperature)

degreesF = 9/5* degreesC + 32;
n = nargout;
if n == 2
    varargout{ 1} = degreesC + 273.15;
end
end
```

converttempii.m

```
function varargout = converttempii(degreesC)
% converttempii converts temperature in degrees C
% to degrees F and maybe also K
% Format: converttempii(C temperature)

n = nargout;
varargout{ 1} = 9/5* degreesC + 32;
if n == 2
    varargout{ 2} = degreesC + 273.15;
end
end
```

10.4 NESTED FUNCTIONS

Just as we have seen that loops can be nested, meaning one inside of another, functions can be nested. The terminology for *nested functions* is that an *outer function* can have within it *inner functions*. When functions are nested, every function must have an **end** statement (much like loops). The general format of a nested function is as follows:

```
outer function header

        body of outer function

        inner function header
            body of inner function
        end % inner function

    more body of outer function

end % outer function
```

The inner function can be in any part of the body of the outer function so there may be parts of the body of the outer function before and after the inner function. There can be multiple inner functions.

The scope of any variable is the workspace of the outermost function in which it is defined and used. This means that a variable defined in the outer function could be used in an inner function (without passing it). A variable defined in the inner function *could* be used in the outer function, but if it is not used in the outer function the scope is just the inner function.

For example, the following function calculates and returns the volume of a cube. Three arguments are passed to it, the length and width of the base of the cube and also the height. The outer function calls a nested function that calculates and returns the area of the base of the cube.

nestedvolume.m

```
function outvol = nestedvolume(len, wid, ht)
% nestedvolume receives the lenght, width, and
% height of a cube and returns the volume; it calls
% a nested function that returns the area of the base
% Format: nestedvolume(length,width,height)

outvol = base * ht;

    function outbase = base
    % returns the area of the base
    outbase = len * wid;
    end % base function

end % nestedvolume function
```

Note

It is not necessary to pass the length and width to the inner function, since the scope of these variables includes the inner function.

An example of calling this function follows:

```
>> v = nestedvolume(3,5,7)
v =
    105
```

Output arguments are different from variables. The scope of an output argument is just the nested function; it cannot be used in the outer function. In this example, *outbase* can only be used in the *base* function; its value, for example, could not be printed from *nestedvolume*. Examples of nested functions will be used in the section on graphical user interfaces.

10.5 RECURSIVE FUNCTIONS

Recursion occurs when something is defined in terms of itself. In programming, a *recursive function* is a function that calls itself. Recursion is used very commonly in programming, although many simple examples (including some shown in this section) are actually not very efficient and can be replaced by iterative methods (loops, or vectorized code in MATLAB). Nontrivial examples go beyond the scope of this book, so the concept of recursion is simply introduced here.

The first example will be of a factorial. Normally, the factorial of an integer n is defined iteratively:

```
n! = 1 * 2 * 3 * ... * n
```

For example, $4! = 1 * 2 * 3 * 4$, or 24.

Another, recursive definition is

```
n! = n * (n - 1)!      general case

1! = 1                 base case
```

This definition is recursive because a factorial is defined in terms of another factorial. There are two parts to any recursive definition: the general (or inductive) case and the base case. We say that in general the factorial of n is defined as n multiplied by the factorial of $(n - 1)$, but the base case is that the factorial of 1 is just 1. The base case stops the recursion.

For example,

```
3! = 3 * 2!
        2! = 2 * 1!
                1! = 1
        = 2
  = 6
```

The way this works is that 3! is defined in terms of another factorial, as 3 * 2!. This expression cannot yet be evaluated, because first we have to find the value of 2!. So, in trying to evaluate the expression 3 * 2!, we are interrupted by the recursive definition. According to the definition, 2! is 2 * 1!.

Again, the expression 2 * 1! cannot yet be evaluated because first we have to find the value of 1!. According to the definition, 1! is 1. Since we now know what 1! is, we can continue with the expression that was just being evaluated; now we know that 2 * 1! is 2 * 1, or 2. Thus, we can now finish the previous expression that was being evaluated; we know that 3 * 2! is 3 * 2, or 6.

This is the way that recursion always works. With recursion, the expressions are put on hold with the interruption of the general case of the recursive definition. This keeps happening until the base case of the recursive definition applies. This finally stops the recursion, and then the expressions that were put on hold are evaluated in the reverse order. In this case, first the evaluation of 2 * 1! was completed, and then 3 * 2!.

There must always be a base case to end the recursion, and the base case must be reached at some point. Otherwise, *infinite recursion* would occur (theoretically, although MATLAB will stop the recursion eventually).

We have already seen the built-in function **factorial** in MATLAB to calculate factorials, and we have seen how to implement the iterative definition using a running product. Now we will instead write a recursive function called *fact*. The function will receive an integer *n*, which we will for simplicity assume is a positive integer, and will calculate *n*! using the recursive definition given previously.

fact.m

```
function facn = fact(n)
% fact recursively finds n!
% Format: fact(n)
if n == 1
    facn = 1;
else
    facn = n * fact(n - 1);
end
end
```

The function calculates one value, using an **if-else** statement to choose between the base and general cases. If the value passed to the function is 1, the function returns 1 since 1! is equal to 1. Otherwise, the general case applies. According to the definition, the factorial of *n*, which is what this function is calculating, is defined as *n* multiplied by the factorial of (*n* – 1). So, the function assigns `n * fact(n - 1)` to the output argument.

How does this work? Exactly the way the example was sketched above for 3!. Let's trace what would happen if the integer 3 is passed to the function:

```
fact(3) tries to assign 3 * fact(2)
                    fact(2) tries to assign 2 * fact(1)
                                        fact(1) assigns 1
                    fact(2) assigns 2
fact(3) assigns 6
```

When the function is first called, 3 is not equal to 1, so the statement

```
facn = n * fact(n - 1);
```

is executed. This will attempt to assign the value of 3 * fact(2) to *facn*, but this expression cannot be evaluated yet and therefore a value cannot be assigned yet because first the value of fact(2) must be found.

Thus, the assignment statement has been interrupted by a recursive call to the *fact* function. The call to the function fact(2) results in an attempt to assign 2 * fact(1), but again this expression cannot yet be evaluated. Next, the call to the function fact(1) results in a complete execution of an assignment statement since it assigns just 1. Once the base case has been reached, the assignment statements that were interrupted can be evaluated, in the reverse order.

Calling the function yields the same result as the built-in **factorial** function, as follows:

```
>> fact(5)
ans =
    120

>> factorial(5)
ans =
    120
```

The recursive factorial function is a very common example of a recursive function. It is somewhat of a lame example, however, since recursion is not necessary to find a factorial. A **for** loop can be used just as well in programming (or, of course, the built-in function in MATLAB).

Another, better, example is of a recursive function that does not return anything, but simply prints. The following function *prtwords* receives a sentence, and prints the words in the sentence in reverse order. The algorithm for the *prtwords* function follows:

- Receive a sentence as an input argument.
- Use **strtok** to break the sentence into the first word and the rest of the sentence.
- If the rest of the sentence is not empty (i.e., if there is more to it), recursively call the *prtwords* function and pass to it the rest of the sentence.
- Print the word.

The function definition follows:

prtwords.m

```
function prtwords(sent)
% prtwords recusively prints the words in a string
% in reverse order
% Format: prtwords(string)

[word, rest] = strtok(sent);
if ~isempty(rest)
    prtwords(rest);
end
disp(word)
end
```

Here is an example of calling the function, passing the sentence "what does this do":

```
>> prtwords('what does this do')
do
this
does
what
```

An outline of what happens when the function is called follows:

```
The function receives 'what does this do'
It breaks it into word = 'what', rest = ' does this do'
Since "rest" is not empty, calls prtwords, passing "rest"

      The function receives ' does this do'
      It breaks it into word = 'does', rest = ' this do'
      Since "rest" is not empty, calls prtwords, passing "rest"

            The function receives ' this do'
            It breaks it into word = 'this', rest = ' do'
            Since "rest" is not empty, calls prtwords, passing "rest"

                  The function receives ' do'
                  It breaks it into word = 'do', rest = ''
                  "rest" is empty so no recursive call
                  Print 'do'

            Print 'this'

      Print 'does'

Print 'what'
```

In this example, the base case is when the rest of the string is empty—in other words, the end of the original sentence has been reached. Every time the function is called, the execution of the function is interrupted by a recursive call to the function, until the base case is reached. When the base case is reached, the entire function can be executed, including printing the word (in the base case, the word "do").

Once that execution of the function is completed, the program returns to the previous version of the function in which the word was "this," and finishes

PRACTICE 10.4

For the following function,

```
recurfn.m
function outvar = recurfn(num)
% Format: recurfn(number)

if num < 0
    outvar = 4;
else
    outvar = 3 + recurfn(num - 1);
end
end
```

what would be returned by the call to the function recurfn(2.3)? Think about it, and then type in the function and test it.

the execution by printing the word "this." This continues; the versions of the function are finished in the reverse order, so the program ends up printing the words from the sentence in the reverse order.

SUMMARY

Common Pitfalls
- Trying to pass just the name of a function to a function function; instead, the function handle must be passed
- Thinking that **nargin** is the number of elements in **varargin** (it may be, but not necessarily; **nargin** is the total number of input arguments)
- Forgetting the base case for a recursive function

Programming Style Guidelines
- Use anonymous functions whenever the function body consists of just a simple expression.
- Store related anonymous functions together in one MAT-file.
- If some inputs and/or outputs will always be passed to/from a function, use standard input arguments/output arguments for them. Use **varargin** and **varargout** only when it is not known ahead of time whether other input/output arguments will be needed.
- Use iteration instead of recursion when possible.

MATLAB Reserved Words
end (for functions)

MATLAB Functions and Commands	
func2str	varargin
str2func	varargout
fplot	nargin
feval	nargout

MATLAB Operator
handle of anonymous functions @

Exercises

1. An approximation for a factorial can be found using Stirling's formula:

$$n! \approx \sqrt{2 \pi n}\left(\frac{n}{e}\right)^{n}$$

Write an anonymous function to implement this.

2. The velocity of sound in air is $49.02 \sqrt{T}$ feet per second where T is the air temperature in degrees Rankine. Write an anonymous function that will calculate this. One argument, the air temperature in degrees R, will be passed to the function and it will return the velocity of sound.

3. The hyperbolic sine for an argument x is defined as

$$hyperbolicsine(x) = (e^x - e^{-x})/2$$

Write an anonymous function to implement this. Compare yours to the built-in function **sinh**.

4. In special relativity, the Lorentz factor is a number that describes the effect of speed on various physical properties when the speed is significant relative to the speed of light. Mathematically, the Lorentz factor is given as

$$\gamma = \frac{1}{\sqrt{1 - \dfrac{v^2}{c^2}}}$$

Write an anonymous function *gamma* that will receive the speed v and calculate the Lorentz factor. Use 3×10^8 m/s for the speed of light, c.

5. Create a set of anonymous functions to do length conversions and store them in a file named *lenconv.mat*. Call each a descriptive name, such as *cmtoinch*, to convert from centimeters to inches.

6. Write a function that will receive data in the form of x and y vectors, and a handle to a plot function, and will produce the plot. For example, a call to the function would look like `wsfn(x,y,@bar)`.

7. Write a function *plot2fnhand* that will receive two function handles as input arguments, and will display in two Figure Windows plots of these functions, with the function names in the titles. The function will create an x vector that ranges from 1 to n (where n is a random integer in the range from 4 to 10). For example, if the function is called as follows:

```
>> plot2fnhand(@sqrt, @exp)
```

and the random integer is 5, the Figure Window 1 would display the **sqrt** function of $x = 1:5$, and the second Figure Window would display **exp(x)** for $x = 1:5$.

8. Write an anonymous function to implement the following quadratic: $3x^2 - 2x + 5$. Then, use **fplot** to plot the function in the range from –6 to 6.

9. Use **feval** as an alternative way to accomplish the following function calls:

```
abs(-4)
size(zeros(4))
```

Use **feval** twice for the second one!

10. There is a built-in function function called **cellfun** that evaluates a function for every element of a cell array. Create a cell array, then call the **cellfun** function, passing the handle of the **length** function and the cell array to determine the length of every element in the cell array.

11. Write a function that will print a random integer. If no arguments are passed to the function, it will print an integer in the range from 1 to 100. If one argument is passed, it is the max and the integer will be in the range from 1 to max. If two arguments are passed, they represent the min and max, and it will print an integer in the range from min to max.

12. The velocity of sound in air is 49.02 $\sqrt{T}$ feet per second where T is the air temperature in degrees Rankine. Write a function to implement this. If just one argument is passed to the function, it is assumed to be the air temperature in degrees Rankine. If, however, two arguments are passed, the two arguments would be first an air temperature and then a character 'f' for Fahrenheit or 'c' for Celsius (so this would then have to be converted to Rankine). *Note:* Degrees R = degrees F + 459.67. Degrees F = 9/5 degrees C + 32.

13. Write a function *areaperim* that will calculate both the area and perimeter of a polygon. The radius r will be passed as an argument to the function. If a second argument is passed to the function, it represents the number of sides n. If, however, only one argument is passed, the function generates a random value for n (an integer in the range from 3 to 8). For a polygon with n sides inscribed in a circle with a radius of r, the area a and perimeter p of the polygon can be found by

$$a = \frac{1}{2}nr^2 \sin\left(\frac{2\pi}{n}\right), \quad p = 2\pi r \sin\left(\frac{\pi}{n}\right)$$

14. Write a function that will receive a variable number of input arguments: the length and width of a rectangle, and possibly also the height of a box that has this rectangle as its base. The function should return the rectangle area if just the length and width are passed, or also the volume if the height is also passed.

15. Write a function that will receive the radius r of a sphere. It will calculate and return the volume of the sphere ($4/3 \pi r^3$). If the function call expects two output arguments, the function will also return the surface area of the sphere ($4\pi r^2$).

16. A basic unit of data storage is the byte (B). One B is equivalent to eight bits. A nibble is equivalent to four bits. Write a function that will receive the number of bytes, and will return the number of bits. If two output arguments are expected, it will also return the number of nibbles.

17. Write a function *arcSector* that receives a radius and an angle in radians of a circular sector. The function will return the area of the sector and, if two output arguments are used when calling the function, the length of the circular arc of the sector. The area of a sector is given as

$$A = \frac{1}{2}r^2\theta$$

and the length of a circular arc is given as

$$l = r\theta$$

The following are some examples of calling the function:

```
>> arcSector(5,pi/4)
ans =
    9.8175
>> [a l] = arcSector(3,pi/4)
a =
    3.5343
l =
    2.3562
```

18. In quantum mechanics, Planck's constant, written as h, is defined as $h = 6.626 * 10^{-34}$ joule-seconds. The Dirac constant hbar is given in terms of Planck's constant:

$$\text{hbar} = \frac{h}{2\pi}$$

Write a function *planck* that will return Planck's constant. If two output arguments are expected, it will also return the Dirac constant.

19. The overall electrical resistance of n resistors in parallel is given as

$$R_T = \left(\frac{1}{R_1} + \frac{1}{R_2} + \frac{1}{R_3} + \ldots + \frac{1}{R_n}\right)^{-1}$$

Write a function *Req* that will receive a variable number of resistance values and will return the equivalent electrical resistance of the resistor network. The following are examples of calling the function:

```
>> Req(100,100)
ans =
    50

>> Req(100,330,1000)
ans =
    71.2743
```

20. Write a function *unwind* that will receive a matrix as an input argument. It will return a row vector created columnwise from the elements in the matrix. If the number of expected output arguments is two, it will also return this as a column vector.

21. A script *ftocmenu* uses the **menu** function to ask the user to choose between output to the screen and to a file. The output is a list of temperature conversions, converting from Fahrenheit to Celsius for values of F ranging from 32 to 62 in steps of 10. If the user chooses File, the script opens a file for writing, calls a function *tempcon* that writes the results to a file (passing the file identifier), and closes the file. Otherwise, it calls the function *tempcon*, passing no arguments, which writes to the screen.

In either case, the function *tempcon* is called by the script. If the file identifier is passed to this function it writes to the file; otherwise, if no arguments are passed, it writes to the screen. The function *tempcon* calls a subfunction that converts one temperature in degrees F to C using the formula: $C = (F-32) * 5/9$. Here is an example of executing the script; in this case, the user chooses the Screen button:

```
>> ftocmenu
32F is 0.0C
42F is 5.6C
52F is 11.1C
62F is 16.7C
>>
```
ftocmenu.m

```
choice = menu('Choose output mode','Screen','File');
if choice == 2
    fid = fopen('yourfilename.dat','w');
    tempcon(fid)
    fclose(fid);
else
    tempcon
end
```

Write the function *tempcon* and its subfunction.

22. The built-in function **clock** returns a vector with six elements representing, in order, the year, month, day, hour, minutes, and seconds. Write a function *whatday* that (using the **clock** function) will always return the current day. If the function call expects two output arguments, it will also return the month. If the function call expects three output arguments, it will also return the year.

23. The built-in function **date** returns a string containing the day, month, and year. Write a function (using the **date** function) that will always return the current day. If the function call expects two output arguments, it will also return the month. If the function call expects three output arguments, it will also return the year.

24. Write a function to calculate the volume of a cone. The volume formula is $V = AH$ where A is the area of the circular base ($A = \pi r^2$ where r is the radius) and H is the height. Use a nested function to calculate A.

25. The two real roots of a quadratic equation $ax^2 + bx + c = 0$ (where a is nonzero) are given by

$$\frac{-b \pm \sqrt{D}}{2 * a}$$

where the discriminant $D = b^2 - 4 * a * c$. Write a function to calculate and return the roots of a quadratic equation. Pass the values of a, b, and c to the function. Use a nested function to calculate the discriminant.

26. A recursive definition of a^n, where a is an integer and n is a non-negative integer, follows:

$$a^n = 1 \quad \text{if } n == 0$$
$$= a * a^{n-1} \quad \text{if } n > 0$$

Write a recursive function called *mypower*, which receives a and n and returns the value of a^n by implementing the previous definition. *Note*: The program should *not* use the ^ operator anywhere; this is to be done recursively instead! Test the function.

27. What does the following function do?

```
function outvar = mystery(x,y)
if y == 1
   outvar = x;
else
   outvar = x + mystery(x,y-1);
end
```

Give one word to describe what this function does with its two arguments.

The Fibonacci numbers is a sequence of numbers 0, 1, 1, 2, 3, 5, 8, 13, 21, 34 . . . . The sequence starts with 0 and 1. All other Fibonacci numbers are obtained by adding the previous two Fibonacci numbers. The higher up in the sequence that you go, the closer the fraction of one Fibonacci number divided by the previous is to the golden ratio. The Fibonacci numbers can be seen in an astonishing number of examples in nature, for example, the arrangement of petals on a sunflower.

28. The Fibonacci numbers is a sequence of numbers F_i:

```
0  1  1  2  3  5  8  13  21  34  ...
```

where F_0 is 0, F_1 is 1, F_2 is 1, F_3 is 2, and so on. A recursive definition is:

$$F_0 = 0$$
$$F_1 = 1$$
$$F_n = F_{n-2} + F_{n-1} \quad \text{if } n > 1$$

Write a recursive function to implement this definition. The function will receive one integer argument n, and it will return one integer value that is the nth Fibonacci number. Note that in this definition there is one general case but two base cases. Then, test the function by printing the first 20 Fibonacci numbers.

29. Use **fgets** to read strings from a file and recursively print them backward.

30. Combinatorial coefficients can be defined recursively as follows:

$$C(n,m) = 1 \qquad\qquad \text{if } m = 0 \text{ or } m = n$$
$$= C(n-1, m-1) + C(n-1, m) \quad \text{otherwise}$$

Write a recursive function to implement this definition.

PART 2

Advanced Topics for Problem Solving with MATLAB

Advanced Plotting Techniques

KEY TERMS

histogram	plot properties	parent/children
stem plot	object	core objects
pie chart	object handle	text box
area plot	graphics primitives	hyetograph
bin	object-oriented	inverse functions
animation	programming	

In Chapter 2, we introduced the use of the function **plot** in the MATLAB® software to get simple, two-dimensional (2D) plots of x and y points represented by two vectors *x* and *y*. We have also seen some functions that allow customization of these plots. In this chapter we will explore other types of plots, ways of customizing plots, and some applications that combine plotting with functions and file input. Additionally, animation, three-dimensional (3D) plots, and graphics' properties will be introduced.

11.1 PLOT FUNCTIONS

So far, we have used **plot** to create two-dimensional plots and **bar** to create bar charts. We have seen how to clear the Figure Window using **clf**, and how to create and number Figure Windows using **figure**. Labeling plots has been accomplished using **xlabel**, **ylabel**, **title**, and **legend**, and we have also seen how to customize the strings passed to these functions using **sprintf**. The **axis** function changes the axes from the defaults that would be taken from the data in the *x* and *y* vectors to the values specified. Finally, the **grid** and **hold** toggle functions print grids or not, or lock the current graph in the Figure Window so that the next plot will be superimposed.

11.1.1 Matrix of plots

Another function that is very useful with all types of plots is **subplot**, which creates a matrix of plots in the current Figure Window. Three arguments are passed to it in the form **subplot(r,c,n)**; where *r* and *c* are the dimensions of the matrix and *n* is the number of the particular plot within this matrix. The plots are numbered rowwise starting in the upper left corner. In many cases, it is useful to create a **subplot** in a <u>for</u> loop so the loop variable can iterate through the integers 1 through *n*.

When the **subplot** function is called in a loop, the first two arguments will always be the same since they give the dimensions of the matrix. The third argument will iterate through the numbers assigned to the elements of the matrix. When the **subplot** function is called, it makes the specified element the "active" plot; then, any plot function can be used complete with formatting such as axis labeling and titles within that element.

For example, the following **subplot** shows the difference, in one Figure Window, between using 10 points and 20 points to plot **sin(x)** between 0 and 2 * π. The **subplot** function creates a *1 × 2* row vector of plots in the Figure Window, so that the two plots are shown side by side. The loop variable *i* iterates through the values 1 and then 2.

The first time through the loop, when *i* has the value 1, 10 * 1 or 10 points are used, and the value of the third argument to the **subplot** function is 1. The second time through the loop, 20 points are used and the third argument to **subplot** is 2. Note that **sprintf** is used to print how many points were used in the plot titles. The resulting Figure Window with both plots is shown in Figure 11.1.

subplotex.m

```
% Demonstrates subplot using a for loop
for i = 1:2
    x = linspace(0,2*pi,10 * i);
    y = sin(x);
    subplot(1,2,i)
    plot(x,y,'ko')
    xlabel('x')
    ylabel('sin(x)')
    title(sprintf('%d Points',10 * i))
end
```

QUICK QUESTION!

What are some options for plotting more than one graph?
Answer: There are several methods, depending on whether you want them in a single Figure Window superim posed (using **hold on**), in a matrix in a single Figure Window (using **subplot**), or in multiple Figure Windows (using **figure(n)**).

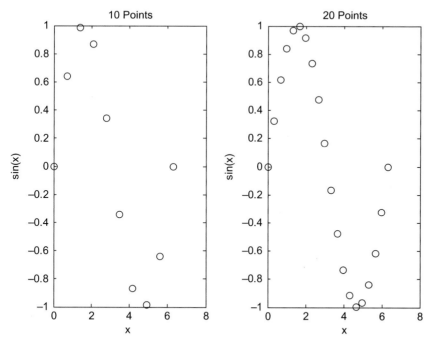

FIGURE 11.1 Subplot to demonstrate a plot using 10 points and 20 points

11.1.2 Plot types

Besides **plot** and **bar**, there are many other plot types such as *histograms, stem plots,* and *pie charts,* as well as other functions that customize graphs.

Described in this section are some of the other plotting functions. The functions **bar, barh, area,** and **stem** essentially display the same data as the **plot** function, but in different forms. The **bar** function draws a bar chart (as we have seen before), **barh** draws a horizontal bar chart, **area** draws the plot as a continuous curve and fills in under the curve that is created, and **stem** draws a stem plot.

For example, the following script creates a Figure Window that uses a *2 × 2* **subplot** to demonstrate these four plot types using the same data points (see Figure 11.2).

subplottypes.m

```
% Subplot to show plot types

year = 2007:2011;
pop = [0.9  1.4  1.7  1.3  1.8];
subplot(2,2,1)
bar(year,pop)
```

Continued

Note

The third argument in the call to the **subplot** function is a single index into the matrix created in the Figure Window; the numbering is rowwise (in contrast to the normal columnwise unwinding that MATLAB uses for matrices).

```
title('bar')
xlabel('Year')
ylabel('Population')
subplot(2,2,2)
barh(year,pop)
title('barh')
xlabel('Year')
ylabel('Population')
subplot(2,2,3)
area(year,pop)
title('area')
xlabel('Year')
ylabel('Population')
subplot(2,2,4)
stem(year,pop)
title('stem')
xlabel('Year')
ylabel('Population')
```

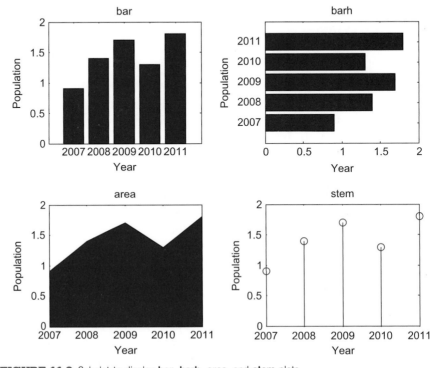

FIGURE 11.2 Subplot to display **bar, barh, area**, and **stem** plots

QUICK QUESTION!

Could we produce the previous **subplot** using a loop?

Answer: Yes, we can store the names of the plots in a cell array. These names are put in the titles, and also concatenated

with the string '(x,y)' and passed to the **eval** function to evaluate the function.

```
loopsubplot.m

% Demonstrates evaluating plot type names in order to
% use the plot functions and put the names in titles

year = 2007:2011;
pop = [0.9  1.4  1.7  1.3  1.8];
titles = {'bar', 'barh', 'area', 'stem'};
for i = 1:4
    subplot(2,2,i)
    eval([titles{i} '(year,pop)'])
    title(titles{i})
    xlabel('Year')
    ylabel('Population')
end
```

For a matrix, the **bar** and **barh** functions will group together the values in each row. For example:

```
>> groupages = [8 19 43 25; 35 44 30 45]
groupages =
     8    19    43    25
    35    44    30    45
>> bar(groupages)
>> xlabel('Group')
>> ylabel('Ages')
```

produces the plot shown in Figure 11.3.

Note that MATLAB groups together the values in the first row and then in the second row. It cycles through colors to distinguish the bars. The 'stack' option will stack rather than grouping the values, so the y value represented by the top of the bar is the sum of the values from that row (shown in Figure 11.4).

```
>> bar(groupages,'stack')
>> xlabel('Group')
>> ylabel('Ages')
```

A *histogram* is a particular type of bar chart that shows the frequency of occurrence of values within a vector. Histograms use what are called *bins* to collect values that are in given ranges. MATLAB has a function to create a histogram, **hist**. Calling the function with the form **hist(vec)** by default takes the values in the vector *vec* and puts them into 10 bins (or, **hist(vec,n)** will put them into *n* bins) and plots this, as shown in Figure 11.5.

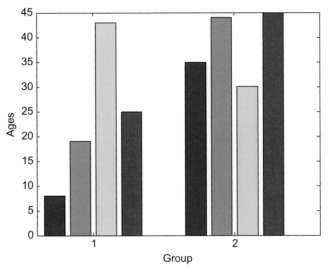

FIGURE 11.3 Data from a matrix in a bar chart

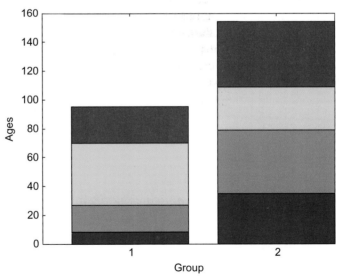

FIGURE 11.4 Stacked bar chart of matrix data

PRACTICE 11.1

Create a file that has two lines with *n* numbers in each. Use **load** to read this into a matrix. Then, use **subplot** to show the **bar** and stacked **bar** charts side by side.

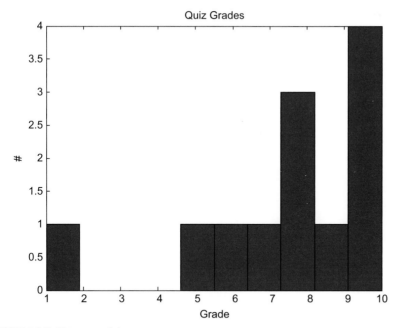

FIGURE 11.5 Histogram of data

```
>> quizzes = [10 8 5 10 10 6 9 7 8 10 1 8];
>> hist(quizzes)
>> xlabel('Grade')
>> ylabel('#')
>> title('Quiz Grades')
```

In this example, the numbers range from 1 to 10 in the vector, and there are 10 bins in the range from 1 to 10. The heights of the bins represent the number of values that fall within that particular bin. The **hist** function actually returns values; the first returned is a vector showing how many of the values from the original vector fall into each of the bins:

```
>> c = hist(quizzes)
c =
     1   0   0   0   1   1   1   3   1   4
```

The bins in a histogram are not necessarily all the same width. Histograms are used for statistical analyses of data; more statistics will be covered in Chapter 13.

MATLAB has a function **pie** that will create a pie chart. Calling the function with the form **pie(vec)** draws a pie chart, using the percentage of each element of *vec* of the whole (the sum). It shows these starting from the top of the circle and going around counterclockwise. For example, the first value in the vector *[11 14 8 3 1]*, 11 is 30% of the sum, 14 is 38% of the sum, and so forth, as shown in Figure 11.6.

```
>> pie([11 14 8 3 1])
```

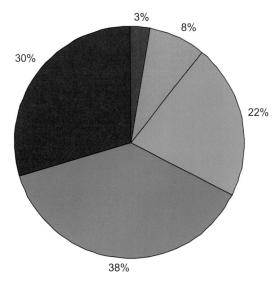

FIGURE 11.6 Pie chart showing percentages

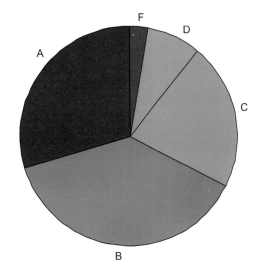

FIGURE 11.7 Pie chart with labels from a cell array

A cell array of labels can also be passed to the **pie** function; these labels will appear instead of the percentages (shown in Figure 11.7).

```
>> pie([11 14 8 3 1], {'A','B','C',...
'D', 'F'})
```

PRACTICE 11.2

A chemistry professor teaches three classes. These are the course numbers and enrollments:

```
CH 101      111
CH 105       52
CH 555       12
```

Use **subplot** to show this information using **pie** charts: the **pie** chart on the left should show the percentage of students in each course, and on the right the course numbers. Put appropriate titles on them.

11.2 ANIMATION

In this section we will examine a couple of ways to *animate* a plot. These are visuals, so the results can't really be shown here; it is necessary to type these into MATLAB to see the results.

We'll start by animating a plot of **sin(x)** with the vectors:

```
>> x = -2*pi : 1/100 : 2*pi;
>> y = sin(x);
```

This results in enough points that we'll be able to see the result using the built-in **comet** function, which shows the plot by first showing the point (x(1), y(1)), and then moving on to the point (x(2), y(2)), and so on, leaving a trail (like a comet!) of all of the previous points.

```
>> comet(x,y)
```

The end result looks similar to the result of **plot(x,y)**.

Another way of animating is to use the built-in function **movie**, which displays recorded movie frames. The frames are captured in a loop using the built-in function **getframe**, and are stored in a matrix. For example, the following script again animates the **sin** function. The **axis** function is used so that MATLAB will use the same set of axes for all frames, and using the **min** and **max** functions on the data vectors *x* and *y* will allow us to see all points. It displays the "movie" once in the **for** loop, and then again when the movie function is called.

sinmovie.m

```
% Shows a movie of the sin function
clear

x = -2*pi: 1/5 : 2*pi;
y = sin(x);
n = length(x);

for i = 1:n
    plot(x(i),y(i),'r*')
    axis([min(x)-1 max(x)+1 min(y)-1 max(y)+1])
    M(i) = getframe;
end
movie(M)
```

11.3 THREE-DIMENSIONAL PLOTS

MATLAB has functions that will display 3D plots. Many of these functions have the same name as the corresponding 2D plot function with a "3" at the end. For example, the 3D line plot function is called **plot3**. Other functions include **bar3**, **bar3h**, **pie3**, **comet3**, and **stem3**.

Vectors representing x, y, and z coordinates are passed to the **plot3** and **stem3** functions. These functions show the points in 3D space. Clicking on the rotate 3D icon and then in the plot allows the user to rotate to see the plot from different angles. Also, using the **grid** function makes it easier to visualize, as shown in Figure 11.8.

```
>> x = 1:5;
>> y = [0 -2 4 11 3];
>> z = 2:2:10;
>> plot3(x,y,z,'k*')
>> grid
>> xlabel('x')
>> ylabel('y')
>> zlabel('z')
>> title('3D Plot')
```

For the **bar3** and **bar3h** functions, *y* and *z* vectors are passed and the function shows 3D bars as shown, for example, for **bar3** in Figure 11.9.

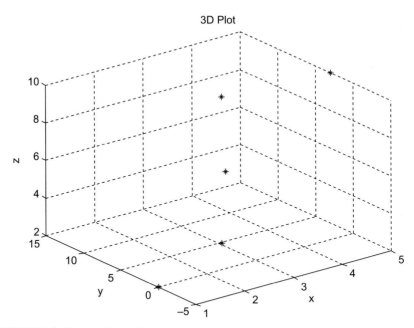

FIGURE 11.8 3D plot with a grid

```
>> y = 1:6;
>> z = [33 11 5 9 22 30];
>> bar3(y,z)
>> xlabel('x')
>> ylabel('y')
>> zlabel('z')
>> title('3D Bar')
```

A matrix can also be passed, such as a 5×5 **spiral** matrix (which "spirals" the integers 1 to 25 or more generally from 1 to n^2 for **spiral(n)**), as shown in Figure 11.10.

```
>> mat = spiral(5)
mat =

    21    22    23    24    25
    20     7     8     9    10
    19     6     1     2    11
    18     5     4     3    12
    17    16    15    14    13
>> bar3(mat)
>> title('3D Spiral')
>> xlabel('x')
>> ylabel('y')
>> zlabel('z')
```

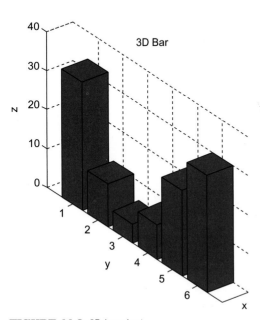

FIGURE 11.9 3D bar chart

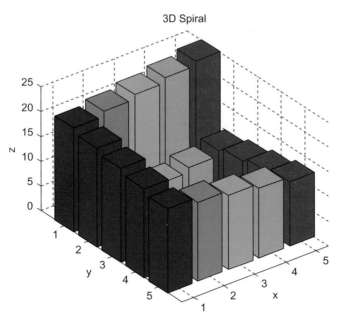

FIGURE 11.10 3D plot of a spiral matrix

Similarly, the **pie3** function shows data from a vector as a 3D pie as shown in Figure 11.11.

```
>> pie3([3 10 5 2])
```

Displaying the result of an animated plot in three dimensions is interesting. For example, try the following using the **comet3** function:

```
>> t = 0:0.001:12*pi;
>> comet3(cos(t), sin(t), t)
```

Other interesting 3D plot types include **mesh** and **surf**. The **mesh** function draws a wireframe mesh of 3D points, whereas the **surf** function uses color to display the parametric surfaces defined by the points. MATLAB has several functions that will create the matrices used for the (x,y,z) coordinates for specified shapes (e.g., **sphere** and **cylinder**).

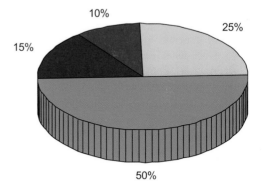

FIGURE 11.11 3D pie chart

For example, passing an integer n to the **sphere** function creates $n + 1 \times n + 1$ matrices for the x, y, and z matrices, which can then be passed to the **mesh** function (Figure 11.12) or the **surf** function (Figure 11.13).

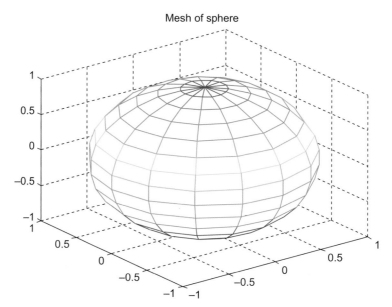

FIGURE 11.12 Mesh plot of sphere

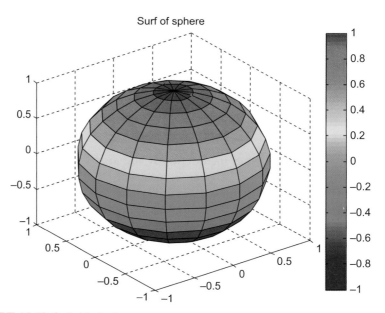

FIGURE 11.13 Surf plot of sphere

```
>> [x,y,z] = sphere(15);
>> size(x)
ans =
    16    16
>> mesh(x,y,z)
>> title('Mesh of sphere')
```

Additionally, the **colorbar** function displays a color bar to the right of the plot, showing the range of colors.

Note: Options for colors will be described in Chapter 14.

```
>> [x,y,z] = sphere(15);
>> surf(x,y,z)
>> title('Surf of sphere')
>> colorbar
```

11.4 CUSTOMIZING PLOTS

There are many ways to customize figures in the Figure Window. Clicking on the Plot Tools icon will bring up the Property Editor and Plot Browser, with many options for modifying the current plot. Additionally, there are *plot properties* that can be modified from the defaults in the plot functions. Using the **help** facility with the function name will show all of the options for that particular plot function.

For example, the **bar** and **barh** functions by default put a "width" of 0.8 between bars. When called as **bar(x,y)**, the width of 0.8 is used. If instead a third argument is passed, it is the width, for example, **barh(x,y,width)**. The following script uses **subplot** to show variations on the width. A width of 0.6 results in more space between the bars. A width of 1 makes the bars touch each other, and with a width greater than 1, the bars actually overlap. The results are shown in Figure 11.14.

```
barwidths.m
```

```
% Subplot to show varying bar widths

year = 2007:2011;
pop = [0.9  1.4  1.7  1.3  1.8];

for i = 1:4
    subplot(1,4,i)
    % width will be 0.6, 0.8, 1, 1.2
    barh(year,pop,0.4 + i*.2)
    title(sprintf('Width = %.1f',0.4 + i*.2))
    xlabel('Population')
    ylabel('Year')
end
```

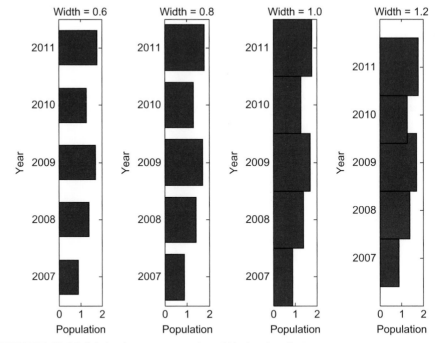

FIGURE 11.14 Subplot demonstrates varying widths in a bar chart

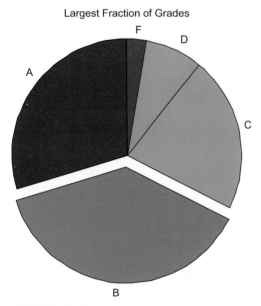

FIGURE 11.15 Exploding pie chart

PRACTICE 11.3

Use **help area** to find out how to change the base level on an **area** chart (from the default of 0).

As another example of customizing plots, pieces of a pie chart can be "exploded" from the rest. In this case, two vectors are passed to the **pie** function: first the data vector, then a **logical** vector; the elements for which the **logical** vector is **true** will be exploded from (separated from) the pie chart.

A third argument—that is, a cell array of labels—can also be passed. The result is seen in Figure 11.15.

```
>> gradenums = [11 14 8 3 1];
>> which = gradenums == max(gradenums)
which =
     0     1     0     0     0
>> pie(gradenums,which,...
{'A','B','C','D','F'})
>> title('Largest Fraction of Grades')
```

11.5 HANDLE GRAPHICS AND PLOT PROPERTIES

MATLAB uses what it calls Handle Graphics® in all of its figures. All figures consist of different *objects*, each of which is assigned a *handle*. The object handle is a unique real number that is used to refer to the object.

Objects include *graphics primitives* such as lines and text, as well as the axes used to orient the objects. The objects are organized hierarchically, and there are properties associated with each object. This is the basis of *object-oriented programming*: objects are organized hierarchically (e.g., a *parent* comes before its *children* in the hierarchy) and this hierarchy has ramifications in terms of the properties; generally children inherit properties from the parents.

The hierarchy in MATLAB, as seen in Help, "Organization of Graphics Objects," can be summarized as follows:

In other words, the Figure Window includes Axes, which are used to orient Core objects (primitives such as **line**, **rectangle**, **text**) and Plot objects (which are used to produce the different plot types, such as bar charts and area plots).

11.5.1 Plot objects and properties

The various plot functions return a handle for the plot object, which can then be stored in a variable. In the following, the **plot** function plots a **sin** function in a Figure Window (as shown in Figure 11.16) and returns a real number, which is the object handle. (Don't try to make sense of the actual number used for the handle!)

This handle will remain valid as long as the object exists.

```
>> x = -2*pi: 1/5 : 2*pi;
>> y = sin(x);
>> h1 = plot(x,y)
h1 =
   159.0142
>> xlabel('x')
>> ylabel('sin(x)')
```

Note
The Figure Window should not be closed, as that would make the object handle invalid since the object wouldn't exist anymore!

Object properties can be displayed using the **get** function, as shown in the following lines of code. This shows properties such as the Color, LineStyle, LineWidth, and so on (and many you will not understand—don't worry about it!).

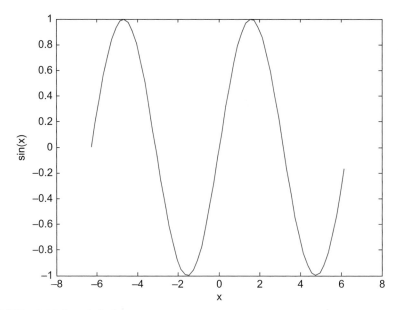

FIGURE 11.16 Plot of **sin** function with default properties

```
>> get(hl)
            DisplayName: ''
             Annotation: [1 x 1 hg.Annotation]
                  Color: [0 0 1]
              EraseMode: 'normal'
              LineStyle: '—'
              LineWidth: 0.5000
                 Marker: 'none'
             MarkerSize: 6
        MarkerEdgeColor: 'auto'
        MarkerFaceColor: 'none'
                  XData: [1 x 63 double]
                  YData: [1 x 63 double]
                  ZData: [1 x 0 double]
           BeingDeleted: 'off'
          ButtonDownFcn: []
               Children: [0 x 1 double]
               Clipping: 'on'
              CreateFcn: []
              DeleteFcn: []
              BusyAction: 'queue'
       HandleVisibility: 'on'
                HitTest: 'on'
          Interruptible: 'on'
               Selected: 'off'
     SelectionHighlight: 'on'
                    Tag: ''
```

```
                Type: 'line'
       UIContextMenu: []
            UserData: []
             Visible: 'on'
              Parent: 158.0131
           XDataMode: 'manual'
         XDataSource: ''
         YDataSource: ''
         ZDataSource: ''
```

A particular property can also be displayed. For example, to determine the line width:

```
>> get(hl,'LineWidth')
ans =
    0.5000
```

The objects, their properties, what the properties mean, and valid values can be found in MATLAB Help. Under the Contents tab, click on Handle Graphics Property Browser. Then, click on Plot Objects; several options can be seen. Click on Lineseries, which is used to create figures using the **plot** function, to see a list of the property names and a brief explanation of each.

For example, the Color property is a vector that stores the color of the line as three separate values for the Red, Green, and Blue intensities, in that order. Each value is in the range from 0 (which means none of that color) to 1. In the previous example, the Color was [0 0 1], which means no red, no green, but full blue—in other words, the line drawn for the **sin** function was blue. More examples of possible values for the Color vector are:

```
[1 0 0] is red
[0 1 0] is green
[0 0 1] is blue
[1 1 1] is white
[0 0 0] is black
[0.5 0.5 0.5] is a shade of grey
```

All of the properties listed by **get** can be changed, using the **set** function. The **set** function is called in the format

```
set(objhandle, 'PropertyName', property value)
```

For example, to change the line width from the default of 0.5 to 1.5:

```
>> set(hl,'LineWidth', 2.5)
```

As long as the Figure Window is still open and this object handle is still valid, the width of the line will be increased.

The properties can also be set in the original function call. For example, the following will set the line width to 2.5 to begin with as seen in Figure 11.17.

```
>> hl = plot(x,y, 'LineWidth', 2.5);
```

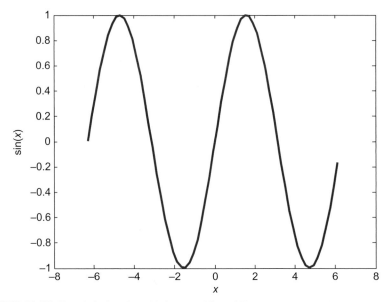

FIGURE 11.17 Plot of **sin** function with increased line width

---

PRACTICE 11.4

Create *x* and *y* vectors, and use the **plot** function to plot the data points represented by these vectors. Store the handle in a variable, and do not close the Figure Window! Use **get** to inspect the properties, and then **set** to change the line width and color.

11.5.2 Core objects

Core Objects in MATLAB are the very basic graphics primitives. A description can be found under the MATLAB help. Under the Contents tab, click on Handle Graphics Objects, and then Core Graphics Objects. The core objects include:

- `line`
- `text`
- `rectangle`
- `patch`
- `image`

These are all built-in functions; **help** can be used to learn how each function is used.

A **line** is a core graphics objects, which is produced by the **plot** function. The following is an example of creating a line object, modifying some properties, and saving the handle in a variable *hl*:

```
>> x = -2*pi: 1/5 : 2*pi;
>> y = sin(x);
```

```
>> hl = line(x,y,'LineWidth', 6, 'Color', [0.5 0.5 0.5])
hl =
   159.0405
```

As seen in Figure 11.18, this draws a reasonably thick grey line for the **sin** function. As before, the handle will be valid as long as the Figure Window is not closed. Some of the properties of this object are:

```
>> get(hl)
      Color = [0.5 0.5 0.5]
      EraseMode = normal
      LineStyle = -
      LineWidth = [ 6]
      Marker = none
      MarkerSize = [ 6]
      MarkerEdgeColor = auto
      MarkerFaceColor = none
      XData = [(1 by 63) double array]
      YData = [(1 by 63) double array]
      ZData = [ ]
          etc.
```

The **text** graphics function allows text to be printed in a Figure Window, including the special characters that are printed using \specchar, where "specchar" is the actual name of the special character. The format of a call to the **text** function is

```
text(x,y,'text string')
```

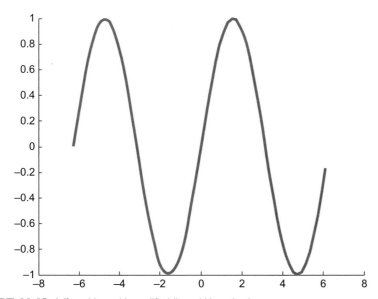

FIGURE 11.18 A **line** object with modified line width and color

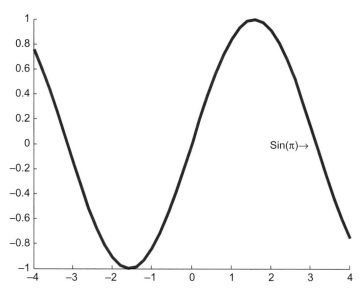

FIGURE 11.19 A **line** object with a text box

where x and y are the coordinates on the graph of the lower left corner of the *text box* in which the text string appears.

To see the options for the special characters, under the Contents tab in Help, click on Handle Graphics Property Browser, then click on Core Objects, and then choose Text. The special characters are shown in a table under the String property. The special characters include letters of the Greek alphabet, arrows, and characters frequently used in equations. For example, Figure 11.19 displays the Greek symbol for **pi** and a right arrow within the text box.

```
>> x = −4:0.2:4;
>> y = sin(x);
>> hp = line(x,y,'LineWidth',3);
>> thand = text(2,0,'Sin(\pi)\rightarrow')
```

Using **get** will display properties of the text box, such as the following:

```
>> get(thand)
    BackgroundColor = none
    Color = [0 0 0]
    EdgeColor = none
    EraseMode = normal
    Editing = off
    Extent = [1.95862 −0.0670554 0.901149 0.110787]
    FontAngle = normal
    FontName = Helvetica
```

```
FontSize = [10]
FontUnits = points
FontWeight = normal
HorizontalAlignment = left
LineStyle = −
LineWidth = [0.5]
Margin = [2]
Position = [2 0 0]
Rotation = [0]
String = Sin(\pi)\rightarrow
Units = data
Interpreter = tex
VerticalAlignment = middle
            etc.
```

Although the Position specified was (2,0), the Extent is the actual extent of the text box, which cannot be seen since the BackgroundColor and EdgeColor are not specified. These can be changed using **set**. For example, the following produces the result shown in Figure 11.20:

```
>> set(thand,'BackgroundColor',[0.8 0.8 0.8],...
         'EdgeColor',[1 0 0])
```

When the Units property has the value of "data," which is the default as shown before, the Extent of the text box is given by a vector `[x y width height]` where x and y are the coordinates of the bottom left corner of the text box.

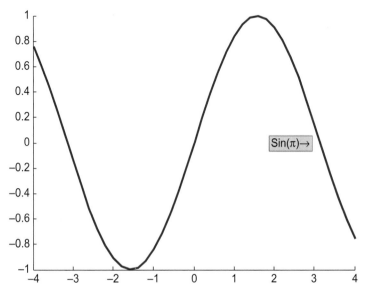

FIGURE 11.20 Text box with a modified edge color and background color

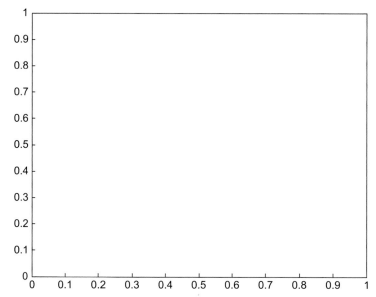

FIGURE 11.21 A **rectangle** object

Another core graphics object is **rectangle**, which can have curvature added to it (!!). Just calling the function **rectangle** without any arguments brings up a Figure Window (shown in Figure 11.21), which at first glance doesn't seem to have anything in it:

```
>> recthand = rectangle;
```

Using the **get** function will display the properties, some of which are excerpted here:

```
>> get(recthand)
    Curvature = [0 0]
    FaceColor = none
    EdgeColor = [0 0 0]
    LineStyle = −
    LineWidth = [0.5]
    Position = [0 0 1 1]
    Type = rectangle
```

The Position of a rectangle is [x y w h] where x and y are the coordinates of the lower left point, w is the width, and h is the height. The default rectangle has a Position of [0 0 1 1]. The default Curvature is [0 0], which means no curvature. The values range from [0 0] (no curvature) to [1 1] (ellipse). A more interesting rectangle object is shown in Figure 11.22.

Note: Properties can be set when calling the **rectangle** function, and also subsequently using the **set** function, as follows:

```
>> rh = rectangle('Position', [0.2, 0.2, 0.5, 0.8],...
   'Curvature',[0.5, 0.5]);
>> axis([0 1.2 0 1.2])
>> set(rh,'Linewidth',3,'LineStyle',':')
```

The **patch** function is used to create a patch graphics object, which is made from two-dimensional polygons. A simple patch in 2D space is defined by specifying the coordinates of three points as shown in Figure 11.23; in this case, the color red is specified for the polygon.

```
>> x = [0 1 0.5];
>> y = [0 0 1];
>> patch(x,y,'r')
```

Patches can also be defined in 3D space. A patch object is defined by both the vertices and the faces of the polygons that connect these vertices. One way of calling this function is patch(fv) where *fv* is a structure variable with fields called *vertices* and *faces*. For example, consider a patch that has four vertices in 3D space, given by the coordinates:

(1)	(0, 0, 0)
(2)	(1, 0, 0)
(3)	(0, 1, 0)
(4)	(0.5, 0.5, 1)

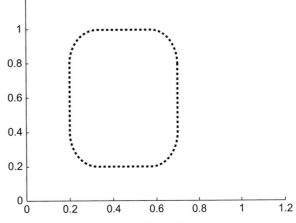

FIGURE 11.22 Rectangle object with curvature

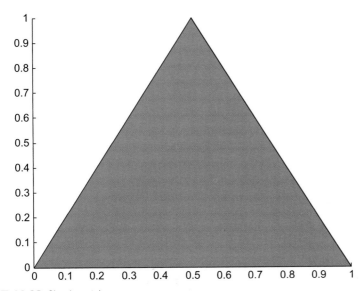

FIGURE 11.23 Simple patch

The order in which the points are given is important, as the faces describe how the vertices are linked. To create these vertices in MATLAB and define faces that connect them, we use a structure variable and then pass it to the **patch** function.

```
polyhedron.vertices = [...
0 0 0
1 0 0
0 1 0
0.5 0.5 1];

polyhedron.faces = [...
1 2 3
1 2 4
1 3 4
2 3 4];

pobj = patch(polyhedron, ...
'FaceColor',[0.8, 0.8, 0.8] ,...
'EdgeColor', 'black');
```

The *polyhedron.vertices* field is a matrix in which each row represents (x,y,z) co-ordinates. The field *polyhedron.faces* defines the faces; for example, the first row in the matrix specifies to draw lines from vertex 1 to vertex 2 to vertex 3 to form the first face. The face color is set to grey and the edge color to black. The figure, as seen in Figure 11.24, shows only two faces. Using the rotate icon on the

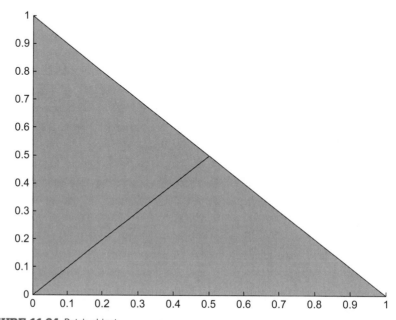

FIGURE 11.24 Patch object

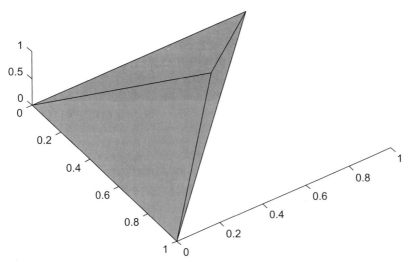

FIGURE 11.25 Rotated **patch** object

Figure Window, the figure can be rotated so that the other edges can be seen, as shown in Figure 11.25.

11.6 PLOT APPLICATIONS

In this section, we will show some examples that integrate plots and many of the other concepts covered to this point in the book. For example, we will have a function that receives an *x* vector, a function handle of a function used to create the *y* vector, and a cell array of plot types as strings and will generate the plots, and we will also show examples of reading data from a file and plotting them.

11.6.1 Plotting from a function

The function below generates a Figure Window (seen in Figure 11.26) that shows different types of plots for the same data. The data are passed as input arguments (as an *x* vector and the handle of a function to create the *y* vector) to the function, as is a cell array with the plot type names. The function generates the Figure Window using the cell array with the plot type names. It creates a function handle for each using the **str2func** function.

```
plotywithcell.m
```
```
function plotxywithcell(x, fnhan, rca)
% plotxywithcell receives an x vector, the handle
% of a function (used to create a y vector), and
% a cell array with plot type names; it creates
% a subplot to show all of these plot types
```

Continued

```
% Format: plotxywithcell(x,fn handle, cell array)

lenrca = length(rca);
y = fnhan(x);
for i = 1:lenrca
    subplot(1,lenrca,i)
    funh = str2func(rca{i});
    funh(x,y)
    title(upper(rca{i}))
    xlabel('x')
    ylabel(func2str(fnhan))
end
end
```

For example, the function could be called as follows:

```
>> anfn = @ (x) x .^ 3;
>> x = 1:2:9;
>> rca = {'bar', 'area', 'plot'};
>> plotxywithcell(x, anfn, rca)
```

The function is general and works for any number of plot types stored in the cell array.

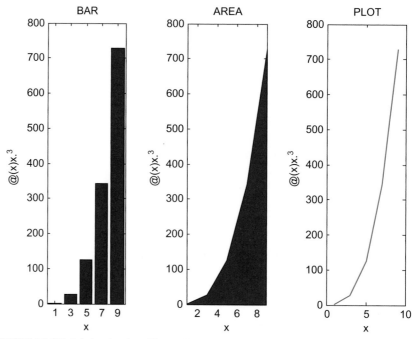

FIGURE 11.26 Subplot showing different file types with their names as titles

11.6.2 Plotting file data

It is often necessary to read data from a file and plot them. Normally, this entails knowing the format of the file. For example, let us assume that a company has two divisions, A and B. Assume that the file *ab11.dat* contains four lines, with the sales figures (in millions) for the two divisions for each quarter of 2011. For example, the file might look like this (and the format will be exactly like this):

```
A5.2B6.4
A3.2B5.5
A4.4B4.3
A4.5B2.2
```

The following script reads in the data and plots the data as bar charts in one Figure Window. The script prints an error message if the file open is not successful or if the file close was not successful. The **axis** command is used to force the x-axis to range from 0 to 3 and the y-axis from 0 to 8, which will result in the axes shown here. The numbers 1 and 2 would show on the x-axis rather than the division labels A and B by default. The **set** function changes the XTickLabel property to use the strings in the cell array as labels on the tick marks on the x-axis; *gca* returns the handle to the axes in the current figure (it stands for "get current axes").

plotdivab.m

```
% Reads sales figures for 2 divisions of a company one
% line at a time as strings, and plots the data

fid = fopen('ab11.dat');
if fid == −1
    disp('File open not successful')
else

    for i = 1:4
        % Every line is of the form A#B#; this separates
        % the characters and converts the #'s to actual
        % numbers
        aline = fgetl(fid);
        aline = aline(2:length(aline));
        [compa rest] = strtok(aline,'B');
        compa = str2num(compa);
        compb = rest(2:length(rest));
        compb = str2num(compb);

        % Data from every line is in a separate subplot
        subplot(1,4,i)
        bar([compa,compb])
        set(gca, 'XTickLabel',{ 'A', 'B'})
        axis([0 3 0 8])
        ylabel('Sales (millions)')
```

Continued

```
      title(sprintf('Quarter %d',i))
  end
  closeresult = fclose(fid);
  if closeresult ~= 0
      disp('File close not successful')
  end
end
```

Running this produces the subplot shown in Figure 11.27.

As another example, a data file called *compsales.dat* stores sales figures (in millions) for divisions in a company. Each line in the file stores the sales number, followed by an abbreviation of the division name, in this format:

```
5.2 X
3.3 A
5.8 P
2.9 Q
```

The script that follows Figure 11.28 uses the **textscan** function to read this information into a cell array, and then uses **subplot** to produce a Figure Window that displays the information in a **bar** chart and in a **pie** chart.

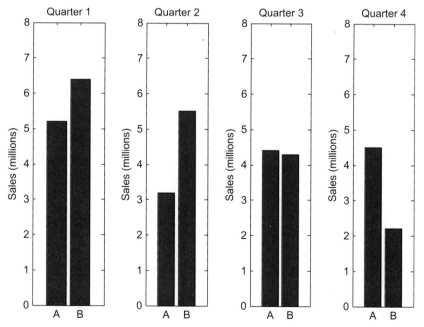

FIGURE 11.27 Subplot with customized x-axis tick labels

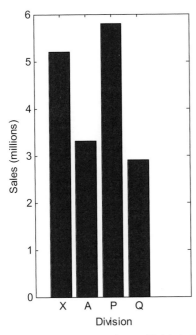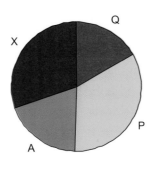

FIGURE 11.28 Bar and pie charts with labels from file data

compsalesbarpie.m

```
% Reads sales figures and plots as a bar chart and a pie chart
fid = fopen('compsales.dat');

if fid == -1
    disp('File open not successful')
else
    % Use textscan to read the numbers and division codes
    % into separate elements in a cell array
    filecell = textscan(fid,'%f %s');
    % plot the bar chart with the division codes on the x ticks
    subplot(1,2,1)
    bar(filecell{1})
    xlabel('Division')
    ylabel('Sales (millions)')
    set(gca, 'XTickLabel', filecell{2})
    % plot the pie chart with the division codes as labels
    subplot(1,2,2)
    pie(filecell{1}, filecell{2})
    title('Sales in millions by division')

    closeresult = fclose(fid);
    if closeresult ~= 0
        disp('File close not successful')
    end
end
```

SUMMARY

Common Pitfalls
- Forgetting that **subplot** numbers the plots rowwise rather than columnwise
- Not realizing that the **subplot** function just creates a matrix within the Figure Window. Each part of this matrix must then be filled with a plot, using any type of plot function
- Closing a Figure Window prematurely—the properties can only be set if the Figure Window is still open!

Programming Style Guidelines
- Always label plots
- Take care to choose the type of plot to highlight the most relevant information

MATLAB Functions and Commands		
subplot	plot3	sphere
barh	bar3	cylinder
area	bar3h	colorbar
stem	pie3	line
hist	comet3	rectangle
pie	stem3	text
comet	spiral	get set
movie	mesh	patch
getframe	surf	image

Exercises

1. Create a data file containing 10 numbers. Write a script that will load the vector from the file, and use **subplot** to do an **area** plot and a **stem** plot with these data in the same Figure Window. (Note: A loop is not needed.) Prompt the user for a title for each plot.

2. Use **subplot** to show the difference between the **sin** and **cos** functions. Create an x vector with 100 linearly spaced points in the range from -2π to 2π, and then two y vectors for **sin(x)** and **cos(x)**. In a 2×1 subplot, use the **plot** function to display them, with appropriate titles.

3. Biomedical engineers are developing an insulin pump for diabetics. To do this, it is important to understand how insulin is cleared from the body after a meal. The concentration of insulin at any time t is described by the equation

$$C = C_0 \, e^{-30t/m}$$

where C_0 is the initial concentration of insulin, t is the time in minutes, and m is the mass of the person in kilograms. Write a script that will graphically show how the weight of the person influences the time for insulin to be cleared from the body.

It will show in a 2×1 **subplot** the concentration of insulin for two subjects— one who weighs 120 lb, and one who weighs 300 lb. For both, the time should increment from 0 to 4 minutes in steps of 0.1 minute, and the initial concentration should be 85. The concentration over time will be shown in each subplot, and the weight of the person should be in the title. The conversion factor is 1 lb = 0.4536 kg. To better compare, use consistent axes for both plots.

4. Write a function *subfnfn* that will receive two function handles as input arguments, and will display in one Figure Window plots of these two functions, with the function names in the titles. Use the default axes. The function will create an x vector that ranges from 1 to n (where n is a random integer in the range from 4 to 10). For example, if the function is called as follows:

```
>> subfnfn(@sqrt, @exp)
```

and if the random integer for n was 9, the Figure Window would look like the image shown in Figure 11.29.

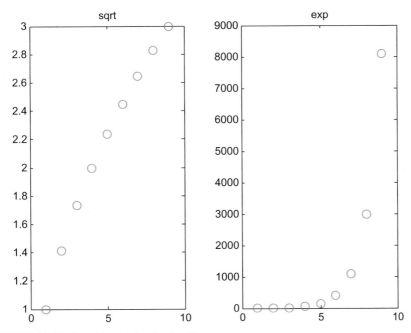

FIGURE 11.29 Subplot using function handles

Massive amounts of temperature data have been accumulated and stored in files. To be able to comb through these data and gain insights into global temperature variations, it is often useful to visualize the information.

5. A file called *avehighs.dat* stores for three locations the average high temperatures for each month for a year (rounded to integers). There are three lines in the file; each stores the location number followed by the 12 temperatures (this format may be assumed). For example, the file might store:

```
432   33 37 42 45 53 72 82 79 66 55 46 41
777   29 33 41 46 52 66 77 88 68 55 48 39
567   55 62 68 72 75 79 83 89 85 80 77 65
```

Write a script that will read these data in and plot the temperatures for the three locations separately in one Figure Window. A **for** loop must be used to accomplish this. For example, if the data are as shown in the previous data block, the Figure Window would appear as in Figure 11.30. The axis labels and titles should be as shown.

6. Sales (in millions) from two different divisions of a company for the four quarters of 2006 are stored in vector variables, such as in the following:

```
div1 = [4.2 3.8 3.7 3.8];
div2 = [2.5 2.7 3.1 3.3];
```

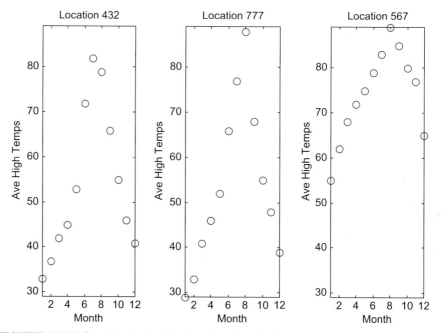

FIGURE 11.30 Subplot to display data from file using a **for** loop

Using **subplot**, show side by side the sales figures for the two divisions. What kind of graph shows this in the best way? Why? In one graph, compare the two divisions. What kind of graph shows this in the best way? Why?

7. Create an x vector that has 30 linearly spaced points in the range from -2π to 2π, and then y as **sin(x)**. Do a **stem** plot of these points, and store the handle in a variable. Use **get** to see the properties of the stem plot, and then **set** to change the face color of the marker.

8. When an object with an initial temperature T is placed in a substance that has a temperature S, according to Newton's law of cooling in t minutes it will reach a temperature T_t using the formula $T_t = S + (T - S)\ e^{(-kt)}$ where k is a constant value that depends on properties of the object. For an initial temperature of 100 and $k = 0.6$, graphically display the resulting temperatures from 1 to 10 minutes for two different surrounding temperatures: 50 and 20. Use the **plot** function to plot two different lines for these surrounding temperatures, and store the handle in a variable. Note that two function handles are actually returned and stored in a vector. Use **set** to change the line width of one of the lines.

9. Write a script that will draw the line $y = x$ between $x = 2$ and $x = 5$, with a random thickness between 1 and 10.

10. In hydrology, *hyetographs* are used to display rainfall intensity during a storm. The intensity could be the amount of rain per hour, recorded every hour for a 24-hour period. Create your own data file to store the intensity in inches per hour every hour for 24 hours. Use a **bar** chart to display the intensities.

11. Write a script that will read x and y data points from a file, and will create an **area** plot with those points. The format of every line in the file is the letter "x," space, the x value, space, the letter "y," space, and the y value. You must assume that the data file is in exactly that format, but you may not assume that the number of lines in the file is known. The number of points will be in the plot title. The script loops until the end of the file is reached, using **fgetl** to read each line as a string. For example, *if* the file contains the following lines,

```
x 0 y 1
x 1.3 y 2.2
x 2.2 y 6
x 3.4 y 7.4
```

when running the script, the result will be as shown in Figure 11.31.

12. A file *houseafford.dat* stores in its three lines years, median incomes, and median home prices for a city. The dollar amounts are in thousands. For example, it might look like this:

```
2004 2005 2006 2007 2008 2009 2010 2011
72 74 74 77 80 83 89 93
250 270 300 310 350 390 410 380
```

Create a file in this format, and then **load** the information into a matrix. Create a horizontal stacked bar chart to display the following information, with an appropriate title. *Note:* Use the 'XData' property to put the years on the axis as shown in Figure 11.32.

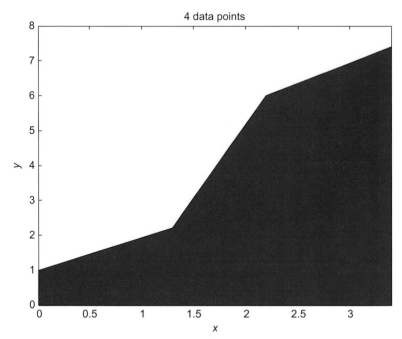

FIGURE 11.31 Area plot produced from *x, y* data read as strings from a file

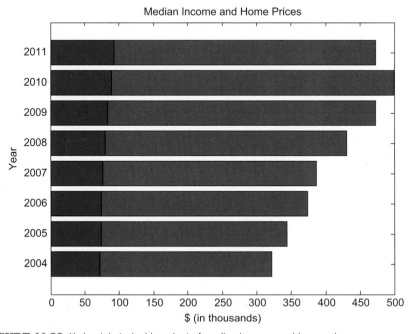

FIGURE 11.32 Horizontal stacked bar chart of median incomes and home prices

13. A file *houseafford.dat* stores in its three lines years, median incomes, and median home prices for a city. The dollar amounts are in thousands. For example, it might look like this:

```
2004 2005 2006 2007 2008 2009 2010 2011
72 74 74 77 80 83 89 93
250 270 300 310 350 390 410 380
```

Create a file in this format, and then **load** the information into a matrix. The ratio of the home price to the income is called the "housing affordability index." Calculate this for every year and plot it. The *x*-axis should show the years (e.g., 2004 to 2011). Store the handle of the plot in a variable and use **get** to see the properties and **set** to change at least one.

14. Do a quick survey of your friends to find out who prefers cheese pizza, pepperoni, or mushroom (no other possibilities; everyone must pick one of those three choices). Draw a pie chart to show the percentage favoring each. Label the pieces of this pizza pie chart!

15. The number of faculty members in each department at a certain college of engineering is:

```
ME 22
BM 45
CE 23
EE 33
```

Experiment with at least three different plot types to graphically depict this information. Make sure that you have appropriate titles, labels, and legends on your plots. Which type(s) work best, and why?

16. The weights of the major components for a given aircraft are important considerations in aircraft design. The components include at the very least the wing, tail, fuselage, and landing gear. Create a data file with values for these weights. Load the data from your file and create a pie chart to show the percentage weight for each component.

17. Create an *x* vector, and then two different vectors (*y* and *z*) based on *x*. Plot them with a legend. Use **help legend** to find out how to position the legend itself on the graph, and experiment with different locations.

18. The wind chill factor (WCF) measures how cold it feels with a given air temperature (T, in degrees Fahrenheit) and wind speed (V, in miles per hour). One formula for WCF is

$$WCF = 35.7 + 0.6\,T - 35.7\,(V^{0.16}) + 0.43\,T\,(V^{0.16})$$

Experiment with different plot types to display the WCF for varying wind speeds and temperatures.

19. Write a script that will plot the **sin** function three times in one Figure Window, using the colors red, green, and blue.

20. Experiment with the **comet** function. Try the example given when **help comet** is entered and then animate your own function using **comet**.

21. Experiment with the **comet3** function. Try the example given when **help comet3** is entered and then animate your own function using **comet3**.

22. Investigate the **scatter** and **scatter3** functions.

23. The exponential and natural log functions are *inverse functions*. What does this mean in terms of the graphs of the functions? Show both functions in one Figure Window and distinguish between them.

24. The electricity generated by wind turbines annually in kilowatt-hours per year is given in a file. The amount of electricity is determined by, among other factors, the diameter of the turbine blade (in feet), and the wind velocity in mph. The file stores on each line the blade diameter, wind velocity, and the approximate electricity generated for the year. For example,

```
5 5 406
5 10 3250
5 15 10970
5 20 26000
10 5 1625
10 10 13000
10 15 43875
10 20 104005
20 5 6500
20 10 52000
20 15 175500
20 20 41600
```

Create this file, and determine how to graphically display these data.

25. In the MATLAB Help, under the Contents tab, click on Functions by Category, then Graphics, then Handle Graphics, and then text to get the MATLAB Function Reference on the function **text** (this is a lot more useful than just typing "help text"!). Read through this, and then on the very bottom click on Text Properties for property descriptions. Create a graph, and then use the text function to put some text on it, including some \specchar commands to increase the font size and to print some Greek letters and symbols.

26. The cost of producing widgets includes an initial setup cost plus an additional cost for each widget, so the total production cost per widget decreases as the number of widgets produced increases. The total revenue is a given dollar amount for each widget sold, so the revenue increases as the number sold increases. The break-even point is the number of widgets produced and sold for which the total production cost is equal to the total revenue. The production cost might be $5000 plus $3.55 per widget, and the widgets might sell for $10 each. Write a script that will find the break-even point using **solve** (see Chapter 15), and then plot the production cost and revenue functions on one graph for 1 to 1000 widgets. Print the break-even point on the graph using **text**.

27. Create a **rectangle** object, and use the **axis** function to change the axes so that you can see the rectangle easily. Change the Position, Curvature, EdgeColor, LineStyle, and LineWidth. Experiment with different values for the Curvature.

28. Write a function that will plot **cos(x)** for *x* values ranging from –pi to pi in steps of 0.1, using black stars (*). It will do this three times across in one Figure Window, with varying line widths. (*Note*: Even if individual points are plotted rather than a solid line, the line width property will change the size of these points.) If no arguments are passed to the function, the line widths will be 1, 2, and 3. If, on the other hand, an argument is passed to the function, it is a multiplier for these values (e.g., if 3 is passed, the line widths will be 3, 6, and 9). The line widths will be printed in the titles on the plots.

29. Write a script that will create the rectangle (shown in Figure 11.33) with a curved rectangle inside it and text inside that. The axes and dimensions of the Figure Window should be as shown here (you should approximate locations based on the axes shown in this figure). The font size for the string is 20. The curvature of the inner rectangle is [0.5, 0.5].

30. Write a script that will display rectangles with varying curvatures and line widths, as shown in Figure 11.34. The script will, in a loop, create a *2 × 2* subplot showing rectangles. In all, both the x and y axes will go from 0 to 1.4. Also, in all, the lower left corner of the rectangle will be at (0.2, 0.2), and the length and width will both be 1. The line width, *i*, is displayed in the title of each plot. The curvature will be [0.2, 0.2] in the first plot, then [0.4, 0.4], [0.6, 0.6], and finally [0.8, 0.8]. Recall that the **subplot** function numbers the elements rowwise.

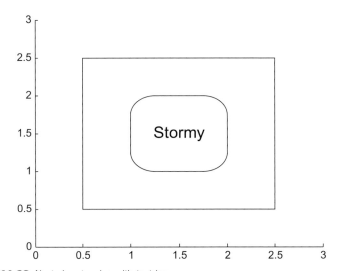

FIGURE 11.33 Nested rectangles with text box

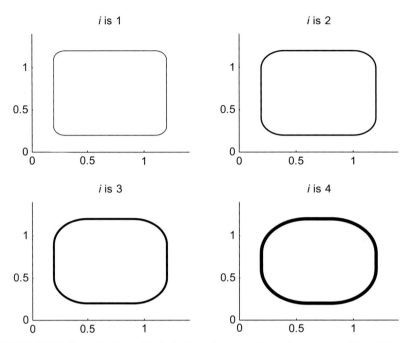

FIGURE 11.34 Example of a script displaying rectangles with varying curves and line widths

31. Write a script that will start with a rounded rectangle. Change both the x and y axes from the default to go from 0 to 3. In a <u>for</u> loop, change the position vector by adding 0.1 to all elements 10 times (this will change the location and size of the rectangle each time). Create a movie consisting of the resulting rectangles. The final result should look like the plot shown in Figure 11.35.

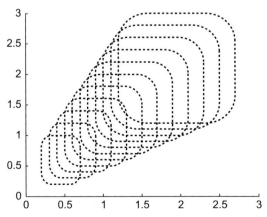

FIGURE 11.35 Curved rectangles produced in a loop

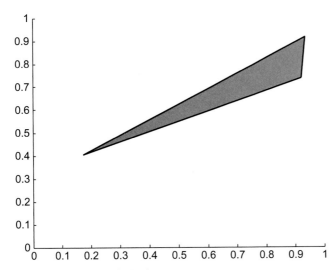

FIGURE 11.36 Patch object with black edge

32. Write a script that will create a two-dimensional **patch** object with just three vertices and one face connecting them. The x and y coordinates of the three vertices will be random real numbers in the range from 0 to 1. The lines used for the edges should be black with a width of 3, and the face should be grey. The axes (both x and *y*) should go from 0 to 1. For example, depending on what the random numbers are, the Figure Window might look like Figure 11.36.

33. Using the **patch** function, create a black box with unit dimensions (so, there will be eight vertices and six faces). Set the edge color to white so that when you rotate the figure, you can see the edges.

34. Write a function *drawpatch* that receives the x and y coordinates of three points as input arguments. If the points are not all on the same straight line, it draws a patch using these three points, and if they are all on the same line, it modifies the coordinates of one of the points and then draws the resulting patch. To test this, it uses two subfunctions. It calls the subfunction *findlin* twice to find the slope and y-intercept of the lines first between point 1 and point 2 and then between point 2 and point 3 (e.g., the values of m and b in the form y = mx + b). It then calls the subfunction *issamelin* to determine whether these are the same line or not. If they are, it modifies point 3. It then draws a patch with a green color for the face and a red edge. Both of the subfunctions use structures (for the points and the lines) (Figure 11.37). For example,

```
>> drawpatch(2,2,4,4,6,1)
```

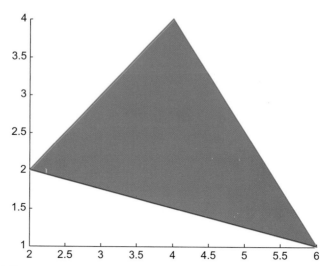

FIGURE 11.37 Patch with red edge

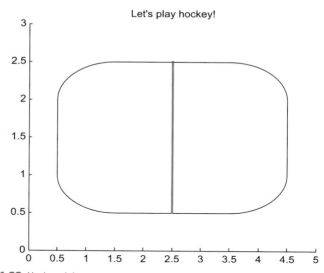

FIGURE 11.38 Hockey rink

35. A hockey rink looks like a rectangle with curvature. Draw a hockey rink, as in Figure 11.38.

36. Use the **cylinder** function to create *x, y,* and *z* matrices and pass them to the **surf** function to get a surface plot. Experiment with different arguments to **cylinder**.

37. Get into the Help, and learn how to do a **contour** plot.

CHAPTER 12

Matrix Representation of Linear Algebraic Equations

KEY TERMS

linear algebraic
  equation
symbolic mathematics
matrix equality
square matrix
main diagonal
diagonal matrix
trace
identity matrix
banded matrix
tridiagonal matrix
lower triangular matrix
upper triangular
  matrix
symmetric matrix

array operations
matrix addition
matrix subtraction
scalar multiplication
array multiplication
matrix multiplication
inner dimensions
outer dimensions
matrix inverse
matrix augmentation
dot product
cross product or outer
  product
coefficients
unknowns

solution set
method of substitution
determinant
Gauss elimination
Gauss-Jordan
  elimination
elementary row
  operations
forward elimination
back substitution
reduced row echelon
  form
skew symmetric

CONTENTS

A *linear algebraic equation* is an equation of the form

$$a_1 x_1 + a_2 x_2 + a_3 x_3 + \ldots + a_n x_n = b$$

Solutions to sets of equations in this form are important in many applications. In the MATLAB® product, to solve systems of equations, there are basically two methods:

- Using a matrix representation
- Using the **solve** function (which is part of Symbolic Math Toolbox™)

In this chapter, we first investigate matrix and vector operations and then use these to solve linear algebraic equations. The use of *symbolic mathematics* including the **solve** function will be covered in Chapter 15.

12.1 MATRIX DEFINITIONS

As we have already seen, a *matrix* can be thought of as a table of values in which there are both rows and columns. The general form of a matrix A (which is sometimes written as [A]) is shown here:

$$A = \begin{bmatrix} a_{11} & a_{12} & \cdots & a_{1n} \\ a_{21} & a_{22} & \cdots & a_{2n} \\ \vdots & \vdots & \vdots & \vdots \\ a_{m1} & a_{m2} & \cdots & a_{mn} \end{bmatrix} = a_{ij} \ i = 1, \ldots, m; \ j = 1, \ldots, n$$

This matrix has m rows and n columns, so the size is $m \times n$.

A *vector* is a special case of a matrix, in which one of the dimensions (either m or n) is 1. A *row vector* is a $1 \times n$ matrix. A *column vector* is an $m \times 1$ matrix. A *scalar* is a special case of a matrix in which both m and n are 1, so it is a single value or a 1×1 matrix.

12.1.1 Matrix properties

In this section we will define some special properties of matrices.

Two matrices are said to be *equal* to each other only if all corresponding elements are equal to each other. For this to be true, their sizes must be the same as well. The definition is [A] = [B] if $a_{ij} = b_{ij}$ for all i, j.

THE PROGRAMMING CONCEPT

To test to see whether two matrices are equal to each other or not, both matrices are passed to a function that will return 1 for **logical true** if they are the same or 0 for **logical false** if not (or if they are not the same size). To write our own function, a flag is first set to 1, for **logical true**. If the two matrices are not the same size, the flag is set to 0 for **false**. Otherwise, using a nested <u>for</u> loop, each element in the first matrix argument *mata* is compared to the corresponding element in *matb*; if they are not the same, the flag is set to 0 for **false**.

myisequal.m

```
function myflag = myisequal(mata,matb)
% myisequal receives two matrices and returns
% logical 1 if they are equal or 0 if not
% Format: myisequal(matrix1, matrix2)

% Assume true until & unless found to be false
myflag = logical(1);
```

```
[r c] = size(mata);
if all(size(mata) ~= size(matb))
    myflag = logical(0);
else
  for i=1:r
      for j = 1:c
         if mata(i,j) ~= matb(i,j)
            myflag = logical(0);
         end
      end
  end
end
end
```

```
>> mata = [2 5 8; 1:3];
>> matb = [2:3:8; 1 2 3];
>> myisequal(mata,matb)
ans =
     1
```

THE EFFICIENT METHOD

In MATLAB, as we have seen, the **isequal** function will also accomplish this:

```
>> isequal(mata,matb)
ans =
     1
```

In addition, the **isequal** function will return logical 0 for **false** if the two matrices are not the same size.

12.1.2 Square matrices

If a matrix has the same number of rows and columns (e.g., if $m == n$), the matrix is *square*. The definitions that follow in this section only apply to square matrices.

The ***main diagonal*** of a square matrix is the set of terms a_{ii} for which the row and column indices are the same, so from the upper left element to the lower right. For example, for the following matrix the set of numbers is 1, 6, 11, and 16.

$$\begin{bmatrix} 1 & 2 & 3 & 4 \\ 5 & 6 & 7 & 8 \\ 9 & 10 & 11 & 12 \\ 13 & 14 & 15 & 16 \end{bmatrix}$$

This is sometimes called just the *diagonal*.

A square matrix is a *diagonal matrix* if all values that are not on the diagonal are 0. The numbers on the diagonal, however, do not have to be all nonzero, although frequently they are. Mathematically, this is written as $a_{ij} = 0$ for $i \sim= j$. The following is an example of a diagonal matrix:

$$\begin{bmatrix} 4 & 0 & 0 \\ 0 & 9 & 0 \\ 0 & 0 & 5 \end{bmatrix}$$

MATLAB has a function **diag** that will return the diagonal of a matrix as a column vector; transposing will result in a row vector instead.

```
>> mymat = reshape(1:16,4,4)'
mymat =
     1     2     3     4
     5     6     7     8
     9    10    11    12
    13    14    15    16

>> diag(mymat)'
ans =
     1     6    11    16
```

The **diag** function can also be used to take a vector of length *n* and create an *n* × *n* square diagonal matrix with the values from the vector on the diagonal:

```
>> v = 1:4;
>> diag(v)
ans =
     1     0     0     0
     0     2     0     0
     0     0     3     0
     0     0     0     4
```

So, the **diag** function can be used two ways: (1) pass a matrix and it returns a vector; or (2) pass a vector and it returns a matrix!

PRACTICE 12.1

Write a function called *isdiagonal* that will return **logical** 1 for **true** if a square matrix is a diagonal matrix, or **logical** 0 for **false** if not.

The *trace* of a square matrix is the sum of all the elements on the diagonal. For example, for the diagonal matrix created using v, it is $1 + 2 + 3 + 4$, or 10.

QUICK QUESTION!

How could we calculate the trace of a square matrix?

Answer: See the following Programming Concept and Efficient Method.

THE PROGRAMMING CONCEPT

To calculate the trace of a square matrix, only one loop is necessary since the only elements in the matrix we are referring to have subscripts (i, i). So, once the size has been determined, the loop variable can iterate from 1 through the number of rows or from 1 through the number of columns (it doesn't matter which, since they have the same value!). The following function calculates and returns the trace of a square matrix, or an empty vector if the matrix argument is not square.

```
mytrace.m

function outsum = mytrace(mymat)
% mytrace calculates the trace of a square matrix
% or an empty vector if the matrix is not square
% Format: mytrace(matrix)

[r c] = size(mymat);
if r ~= c
    outsum = [];
else
    outsum = 0;
    for i = 1:r
        outsum = outsum + mymat(i,i);
    end
end
end
```

```
>> mymat = reshape(1:16,4,4)'
mymat =
     1    2    3    4
     5    6    7    8
     9   10   11   12
    13   14   15   16

>> mytrace(mymat)
ans =
    34
```

THE EFFICIENT METHOD

In MATLAB, there is a built-in function **trace** to calculate the trace of a square matrix:

```
>> trace(mymat)
ans =
    34
```

A square matrix is an *identity* matrix, called [I], if $a_{ij} = 1$ for $i == j$ and $a_{ij} = 0$ for $i \sim= j$. In other words, all of the numbers on the diagonal are 1 and all others are 0. Following is a 3×3 identity matrix:

$$\begin{bmatrix} 1 & 0 & 0 \\ 0 & 1 & 0 \\ 0 & 0 & 1 \end{bmatrix}$$

Note that any identity matrix is a special case of a diagonal matrix.

Identity matrices are very important and useful. MATLAB has a built-in function **eye** that will create an $n \times n$ identity matrix, given the value of *n*:

```
>> eye(5)
ans =
    1    0    0    0    0
    0    1    0    0    0
    0    0    1    0    0
    0    0    0    1    0
    0    0    0    0    1
```

Note

i is built into MATLAB as the square root of −1, so another name is used for the function that creates an identity matrix: **eye**, which sounds like "i" (... get it?).

Several special cases of matrices are related to diagonal matrices.

A *banded matrix* is a matrix of all 0s, with the exception of the main diagonal and other diagonals next to (above and below) the main. For example, the following matrix has 0s except for the band of three diagonals; this is a particular kind of banded matrix called a *tridiagonal matrix*.

$$\begin{bmatrix} 1 & 2 & 0 & 0 \\ 5 & 6 & 7 & 0 \\ 0 & 10 & 11 & 12 \\ 0 & 0 & 15 & 16 \end{bmatrix}$$

A *lower triangular matrix* has all 0s above the main diagonal. For example,

$$\begin{bmatrix} 1 & 0 & 0 & 0 \\ 5 & 6 & 0 & 0 \\ 9 & 10 & 11 & 0 \\ 13 & 14 & 15 & 16 \end{bmatrix}$$

An *upper triangular matrix* has all 0s below the main diagonal. For example,

$$\begin{bmatrix} 1 & 2 & 3 & 4 \\ 0 & 6 & 7 & 8 \\ 0 & 0 & 11 & 12 \\ 0 & 0 & 0 & 16 \end{bmatrix}$$

It is possible for there to be 0s on the diagonal and in the lower part or upper part and still be a lower or upper triangular matrix, respectively.

THE EFFICIENT METHOD

MATLAB has functions **triu** and **tril** that will take a matrix and make it into an upper triangular or lower triangular matrix by replacing the appropriate elements with 0s.

```
>> mymat
mymat =
        1       2       3       4
        5       6       7       8
        9      10      11      12
       13      14      15      16

>> triu(mymat)
ans =
        1       2       3       4
        0       6       7       8
        0       0      11      12
        0       0       0      16

>> tril(mymat)
ans =
        1       0       0       0
        5       6       0       0
        9      10      11       0
       13      14      15      16
```

THE PROGRAMMING CONCEPT

Nested loops would be required to accomplish this without using **triu** or **tril**.

A square matrix is *symmetric* if $a_{ij} = a_{ji}$ for all i, j. In other words, all of the values opposite the diagonal from each other must be equal to each other. In this example, there are three pairs of values opposite the diagonals, all of which are equal (the 2s, 9s, and 4s).

$$\begin{bmatrix} 1 & 2 & 9 \\ 2 & 5 & 4 \\ 9 & 4 & 6 \end{bmatrix}$$

PRACTICE 12.2

For the following matrices:

$$\begin{matrix} A & B & C \end{matrix}$$
$$\begin{bmatrix} 4 & 3 \\ 0 & 1 \end{bmatrix} \begin{bmatrix} 1 & 2 & 3 \\ 4 & 5 & 6 \end{bmatrix} \begin{bmatrix} 1 & 4 & 2 \\ 4 & 0 & 3 \\ 2 & 3 & 6 \end{bmatrix}$$

Which are equal? Which are square?
For all square matrices:
- Calculate the trace.
- Which are symmetric?
- Which are diagonal?
- Which are lower triangular?
- Which are upper triangular?

12.1.3 Array operations

As we have seen in Chapter 5, operators that are applied term by term or element by element, implying that the matrices must be the same size, are sometimes referred to as *array operations*. These include addition, subtraction, multiplication, division, and exponentiation.

Matrix addition means adding two matrices element by element, which means they must be of the same size. In mathematical terms, this is written $c_{ij} = a_{ij} + b_{ij}$. In MATLAB, this is accomplished with the $+$ operator. Similar to matrix addition, *matrix subtraction* means to subtract term by term, so in mathematical terms, $c_{ij} = a_{ij} - b_{ij}$. This would also be accomplished using a nested **for** loop in most languages, or by using the – (minus) operator in MATLAB.

Scalar multiplication means to multiply every element by a scalar (a number). This would also be accomplished using a nested **for** loop in most languages, or by using the * operator in MATLAB.

To multiply matrices element by element (which is *not* matrix multiplication!) in MATLAB, the .* operator is used, and the matrices must have the same dimensions. This is called *array multiplication* since it is an array operation (term by term).

```
>> A = [1:3;4:6];
>> B = [100 10 1; 10 100 1];
>> C = A .* B
C =
      100     20      3
       40    500      6
```

Array division (dividing term by term or element by element) is accomplished using the ./ or .\ operators, and array exponentiation (which is a form of array multiplication) is done with the .^ operator.

12.1.4 Matrix multiplication

Matrix multiplication does *not* mean multiplying term by term; it is not an array operation. Matrix multiplication has a very specific meaning. First of all, to multiply a matrix A by a matrix B to result in a matrix C, the number of columns of A must be the same as the number of rows of B. If the matrix A has dimensions $m \times n$, that means that matrix B must have dimensions $n \times$ *something*; we'll call it p.

We say that the ***inner dimensions*** must be the same. The resulting matrix C has the same number of rows as A and the same number of columns as B (i.e., the ***outer dimensions*** $m \times p$). In mathematical notation,

$$[A]_{m \times n}[B]_{n \times p} = [C]_{m \times p}$$

This only defines the size of C.

The elements of the matrix C are defined as the sum of products of corresponding elements in the rows of A and columns of B, or in other words,

$$c_{ij} = \sum_{k=1}^{n} a_{ik}b_{kj}$$

In the following example, A is 2×3 and B is 3×4; the inner dimensions are both 3, so the matrix multiplication is possible. C will have as its size the outer dimensions 2×4. The elements in C are obtained using the summation just described. The first row of C is obtained using the first row of A and in succession the columns of B. For example, C(1,1) is $3*1 + 8*4 + 0*0$ or 35. C(1,2) is $3*2 + 8*5 + 0*2$ or 46.

$$
\begin{matrix}
A \\
\begin{bmatrix} 3 & 8 & 0 \\ 1 & 2 & 5 \end{bmatrix}
\end{matrix}
*
\begin{matrix}
B \\
\begin{bmatrix} 1 & 2 & 3 & 1 \\ 4 & 5 & 1 & 2 \\ 0 & 2 & 3 & 0 \end{bmatrix}
\end{matrix}
=
\begin{matrix}
C \\
\begin{bmatrix} 35 & 46 & 17 & 19 \\ 9 & 22 & 20 & 5 \end{bmatrix}
\end{matrix}
$$

THE PROGRAMMING CONCEPT

To multiply two matrices together, three nested **for** loops are required. The two outer loops iterate through the rows and columns of C, which is $m \times p$. For each element in C (c_{ij}), the inner loop sums $a_{ik} * b_{kj}$ for values of k from 1 through n. The following is a script that will accomplish this:

```
mymatmult.m
```

```
% This script demonstrates matrix multiplication
A = [3 8 0; 1 2 5];
B = [1 2 3 1; 4 5 1 2; 0 2 3 0];

[m n] = size(A);
[nb p] = size(B);
if n ~= nb
  disp('Cannot perform this matrix multiplication')
else
  % Preallocate C
  C = zeros(m,p);
  % Outer 2 loops iterate through the elements in C
  %     which has dimensions m by p
  for i = 1:m
     for j = 1:p
        % Inner loop performs the sum for each
        % element in C
        mysum = 0;
        for k = 1:n
            mysum = mysum + A(i,k) * B(k,j);
        end
        C(i,j) = mysum;
     end
  end
  % display C
  C
end
```

```
>> mymatmult
C =
     35    46    17    19
      9    22    20     5
```

THE EFFICIENT METHOD

In MATLAB, the * operator will perform this matrix multiplication:

```
>> A = [3 8 0; 1 2 5];
>> B = [1 2 3 1; 4 5 1 2; 0 2 3 0];
>> C = A * B
C =
     35    46    17    19
      9    22    20     5
```

QUICK QUESTION!

What happens if a matrix M is multiplied by an identity matrix (of the appropriate size)?

Answer: For the size to be appropriate, the dimensions of the identity matrix would be the same as the number of columns of M. The result of the multiplication will always be the original matrix M (thus, it is similar to multiplying a scalar by 1).

```
>> A = [3 8 0; 1 2 5]
A =
     3    8    0
     1    2    5

>> A * eye (3)
ans =
     3    8    0
     1    2    5
```

```
>> B = [1 2 3 1; 4 5 1 2; 0 2 3 0]
B =
     1    2    3    1
     4    5    1    2
     0    2    3    0

>> [r c] = size (B);
>> B * eye (c)
ans =
     1    2    3    1
     4    5    1    2
     0    2    3    0
```

PRACTICE 12.3

Multiply these two matrices by hand and then verify the result in MATLAB.

$$
\overset{A}{\begin{bmatrix} 1 & 3 \\ 4 & 2 \end{bmatrix}} \quad \overset{B}{\begin{bmatrix} 2 & 1 \\ -1 & 3 \end{bmatrix}}
$$

12.1.5 Matrix operations

There are several common operations on matrices, some of which we have seen already. These include matrix transpose, matrix augmentation, and array operations.

A *matrix transpose* interchanges the rows and columns of a matrix. For a matrix A, its transpose is written A^T. For example, if

$$
A = \begin{bmatrix} 1 & 2 & 3 \\ 4 & 5 & 6 \end{bmatrix}
$$

then

$$
A^T = \begin{bmatrix} 1 & 4 \\ 2 & 5 \\ 3 & 6 \end{bmatrix}
$$

In MATLAB, as we have seen, there is a built-in transpose operator—the apostrophe.

If the result of multiplying a matrix A by another matrix is the identity matrix I, then the second matrix is the *inverse* of matrix A. The inverse of a matrix A is written as A^{-1}, so

$$[A][A^{-1}] = [I]$$

How to actually compute the inverse A^{-1} of a matrix by hand is not so easy. MATLAB, however, has a function **inv** to compute a matrix inverse. For example, here a matrix is created, its inverse is found, and then it is multiplied by the original matrix to verify that the product is in fact the identity matrix:

```
>> a = [1 2; 2 2]
a =
     1     2
     2     2

>> ainv = inv(a)
ainv =
    -1.0000    1.0000
     1.0000   -0.5000

>> a * ainv
ans =
     1     0
     0     1
```

Matrix augmentation means adding column(s) to the original matrix. For example, the matrix A

$$A = \begin{bmatrix} 1 & 3 & 7 \\ 2 & 5 & 4 \\ 9 & 8 & 6 \end{bmatrix}$$

might be augmented with a *3 × 3* identity matrix:

$$\left[\begin{array}{ccc|ccc} 1 & 3 & 7 & 1 & 0 & 0 \\ 2 & 5 & 4 & 0 & 1 & 0 \\ 9 & 8 & 6 & 0 & 0 & 1 \end{array} \right]$$

Sometimes in mathematics the vertical line is shown to indicate that the matrix has been augmented. In MATLAB, matrix augmentation can be accomplished using square brackets to concatenate the two matrices. The square matrix *a* is concatenated with an identity matrix that has the same size as the matrix *a*:

```
>> a = [1 3 7; 2 5 4; 9 8 6]
a =
     1     3     7
     2     5     4
     9     8     6
```

```
>> [a eye(size(a))]
ans =
     1     3     7     1     0     0
     2     5     4     0     1     0
     9     8     6     0     0     1
```

Of course, as we have seen already, it is more efficient to preallocate the matrix to the correct dimensions to begin with. Particularly for large matrices, augmenting a matrix in this fashion is inefficient.

12.1.6 Vector operations

Since vectors are just special cases of matrices, the matrix operations previously described (addition, subtraction, scalar multiplication, multiplication, transpose) work on vectors as well, as long as the dimensions are correct.

For vectors, we have already seen that the transpose of a row vector is a column vector, and the transpose of a column vector is a row vector.

To multiply vectors, they must have the same number of elements, but one must be a row vector and the other a column vector. For example, for a column vector c and row vector r:

$$c = \begin{bmatrix} 5 \\ 3 \\ 7 \\ 1 \end{bmatrix} \quad r = [6\ 2\ 3\ 4]$$

Note that r is 1×4, and c is 4×1. Recall that to multiply two matrices,

$$[A]_{m \times n}[B]_{n \times p} = [C]_{m \times p}$$

so

$$[r]_{1 \times 4}[c]_{4 \times 1} = [s]_{1 \times 1}$$

or, in other words, a scalar:

$$[6\ 2\ 3\ 4]\begin{bmatrix} 5 \\ 3 \\ 7 \\ 1 \end{bmatrix} = 6*5 + 2*3 + 3*7 + 4*1 = 61$$

whereas $[c]_{4 \times 1}[r]_{1 \times 4} = [M]_{4 \times 4}$, or in other words, a 4×4 matrix:

$$\begin{bmatrix} 5 \\ 3 \\ 7 \\ 1 \end{bmatrix}[6\ 2\ 3\ 4] = \begin{bmatrix} 30 & 10 & 15 & 20 \\ 18 & 6 & 9 & 12 \\ 42 & 14 & 21 & 28 \\ 6 & 2 & 3 & 4 \end{bmatrix}$$

In MATLAB, these operations are accomplished using the * operator, which is the matrix multiplication operator. First, the column vector c and row vector r are created.

```
>> c = [5 3 7 1]';
>> r = [6 2 3 4];
>> r * c
ans =
    61
```

```
>> c * r
ans =
    30    10    15    20
    18     6     9    12
    42    14    21    28
     6     2     3     4
```

Dot product and cross product

There are also operations specific to vectors: the *dot product* and *cross product*. The *dot product*, or *inner product*, of two vectors a and b is written as a • b and is defined as

$$a_1b_1 + a_2b_2 + a_3b_3 + \ldots + a_nb_n = \sum_{i=1}^{n} a_ib_i$$

In other words, this is like matrix multiplication when multiplying a row vector a by a column vector b; the result is a scalar. This can be accomplished using the * operator and transposing the second vector, or by using the **dot** function in MATLAB:

```
>> vec1 = [4 2 5 1];
>> vec2 = [3 6 1 2];
>> vec1 * vec2'
ans =
    31
```

```
>> dot(vec1,vec2)
ans =
    31
```

The *cross product* or *outer product* a × b of two vectors a and b is defined only when both a and b are vectors in three-dimensional space, which means that both must have three elements. It can be defined as a matrix multiplication of a matrix composed from the elements from a in a particular manner shown here and the column vector b.

$$a \times b = \begin{bmatrix} 0 & -a_3 & a_2 \\ a_3 & 0 & -a_1 \\ -a_2 & a_1 & 0 \end{bmatrix} \begin{bmatrix} b_1 \\ b_2 \\ b_3 \end{bmatrix} = [a_2b_3 - a_3b_2, a_3b_1 - a_1b_3, a_1b_2 - a_2b_1]$$

MATLAB has a built-in function **cross** to accomplish this.

```
>> vec1 = [4 2 5];
>> vec2 = [3 6 1];
>> cross(vec1,vec2)
ans =
   -28   11   18
```

12.2 MATRIX SOLUTIONS TO SYSTEMS OF LINEAR ALGEBRAIC EQUATIONS

A *linear algebraic equation* is an equation of the form

$$a_1x_1 + a_2x_2 + a_3x_3 + \ldots + a_nx_n = b$$

where the a's are constant *coefficients*, the x's are the **unknowns**, and b is a *constant*. A solution is a sequence of numbers that satisfy the equation. For example,

$$4x_1 + 5x_2 - 2x_3 = 16$$

is such an equation in which there are three unknowns: x_1, x_2, and x_3. One solution to this equation is $x_1 = 3$, $x_2 = 4$, and $x_3 = 8$, since 4 * 3 + 5 * 4 − 2 * 8 is equal to 16.

A system of linear algebraic equations is a set of equations of the form:

$$a_{11}x_1 + a_{12}x_2 + a_{13}x_3 + \ldots + a_{1n}x_n = b_1$$
$$a_{21}x_1 + a_{22}x_2 + a_{23}x_3 + \ldots + a_{2n}x_n = b_2$$
$$a_{31}x_1 + a_{32}x_2 + a_{33}x_3 + \ldots + a_{3n}x_n = b_3$$
$$\vdots \quad \vdots \quad \vdots \quad \quad \vdots \quad \vdots$$
$$a_{m1}x_1 + a_{m2}x_2 + a_{m3}x_3 + \ldots + a_{mn}x_n = b_m$$

This is called an *m* × *n* system of equations; there are m equations and n unknowns.

Because of the way that matrix multiplication works, these equations can be represented in matrix form as A x = b where A is a matrix of the coefficients, x is a column vector of the unknowns, and b is a column vector of the constants from the right side of the equations:

$$
\begin{matrix}
& \mathbf{A} & & & & \mathbf{x} & = & \mathbf{b} \\
\begin{bmatrix}
a_{11} & a_{12} & a_{13} & \cdots & a_{1n} \\
a_{21} & a_{22} & a_{23} & \cdots & a_{2n} \\
a_{31} & a_{32} & a_{33} & \cdots & a_{3n} \\
\vdots & \vdots & \vdots & \vdots & \vdots \\
a_{m1} & a_{m2} & a_{m3} & \cdots & a_{mn}
\end{bmatrix}
&
\begin{bmatrix}
x_1 \\ x_2 \\ x_3 \\ \vdots \\ x_n
\end{bmatrix}
& = &
\begin{bmatrix}
b_1 \\ b_2 \\ b_3 \\ \vdots \\ b_m
\end{bmatrix}
\end{matrix}
$$

A *solution set* is the set of all possible solutions to the system of equations (all sets of values for the unknowns that solve the equations). All systems of linear equations have either of the following:

■ no solutions
■ one solution
■ infinitely many solutions

One of the main concepts of the subject of *linear algebra* is the different methods of solving (or, attempting to!) systems of linear algebraic equations. MATLAB has many functions that assist in this process.

Once the system of equations has been written in matrix form, what we want is to solve the equation $Ax = b$ for the unknowns x. To do this, we need to isolate x on one side of the equation. If we were working with scalars, we would divide both sides of the equation by A. In fact, with MATLAB we can use the **divided into** operator to do this. However, most languages cannot do this with matrices, so we instead multiply both sides of the equation by the inverse of the coefficient matrix A:

$$A^{-1} A x = A^{-1} b$$

Then, because multiplying a matrix by its inverse results in the identity matrix I, and because multiplying any matrix by I results in the original matrix, we have:

$$I x = A^{-1} b$$

or

$$x = A^{-1} b$$

This means that the column vector of unknowns x is found as the inverse of matrix A multiplied by the column vector b. So, if we can find the inverse of A, we can solve for the unknowns in x.

For example, consider the following three equations with three unknowns—x_1, x_2, and x_3:

$$4x_1 - 2x_2 + 1x_3 = 7$$
$$1x_1 + 1x_2 + 5x_3 = 10$$
$$-2x_1 + 3x_2 - 1x_3 = 2$$

We write this in the form $Ax = b$ where A is a matrix of the coefficients, x is a column vector of the unknowns x_i, and b is a column vector of the values on the right side of the equations:

$$
\begin{matrix} A & x & b \end{matrix}
$$
$$
\begin{bmatrix} 4 & -2 & 1 \\ 1 & 1 & 5 \\ -2 & 3 & -1 \end{bmatrix}
\begin{bmatrix} x_1 \\ x_2 \\ x_3 \end{bmatrix}
=
\begin{bmatrix} 7 \\ 10 \\ 2 \end{bmatrix}
$$

The solution is then $x = A^{-1} b$. In MATLAB there are two simple ways to solve this. The built-in function **inv** can be used to get the inverse of A and then we multiply this by b, or we can use the divided into operator.

```
>> A = [4 −2 1; 1 1 5; −2 3 −1];
>> b = [7;10;2];
>> x = inv(A) * b
x =
    3.0244
    2.9512
    0.8049

>> x = A\b
x =
    3.0244
    2.9512
    0.8049
```

12.2.1 Solving 2 × 2 systems of equations

Although this may seem easy in MATLAB, in general, finding solutions to systems of equations is not. However, 2×2 systems are fairly straightforward, and there are several methods of solution for these systems for which MATLAB has built-in functions.

Consider the following 2×2 system of equations:

$$x_1 + 2x_2 = 2$$
$$2x_1 + 2x_2 = 6$$

First, to visualize the solution, we will change both equations to the equation of a straight line by writing each in the form $y = mx + b$ (by changing x_1 to x and x_2 to y):

$$x + 2y = 2 \quad \rightarrow \quad 2y = -x + 2 \quad \rightarrow \quad y = -0.5x + 1$$
$$2x + 2y = 6 \quad \rightarrow \quad 2y = -2x + 6 \quad \rightarrow \quad y = -x + 3$$

In MATLAB we can plot these lines using a script; the results are seen in Figure 12.1.

plot2by2.m

```
% Plot a 2 by 2 system as straight lines
x = −2:5;
y1 = −0.5 * x + 1;
y2 = −x + 3;
plot(x,y1,x,y2)
axis([ −2 5 −4 6] )
xlabel('x')
ylabel('y')
title('Visualize 2 x 2 system')
```

The intersection of the lines is the point $(4, -1)$. In other words, $x = 4$ and $y = -1$. Changing back to x_1 and x_2, we have $x_1 = 4$ and $x_2 = -1$. This allows us to visualize the solution.

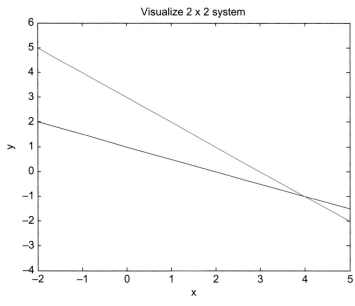

Visualize 2 x 2 system

FIGURE 12.1 Visualizing
2 × 2 systems of equations
as straight lines

This system of equations in matrix form is:

$$\begin{matrix} A & x & b \end{matrix}$$

$$\begin{bmatrix} 1 & 2 \\ 2 & 2 \end{bmatrix} \begin{bmatrix} x_1 \\ x_2 \end{bmatrix} = \begin{bmatrix} 2 \\ 6 \end{bmatrix}$$

We have already seen that the solution is $x = A^{-1}b$, so we can solve this if we can find the inverse of A. One method of finding the inverse for a 2 × 2 matrix involves calculating the **determinant** D.

For a 2 × 2 matrix

$$A = \begin{bmatrix} a_{11} & a_{12} \\ a_{21} & a_{22} \end{bmatrix}$$

the determinant D is defined as:

$$D = \begin{vmatrix} a_{11} & a_{12} \\ a_{21} & a_{22} \end{vmatrix} = a_{11}a_{22} - a_{12}a_{21}$$

It is written using vertical lines around the coefficients of the matrix, and is defined as the product of the values on the diagonal minus the product of the other two numbers.

For a 2 × 2 matrix, the matrix inverse is defined in terms of D as

$$A^{-1} = \frac{1}{D} \begin{bmatrix} a_{22} & -a_{12} \\ -a_{21} & a_{11} \end{bmatrix}$$

Note

If the determinant D is 0, it will not be possible to find the inverse of the matrix A.

The inverse is therefore the result of multiplying the scalar 1/D by every element in the previous matrix. Note that this is not the matrix A, but is determined using the elements from A in the following manner: the values on the diagonal are reversed, and the negation operator is used on the other two values.

For our coefficient matrix,

$$A = \begin{bmatrix} 1 & 2 \\ 2 & 2 \end{bmatrix}, \qquad D = \begin{vmatrix} 1 & 2 \\ 2 & 2 \end{vmatrix} = 1*2 - 2*2 \text{ or } -2$$

so

$$A^{-1} = \frac{1}{1*2 - 2*2} \begin{bmatrix} 2 & -2 \\ -2 & 1 \end{bmatrix} = \frac{1}{-2} \begin{bmatrix} 2 & -2 \\ -2 & 1 \end{bmatrix} = \begin{bmatrix} -1 & 1 \\ 1 & -\frac{1}{2} \end{bmatrix}$$

and

$$\begin{bmatrix} x_1 \\ x_2 \end{bmatrix} = \begin{bmatrix} -1 & 1 \\ 1 & -\frac{1}{2} \end{bmatrix} \begin{bmatrix} 2 \\ 6 \end{bmatrix}$$

The unknowns are found by performing this matrix multiplication. Consequently,

$$x_1 = -1 * 2 + 1 * 6 = 4$$
$$x_2 = 1 * 2 + (-1/2) * 6 = -1$$

This, of course, is the same solution as found by the intersection of the two lines.

To do this in MATLAB, we would first create the coefficient matrix variable a and column vector b.

```
>> a = [1 2; 2 2];
>> b = [2;6];
```

THE PROGRAMMING CONCEPT

For 2×2 matrices, the determinant and inverse are found using simple expressions.

```
>> deta = a(1,1) * a(2,2) - a(1,2) * a(2,1)
deta =
     -2

>> inva = (1/deta) * [a(2,2) -a(1,2); -a(2,1) a(1,1)]
inva =
    -1.0000    1.0000
     1.0000   -0.5000
```

THE EFFICIENT METHOD

We have already seen that MATLAB has a built-in function, **inv**, to find a matrix inverse. It also has a built-in function **det** to find a determinant:

```
>> det(a)
ans =
     -2

>> inv(a)
ans =
    -1.0000    1.0000
     1.0000   -0.5000
```

PRACTICE 12.4

For the following 2×2 system of equations:

$$x_1 + 3x_2 = 2$$
$$2x_1 + 4x_2 = 5$$

Do the following on paper:

■ Write the equations in matrix form $Ax = b$.
■ Solve by finding the inverse A^{-1} and then $x = A^{-1} b$.

Now, get into MATLAB and check your answers.

12.2.2 Gauss and Gauss-Jordan elimination

For *2 × 2* systems of equations, there are solution methods that are well-defined and simple. However, for larger systems of equations, finding solutions is frequently not as straightforward.

Two related methods of solving systems of linear equations will be described here: *Gauss elimination* and *Gauss-Jordan elimination*. They are both based on the observation that systems of equations are equivalent if they have the same solution set. Also, performing simple operations on the rows of a matrix, called *elementary row operations* (EROs), result in equivalent systems. These fall into three categories:

1. *Scaling:* This changes a row by multiplying it by a nonzero scalar $sr_i \rightarrow r_i$.
2. *Interchange rows:* For example, interchanging rows r_i and r_j is written as

$$r_i \longleftrightarrow r_j$$

3. *Replacement:* Replace a row by adding it to (or subtracting from it) a multiple of another row. For a given row r_i, this is written as

$$r_i + -sr_j \rightarrow r_i$$

Note that when replacing row r_i, nothing is multiplied by it. Instead, row r_j is multiplied by a scalar s (which could be a fraction) and that is added to or subtracted from row r_i. For example, for the matrix

$$\begin{bmatrix} 4 & 2 & 3 \\ 1 & 4 & 0 \\ 2 & 5 & 3 \end{bmatrix}$$

an example of interchanging rows would be $r_1 \longleftrightarrow r_3$, which would yield:

$$\begin{bmatrix} 4 & 2 & 3 \\ 1 & 4 & 0 \\ 2 & 5 & 3 \end{bmatrix} r_1 \longleftrightarrow r_3 \begin{bmatrix} 2 & 5 & 3 \\ 1 & 4 & 0 \\ 4 & 2 & 3 \end{bmatrix}$$

Now, starting with this matrix, an example of scaling would be $2r_2 \rightarrow r_2$, which means all elements in row 2 are multiplied by 2. This yields:

$$\begin{bmatrix} 2 & 5 & 3 \\ 1 & 4 & 0 \\ 4 & 2 & 3 \end{bmatrix} 2r_2 \rightarrow r_2 \begin{bmatrix} 2 & 5 & 3 \\ 2 & 8 & 0 \\ 4 & 2 & 3 \end{bmatrix}$$

Now, starting with this matrix, an example of a replacement would be $r_3 - 2r_2 \rightarrow r_3$. Element by element, row 3 is replaced by the element in row 3 minus 2 *, the corresponding element in row 2. This yields:

$$\begin{bmatrix} 2 & 5 & 3 \\ 2 & 8 & 0 \\ 4 & 2 & 3 \end{bmatrix} r_3 - 2r_2 \rightarrow r_3 \begin{bmatrix} 2 & 5 & 3 \\ 2 & 8 & 0 \\ 0 & -14 & 3 \end{bmatrix}$$

PRACTICE 12.5

Show the result of each of the following EROs:

$$\begin{bmatrix} 4 & 2 & 3 \\ 1 & 4 & 0 \\ 2 & 5 & 3 \end{bmatrix} r_1 \longleftrightarrow r_2$$

$$\begin{bmatrix} 4 & 2 & 3 \\ 1 & 4 & 0 \\ 2 & 5 & 3 \end{bmatrix} r_2 - \tfrac{1}{4} r_1 \rightarrow r_2$$

$$\begin{bmatrix} 4 & 2 & 3 \\ 1 & 4 & 0 \\ 2 & 5 & 3 \end{bmatrix} \tfrac{1}{2} r_1 \rightarrow r_1$$

Both the Gauss and Gauss-Jordan methods begin with the matrix form $Ax = b$ of a system of equations, and then augment the coefficient matrix A with the column vector b.

Gauss elimination
The Gauss elimination method consists of:

- creating the augmented matrix [A|b]
- applying EROs to this augmented matrix to get an upper triangular form (this is called *forward elimination*)
- *back substitution* to solve

For example, for a *2 × 2* system, the augmented matrix would be:

$$\begin{bmatrix} a_{11} & a_{12} & b_1 \\ a_{21} & a_{22} & b_2 \end{bmatrix}$$

Then, elementary row operations are applied to get the augmented matrix into an upper triangular form (i.e., the square part of the matrix on the left is in upper triangular form):

$$\begin{bmatrix} a'_{11} & a'_{12} & b'_1 \\ 0 & a'_{22} & b'_2 \end{bmatrix}$$

So, the goal is simply to replace a_{21} with 0. Here, the primes indicate that the values (may) have been changed.

Putting this back into the equation form yields

$$\begin{bmatrix} a'_{11} & a'_{12} \\ 0 & a'_{22} \end{bmatrix} \begin{bmatrix} x_1 \\ x_2 \end{bmatrix} = \begin{bmatrix} b'_1 \\ b'_2 \end{bmatrix}$$

Performing this matrix multiplication for each row results in:

$a'_{11} x_1 + a'_{12} x_2 = b'_1$

$a'_{22} x_2 = b'_2$

So, the solution is

$x_2 = b'_2 / a'_{22}$

$x_1 = (b'_1 - a'_{12} x_2) / a'_{11}$

Similarly, for a 3×3 system, the augmented matrix is reduced to upper triangular form:

$$\begin{bmatrix} a_{11} & a_{12} & a_{13} & b_1 \\ a_{21} & a_{22} & a_{23} & b_2 \\ a_{31} & a_{32} & a_{33} & b_3 \end{bmatrix} \rightarrow \begin{bmatrix} a'_{11} & a'_{12} & a'_{13} & b'_1 \\ 0 & a'_{22} & a'_{23} & b'_2 \\ 0 & 0 & a'_{33} & b'_3 \end{bmatrix}$$

(This will be done systematically by first getting a 0 in the a_{21} position, then a_{31}, and finally a_{32}.) Then, the solution will be:

$x_3 = b_3' / a_{33}'$

$x_2 = (b_2' - a_{23}'x_3) / a_{22}'$

$x_1 = (b_1' - a_{13}'x_3 - a_{12}'x_2) / a_{11}'$

Note that we find the last unknown, x_3, first, then the second unknown, and then the first unknown. This is why it is called back substitution.

As an example, consider the following 2×2 system of equations:

$x_1 + 2x_2 = 2$
$2x_1 + 2x_2 = 6$

As a matrix equation $Ax = b$, this is:

$$\begin{bmatrix} 1 & 2 \\ 2 & 2 \end{bmatrix} \begin{bmatrix} x_1 \\ x_2 \end{bmatrix} = \begin{bmatrix} 2 \\ 6 \end{bmatrix}$$

The first step is to augment the coefficient matrix A with b to get an augmented matrix [A|b]:

$$\begin{bmatrix} 1 & 2 & 2 \\ 2 & 2 & 6 \end{bmatrix}$$

For forward elimination, we want to get a 0 in the a_{21} position. To accomplish this, we can modify the second line in the matrix by subtracting from it 2 * the first row.

The way we would write this ERO follows:

$$\begin{bmatrix} 1 & 2 & 2 \\ 2 & 2 & 6 \end{bmatrix} r_2 - 2r_1 \rightarrow r_2 \begin{bmatrix} 1 & 2 & 2 \\ 0 & -2 & 2 \end{bmatrix}$$

Now, putting it back in matrix equation form:

$$\begin{bmatrix} 1 & 2 \\ 0 & -2 \end{bmatrix} \begin{bmatrix} x_1 \\ x_2 \end{bmatrix} = \begin{bmatrix} 2 \\ 2 \end{bmatrix}$$

says that the second equation is now $-2x_2 = 2$, so $x_2 = -1$. Plugging into the first equation,

$$x_1 + 2(-1) = 2, \text{ so } x_1 = 4$$

This is back substitution.

Gauss-Jordan elimination

The Gauss-Jordan elimination method starts the same way that the Gauss elimination method does, but then instead of back substitution, the elimination continues. The Gauss-Jordan method consists of:

- Creating the augmented matrix [A|b]
- Forward elimination by applying EROs to get an upper triangular form
- Back elimination to a diagonal form that yields the solution

For a 2×2 system, this method would yield

$$\begin{bmatrix} a_{11} & a_{12} & b_1 \\ a_{21} & a_{22} & b_2 \end{bmatrix} \rightarrow \begin{bmatrix} a'_{11} & 0 & b'_1 \\ 0 & a'_{22} & b'_2 \end{bmatrix}$$

and for a 3×3 system,

$$\begin{bmatrix} a_{11} & a_{12} & a_{13} & b_1 \\ a_{21} & a_{22} & a_{23} & b_2 \\ a_{31} & a_{32} & a_{33} & b_3 \end{bmatrix} \rightarrow \begin{bmatrix} a'_{11} & 0 & 0 & b'_1 \\ 0 & a'_{22} & 0 & b'_2 \\ 0 & 0 & a'_{33} & b'_3 \end{bmatrix}$$

Note that the resulting diagonal form does not include the right-most column. For example, for the 2×2 system, forward elimination yielded the matrix:

$$\begin{bmatrix} 1 & 2 & 2 \\ 0 & -2 & 2 \end{bmatrix}$$

Now, to continue with back elimination, we need a 0 in the a_{12} position:

$$\begin{bmatrix} 1 & 2 & 2 \\ 0 & -2 & 2 \end{bmatrix} r_1 + r_2 \rightarrow r_1 \begin{bmatrix} 1 & 0 & 4 \\ 0 & -2 & 2 \end{bmatrix}$$

So, the solution is $x_1 = 4$; $-2x_2 = 2$ or $x_2 = -1$.

Here is an example of a 3×3 system:

$$\begin{aligned} x_1 + 3x_2 \quad\quad &= 1 \\ 2x_1 + \quad x_2 + 3x_3 &= 6 \\ 4x_1 + 2x_2 + 3x_3 &= 3 \end{aligned}$$

In matrix form, the augmented matrix [A|b] is

$$\begin{bmatrix} 1 & 3 & 0 & 1 \\ 2 & 1 & 3 & 6 \\ 4 & 2 & 3 & 3 \end{bmatrix}$$

For forward substitution (done systematically by first getting a 0 in the a_{21} position, then a_{31}, and finally a_{32}):

$$\begin{bmatrix} 1 & 3 & 0 & 1 \\ 2 & 1 & 3 & 6 \\ 4 & 2 & 3 & 3 \end{bmatrix} r_2 - 2r_1 \rightarrow r_2 \begin{bmatrix} 1 & 3 & 0 & 1 \\ 0 & -5 & 3 & 4 \\ 4 & 2 & 3 & 3 \end{bmatrix} r_3 - 4r_1 \rightarrow r_3 \begin{bmatrix} 1 & 3 & 0 & 1 \\ 0 & -5 & 3 & 4 \\ 0 & -10 & 3 & -1 \end{bmatrix}$$

$$r_3 - 2r_2 \rightarrow r_3 \begin{bmatrix} 1 & 3 & 0 & 1 \\ 0 & -5 & 3 & 4 \\ 0 & 0 & -3 & -9 \end{bmatrix}$$

For the Gauss method, this is followed by back substitution:

$$\begin{aligned} -3x_3 &= -9 \\ x_3 &= 3 \end{aligned}$$

$$\begin{aligned} -5x_2 + 3\,(3) &= 4 \\ -5x_2 &= -5 \\ x_2 &= 1 \end{aligned}$$

$$\begin{aligned} x_1 + 3\,(1) &= 1 \\ x_1 &= -2 \end{aligned}$$

For the Gauss-Jordan method, this is instead followed by back elimination:

$$\begin{bmatrix} 1 & 3 & 0 & 1 \\ 0 & -5 & 3 & 4 \\ 0 & 0 & -3 & -9 \end{bmatrix} r_2 + r_3 \rightarrow r_2 \begin{bmatrix} 1 & 3 & 0 & 1 \\ 0 & -5 & 0 & -5 \\ 0 & 0 & -3 & -9 \end{bmatrix}$$

$$r_1 + 3/5r_2 \rightarrow r_1 \begin{bmatrix} 1 & 0 & 0 & -2 \\ 0 & -5 & 0 & -5 \\ 0 & 0 & -3 & -9 \end{bmatrix}$$

Thus,

$$x_1 = -2$$

$$-5x_2 = -5$$
$$x_2 = 1$$

$$-3x_3 = -9$$
$$x_3 = 3$$

Here's an example of performing these substitutions using MATLAB:

```
>> a = [1 3 0; 2 1 3; 4 2 3]
a =
     1    3    0
     2    1    3
     4    2    3

>> b = [1 6 3]'
b =
     1
     6
     3

>> ab = [a b]
ab =
     1    3    0    1
     2    1    3    6
     4    2    3    3

>> ab(2,:) = ab(2,:) - 2 * ab(1,:)
ab =
     1    3    0    1
     0   -5    3    4
     4    2    3    3

>> ab(3,:) = ab(3,:) - 4 * ab(1,:)
ab =
     1    3    0    1
     0   -5    3    4
     0  -10    3   -1

>> ab(3,:) = ab(3,:) - 2 * ab(2,:)
ab =
     1    3    0    1
     0   -5    3    4
     0    0   -3   -9

>> ab(2,:) = ab(2,:) + ab(3,:)
ab =
     1    3    0    1
     0   -5    0   -5
     0    0   -3   -9
```

```
>> ab(1,:) = ab(1,:) + 3/5 * ab(2,:)
ab =
    1    0    0   -2
    0   -5    0   -5
    0    0   -3   -9
```

12.2.3 Reduced row echelon form

The Gauss-Jordan method results in a diagonal form; for example, for a 3×3 system:

$$
\begin{bmatrix}
a_{11} & a_{12} & a_{13} & b_1 \\
a_{21} & a_{22} & a_{23} & b_2 \\
a_{31} & a_{32} & a_{33} & b_3
\end{bmatrix}
\rightarrow
\begin{bmatrix}
a'_{11} & 0 & 0 & b'_1 \\
0 & a'_{22} & 0 & b'_2 \\
0 & 0 & a'_{33} & b'_3
\end{bmatrix}
$$

Reduced row echelon form takes this one step further to result in all 1s rather than the a's, so that the column of b's is the solution. All that is necessary to accomplish this is to scale each row.

$$
\begin{bmatrix}
a_{11} & a_{12} & a_{13} & b_1 \\
a_{21} & a_{22} & a_{23} & b_2 \\
a_{31} & a_{32} & a_{33} & b_3
\end{bmatrix}
\rightarrow
\begin{bmatrix}
1 & 0 & 0 & b'_1 \\
0 & 1 & 0 & b'_2 \\
0 & 0 & 1 & b'_3
\end{bmatrix}
$$

In other words, we are reducing [A|b] to [I|b']. MATLAB has a built-in function to do this, called **rref**. For example, for the previous example:

```
>> a = [1 3 0; 2 1 3; 4 2 3];
>> b = [1 6 3]';
>> ab = [a b];
>> rref(ab)
ans =
    1    0    0   -2
    0    1    0    1
    0    0    1    3
```

The solution is found from the last column, so $x_1 = -2$, $x_2 = 1$, and $x_3 = 3$. To get this in a column vector in MATLAB:

```
>> x = ans(:,end)
x =
   -2
    1
    3
```

12.2.4 Finding a matrix inverse by reducing an augmented matrix

For a system of equations larger than a 2×2 system, one method of finding the inverse of a matrix A mathematically involves augmenting the matrix with an identity matrix of the same size, and then reducing it. The algorithm is:

- Augment the matrix with I: [A|I].
- Reduce it to the form [I|X]; X will be A^{-1}.

For example, in MATLAB we can start with a matrix, augment it with an identity matrix, and then use the **rref** function to reduce it.

```
>> a = [1 3 0; 2 1 3; 4 2 3];
>> rref([a eye(size(a))])
ans =
    1.0000         0         0   -0.2000   -0.6000    0.6000
         0    1.0000         0    0.4000    0.2000   -0.2000
         0         0    1.0000         0    0.6667   -0.3333
```

In MATLAB, the **inv** function can be used to verify the result.

```
>> inv(a)
ans =
   -0.2000   -0.6000    0.6000
    0.4000    0.2000   -0.2000
         0    0.6667   -0.3333
```

SUMMARY

Common Pitfalls
- Confusing matrix multiplication and array multiplication. Array operations, including multiplication, division, and exponentiation, are performed term by term (so the arrays must have the same size); the operators are .*, ./, .\, and .^. For matrix multiplication to be possible, the inner dimensions must agree and the operator is *.
- Forgetting that to augment one matrix with another, the number of rows must be the same in each.

Programming Style Guidelines
- Use vectorized code when performing matrix operations.

MATLAB Functions and Commands		
diag	tril	det
trace	inv	rref
eye	dot	magic
triu	cross	pascal

MATLAB Operators
* matrix multiplication
\ matrix division

Exercises

1. For the following matrices A, B, and C:

$$A = \begin{bmatrix} 1 & 4 \\ 3 & 2 \end{bmatrix} \quad B = \begin{bmatrix} 2 & 1 & 3 \\ 1 & 5 & 6 \\ 3 & 6 & 0 \end{bmatrix} \quad C = \begin{bmatrix} 3 & 2 & 5 \\ 4 & 1 & 2 \end{bmatrix}$$

   ▪ Which are symmetric?
   ▪ For all square matrices, give their trace.
   ▪ Give the result of 3 * A.
   ▪ Give the result of A * C.
   ▪ Are there any other matrix multiplications that can be performed? If so, list them.

2. For the following vectors and matrices A, B, and C:

$$A = \begin{bmatrix} 4 & 1 & -1 \\ 2 & 3 & 0 \end{bmatrix} \qquad B = [1 \quad 4] \qquad C = \begin{bmatrix} 2 \\ 3 \end{bmatrix}$$

   Perform the following operations, if possible. If not, just say it can't be done!

```
A * B
B * C
C * B
B + C^T
```

3. Given the following matrices:

$$A = \begin{bmatrix} 3 & 2 & 1 \\ 0 & 5 & 2 \\ 1 & 0 & 3 \end{bmatrix} \qquad B = \begin{bmatrix} 2 \\ 1 \\ 3 \end{bmatrix} \qquad I = \begin{bmatrix} 1 & 0 & 0 \\ 0 & 1 & 0 \\ 0 & 0 & 1 \end{bmatrix}$$

   Perform the following MATLAB operations, if they can be done. If not, explain why.

```
A * B
B * A
I + A
A .* I
trace(A)
```

4. Write a function *issquare* that will receive an array argument, and will return **logical** 1 for **true** if it is a square matrix, or **logical** 0 for **false** if it is not.

5. What is the value of the trace of an $n \times n$ identity matrix?

6. Write a function *mymatdiag* that will receive a matrix argument, and will return a vector consisting of the main diagonal (without using the built-in **diag** function). *Note:* This is only possible if the argument is a square matrix, so the function should first check this by calling the *issquare* function from Exercise 4. If the argument is a square matrix, the function will return the diagonal; otherwise, it will return an empty vector.

7. Write a function that will receive a square matrix as an input argument, and will return a row vector containing the diagonal of the matrix. If the function is called with a vector of two variables on the left side of the assignment, the function will also return the trace of the matrix. (*Note:* It will only return the trace if there are two variables on the left side of the assignment.) You may assume that the matrix is square. The function must preallocate the diagonal vector to the correct size.

8. Write a function *randdiag* that will return an $n \times n$ diagonal matrix, with random integers each in the range from *low* to *high* on the diagonal. Three arguments are passed to the function, in the following order: the value of n, low, and high.

9. Write a function *myeye* to return an $n \times n$ identity matrix (without using **eye**).

10. Write a function *myupp* that will receive an integer argument n, and will return an $n \times n$ upper triangular matrix of random integers.

11. When using the Gauss elimination to solve a set of algebraic equations, the solution can be obtained through back substitution when the corresponding matrix is in its upper triangular form. Write a function *istriu* that receives a matrix variable and returns a logical 1 if the matrix is in upper triangular form, or a logical 0 if not. Do this problem two ways: using loops and using built-in functions.

12. Write a function to determine whether a square matrix is a diagonal matrix. This function will return **logical** 1 for **true** if it is, or **logical** 0 if not.

13. Write a function *mymatsub* that will receive two matrix arguments and will return the result of subtracting the elements in one matrix from another (by looping and subtracting term by term). If it is not possible to subtract, return an empty matrix.

14. Write a function to receive a matrix and return its transpose (for more programming practice, do not use the built-in operator for the transpose).

15. We have already seen the **zeros** function, which returns a matrix of all 0s. Similarly, there is a function **ones** that returns a matrix of all 1s. (*Note:* No, there aren't functions called twos, threes, fifteens—just **ones** and **zeros**!) However, write a *fives* function that will receive two arguments for the number of rows and columns and will return a matrix with that size of all 5s.

16. The function **magic(n)** returns an $n \times n$ magic matrix, which is a matrix for which the sum of all rows, columns, and the diagonal are the same. Investigate this built-in function.

17. The function **pascal(n)** returns an $n \times n$ matrix made from Pascal's triangle. Investigate this built-in function, and then write your own.

18. Rewrite the following system of equations in matrix form:

$$
\begin{aligned}
4x_1 - x_2 + 3x_4 &= 10 \\
-2x_1 + 3x_2 + x_3 - 5x_4 &= -3 \\
x_1 + x_2 - x_3 + 2x_4 &= 2 \\
3x_1 + 2x_2 - 4x_3 &= 4
\end{aligned}
$$

Set it up in MATLAB and use any method to solve.

19. For the following 2×2 system of equations:

$$-3x_1 + x_2 = -4$$
$$-6x_1 + 2x_2 = 4$$

- In MATLAB, rewrite the equations as equations of straight lines and plot them to find the intersection.
- Solve for one of the unknowns and then substitute into the other equation to solve for the other unknown.
- Find the determinant D.
- How many solutions are there? One? None? Infinite?

20. For the following 2×2 system of equations:

$$-3x_1 + x_2 = 2$$
$$-6x_1 + 2x_2 = 4$$

- Rewrite the equations as equations of straight lines and plot them to find the intersection.
- Solve for one of the unknowns and then substitute into the other equation to solve for the other unknown.
- Find the determinant D.
- How many solutions are there? One? None? Infinite?

21. Write a function to return the determinant of a 2×2 matrix.

22. For a 2×2 system of equations, Cramer's rule states that the unknowns x are fractions of determinants. The numerator is found by replacing the column of coefficients of the unknown by constants b. Thus,

$$x_1 = \frac{\begin{vmatrix} b_1 & a_{12} \\ b_2 & a_{22} \end{vmatrix}}{D} \quad \text{and} \quad x_2 = \frac{\begin{vmatrix} a_{11} & b_1 \\ a_{21} & b_2 \end{vmatrix}}{D}$$

Use Cramer's rule to solve the following 2×2 system of equations:

$$-3x_1 + 2x_2 = -1$$
$$4x_1 - 2x_2 = -2$$

23. Write a function to implement Cramer's rule (see Exercise 22).

24. Write a function to return the inverse of a 2×2 matrix.

25. Given the following 2×2 system of equations:

$$3x_1 + x_2 = 2$$
$$2x_1 \quad\quad = 4$$

Use all methods presented in the text to solve it, and to visualize the solution. Do all of the math by hand, and then also use MATLAB.

26. ERO practice: Show the result of each of the following EROs:

$$\begin{bmatrix} 4 & 2 & 3 \\ 1 & 4 & 0 \\ 2 & 5 & 3 \end{bmatrix} \frac{1}{4} r_1 \rightarrow r_1$$

$$\begin{bmatrix} 4 & 2 & 3 \\ 1 & 4 & 0 \\ 2 & 5 & 3 \end{bmatrix} r_2 \longleftrightarrow r_3$$

$$\begin{bmatrix} 4 & 2 & 3 \\ 1 & 4 & 0 \\ 2 & 5 & 3 \end{bmatrix} r_3 - 2r_2 \rightarrow r_3$$

27. For the following 2×2 system of equations:

$$3x_1 + 2x_2 = 4$$
$$x_1 = 2$$

- Write it in matrix form.
- Using the method for 2×2 systems, find the determinant D.
- Use D to find the inverse of A.
- Use the Gauss elimination method to find the solution.
- Use the Gauss-Jordan method to find the solution.
- Check your work in MATLAB.

28. For the following set of equations:

$$2x_1 + 2x_2 + x_3 = 2$$
$$x_2 + 2x_3 = 1$$
$$x_1 + x_2 + 3x_3 = 3$$

- Put the set in the augmented matrix [A|b].
- Solve using Gauss.
- Solve using Gauss-Jordan.
- In MATLAB, create the matrix A and column vector b. Find the inverse and determinant of A. Solve for x.

29. Given the following system of equations:

$$x_1 - 2x_2 + x_3 = 2$$
$$2x_1 - 5x_2 + 3x_3 = 6$$
$$x_1 + 2x_2 + 2x_3 = 4$$
$$2x_1 + 3x_3 = 6$$

Write this in matrix form and use either Gauss or Gauss-Jordan to solve it. Check your answer using MATLAB.

30. Write a function that will augment a matrix with an identity matrix of the appropriate dimensions, without using any built-in functions (except **size**). This function will receive a matrix argument, and will return the augmented matrix.

31. Write a function *myrrefinv* that will receive a square matrix A as an argument, and will return the inverse of A. The function cannot use the built-in **inv** function; instead, it must augment the matrix with I and use **rref** to reduce it to the form $[I|A^{-1}]$. Examples of calling it are:

```
>> a = [4 3 2; 1 5 3; 1 2 3]
a =
    4    3    2
    1    5    3
    1    2    3
>> inv(a)
ans =
   0.3000  -0.1667  -0.0333
        0   0.3333  -0.3333
  -0.1000  -0.1667   0.5667
>> disp(myrrefinv(a))
   0.3000  -0.1667  -0.0333
        0   0.3333  -0.3333
  -0.1000  -0.1667   0.5667
```

32. For the following set of equations:

$$2x_1 + 2x_2 + x_3 = 2$$
$$x_2 + 2x_3 = 1$$
$$x_1 + x_2 + 3x_3 = 3$$

   ▪ In MATLAB, create the coefficient matrix A and vector b. Solve for x using the inverse, using the built-in function.
   ▪ Create the augmented matrix [A|b] and solve using the **rref** function.

33. Analyzing electric circuits can be accomplished by solving sets of equations. For a particular circuit, the voltages V_1, V_2, and V_3 are found through the system:

$$V_1 = 5$$
$$-6V_1 + 10V_2 - 3V_3 = 0$$
$$-V_2 + 51V_3 = 0$$

Put these equations in matrix form and solve in MATLAB.

Some operations are easier to do if a matrix (in particular, if it is really large) is partitioned into blocks. Partitioning into blocks also allows utilization of grid computing or parallel computing, where the operations are spread over a grid of computers. For example, if

$$A = \begin{bmatrix} 1 & -3 & 2 & 4 \\ 2 & 5 & 0 & 1 \\ -2 & 1 & 5 & -3 \\ -1 & 3 & 1 & 2 \end{bmatrix}$$

it can be partitioned into

$$\begin{bmatrix} A_{11} & A_{12} \\ A_{21} & A_{22} \end{bmatrix}$$

where

$$A_{11} = \begin{bmatrix} 1 & -3 \\ 2 & 5 \end{bmatrix}, A_{12} = \begin{bmatrix} 2 & 4 \\ 0 & 1 \end{bmatrix}, A_{21} = \begin{bmatrix} -2 & 1 \\ -1 & 3 \end{bmatrix}, A_{22} = \begin{bmatrix} 5 & -3 \\ 1 & 2 \end{bmatrix}$$

If B *is the same size,*

$$B = \begin{bmatrix} 2 & 1 & -3 & 0 \\ 1 & 4 & 2 & -1 \\ 0 & -1 & 5 & -2 \\ 1 & 0 & 3 & 2 \end{bmatrix}$$

partition it into

$$\begin{bmatrix} B_{11} & B_{12} \\ B_{21} & B_{22} \end{bmatrix}$$

34. Create the matrices A and B, and partition them in MATLAB. Show that matrix addition, matrix subtraction, and scalar multiplication can be performed block by block, and concatenated for the overall result.

35. For matrix multiplication, use the following blocks:

$$A * B = \begin{bmatrix} A_{11} & A_{12} \\ A_{21} & A_{22} \end{bmatrix} \begin{bmatrix} B_{11} & B_{12} \\ B_{21} & B_{22} \end{bmatrix} = \begin{bmatrix} A_{11}B_{11} + A_{12}B_{21} & A_{11}B_{12} + A_{12}B_{22} \\ A_{21}B_{11} + A_{22}B_{21} & A_{21}B_{12} + A_{22}B_{22} \end{bmatrix}$$

Perform this in MATLAB for the given matrices.

36. We have seen that a square matrix is **symmetric** if $a_{ij} = a_{ji}$ for all i, j. We say that a square matrix is **skew symmetric** if $a_{ij} = -a_{ji}$ for all i, j. Note that this means that all of the values on the diagonal must be 0. Write a function that will receive a square matrix as an input argument, and will return **logical** 1 for **true** if the matrix is skew symmetric or **logical** 0 for **false** if not.

Basic Statistics, Sets, Sorting, and Indexing

CONTENTS

There are a lot of statistical analyses that can be performed on data sets. In the MATLAB® software, the statistical functions are in the data analysis help topic called **datafun**.

In general, we will write a data set of n values as

```
x = { x_1, x_2, x_3, x_4, ..., x_n}
```

In MATLAB, this will generally be represented as a row vector called x.

Statistics can be used to characterize properties of a data set. For example, consider a set of exam grades {33, 75, 77, 82, 83, 85, 85, 91, 100}. What is a "normal," "expected," or "average" exam grade? There are several ways that this could be interpreted. Perhaps the most common is the **mean** grade, which is found by summing the grades and dividing by the number of them (the result of that would be 79). Another way of interpreting that would be the grade found the most often, which would be 85. Also, the value in the middle of the list, 83, could be used. Another property that is useful to know is how spread out the data values are within the data set.

This section will cover some simple statistics, as well as set operations that can be performed on data sets. Some statistical functions require that the data set be *sorted*, so sorting will also be covered. Using *index vectors* is a way of representing the data in order, without physically sorting the data set. Finally, *searching* for values within a data set or a database is useful, so some basic searching techniques will be explained.

13.1 STATISTICAL FUNCTIONS

MATLAB has built-in functions for many statistics; the simplest of which we have already seen (e.g., **min** and **max** to find the minimum or maximum value in a data set).

```
>> x = [9 10 10 9 8 7 3 10 9 8 5 10];
>> min(x)
ans =
     3

>> max(x)
ans =
    10
```

Both of these functions also return the index of the smallest or largest value; if there is more than one occurrence, it returns the first. For example, in the following data set 10 is the largest value; it is found in three elements in the vector but the index returned is the first element in which it is found (which is 2):

```
>> x = [9 10 10 9 8 7 3 10 9 8 5 10];
>> [maxval, maxind] = max(x)
maxval =
    10
maxind =
     2
```

For matrices, the **min** and **max** functions operate columnwise by default:

```
>> mat = [9 10 17 5; 19 9 11 14]
mat =
     9    10    17     5
    19     9    11    14

>> [minval, minind] = min(mat)
minval =
     9     9    11     5

minind =
     1     2     2     1
```

To find the minimum (or maximum) for each row, the dimension of 2 (which is how MATLAB refers to rows) can be specified as the third argument to the **min** (or **max**) function. The second argument *must* be an empty vector:

```
>> min(mat,[],2)
ans =
     5
     9
```

These functions can also compare vectors or matrices (with the same dimensions) and return the minimum (or maximum) values from corresponding elements. For example, the following iterates through all elements in the two vectors, comparing corresponding elements, and returning the minimum for each:

```
>> x = [3 5 8 2 11];
>> y = [2 6 4 5 10];
>> min(x,y)
ans =
     2     5     4     2    10
```

Some of the other functions in the **datafun** help topic that have been described already include **sum**, **prod**, **cumsum**, **cumprod**, and **hist**. Other statistical operations, and the functions that perform them in MATLAB, will be described in the rest of this section.

13.1.1 Mean

The *arithmetic mean* of a data set is what is usually called the *average* of the values, or in other words, the sum of the values divided by the number of values in the data set. Mathematically, we would write this as

$$\frac{\sum_{i=1}^{n} x_i}{n}$$

THE PROGRAMMING CONCEPT

Calculating a mean, or average, would normally be interpreted as looping through the elements of a vector, adding them together, and then dividing by the number of elements:

mymean.m

```
function outv = mymean(vec)
% mymean returns the mean of a vector
% Format: mymean(vector)

mysum = 0;
```

Continued

THE PROGRAMMING CONCEPT—CONT'D

```
for i = 1:length(vec)
    mysum = mysum + vec(i);
end
outv = mysum/length(vec);
end
```

```
>> x = [9 10 10 9 8 7 3 10 9 8 5 10];
>> mymean(x)
ans =
    8.1667
```

THE EFFICIENT METHOD

There is a built-in function, **mean**, in MATLAB to accomplish this:

```
>> mean(x)
ans =
    8.1667
```

For a matrix, the **mean** function operates columnwise. To find the mean of each row, the dimension of 2 is passed as the second argument to the function, as is the case with the functions **sum**, **prod**, **cumsum**, and **cumprod** (the [] as a middle argument is not necessary for these functions like it is for **min** and **max**).

```
>> mat = [8 9 3; 10 2 3; 6 10 9]
mat =

     8     9     3
    10     2     3
     6    10     9

>> mean(mat)
ans =
     8     7     5

>> mean(mat,2)
ans =
    6.6667
    5.0000
    8.3333
```

Sometimes a value that is much larger or smaller than the rest of the data (called an *outlier*) can throw off the mean. For example, in the following all of the numbers in the data set are in the range from 3 to 10, with the exception of the 100 in the middle. Because of this outlier, the mean of the values in this vector is actually larger than any of the other values in the vector.

```
>> xwithbig = [9 10 10 9 8 100 7 3 10 9 8 5 10];
>> mean(xwithbig)
ans =
   15.2308
```

Typically, an outlier like this represents an error of some kind, perhaps in the data collection. To handle this, sometimes the minimum and maximum values from a data set are discarded before the mean is computed. In this example, a **logical** vector indicating which elements are neither the largest nor smallest values is used to index into the original data set, resulting in removing the minimum and the maximum.

```
>> xwithbig = [9 10 10 9 8 100 7 3 10 9 8 5 10];
>> length(xwithbig)
ans =
   13

>> newx = xwithbig(xwithbig ~= min(xwithbig) & ...
              xwithbig ~= max(xwithbig))
newx =
   9   10   10   9   8   7   10   9   8   5   10

>> length(newx)
ans =
   11
```

Instead of just removing the minimum and maximum values, sometimes the largest and smallest 1% or 2% of values are removed, especially if the data set is very large.

There are several other means that can be computed. The *harmonic mean* of the *n* values in a vector or data set *x* is defined as

$$\frac{n}{\dfrac{1}{x_1} + \dfrac{1}{x_2} + \dfrac{1}{x_3} + \cdots \dfrac{1}{x_n}}$$

This could be implemented in an anonymous function using the built-in **sum** function. For example, the following anonymous function calculates this, and stores the handle in a variable called *harmhand*.

```
>> harmhand = @ (x) length(x) / sum(1 ./ x);
>> x = [9 10 10 9 8 7 3 10 9 8 5 10];
>> harmhand(x)
ans =
   7.2310
```

Note

Statistics Toolbox™ has functions for these means, called **harmmean** and **geomean**, as well as a function **trimmean** that trims the highest and lowest 2% of data values.

The *geometric mean* of the n values in a vector x is defined as the nth root of the product of the data set values.

$$\sqrt[n]{x_1{}^*x_2{}^*x_3{}^*\ldots{}^*x_n}$$

The following anonymous function implements this definition, using **prod**:

```
>> geomhand = @ (x) prod(x)^(1/length(x));
>> geomhand(x)
ans =
    7.7775
```

13.1.2 Variance and standard deviation

The *standard deviation* and *variance* are ways of determining the spread of the data. The variance is usually defined in terms of the arithmetic mean as:

$$var = \frac{\sum_{i=1}^{n}(x_i - mean)^2}{n-1}$$

Sometimes, the denominator is defined as n rather than $n - 1$. The default definition in MATLAB uses $n - 1$ for the denominator, so we will use that definition here.

For example, for the vector [8, 7, 5, 4, 6], there are $n = 5$ values so $n - 1$ is 4. Also, the mean of this data set is 6. The variance would be

$$var = \frac{(8-6)^2 + (7-6)^2 + (5-6)^2 + (4-6)^2 + (6-6)^2}{4}$$

$$= \frac{4+1+1+4+0}{4} = 2.5$$

The built-in function to calculate the variance is called **var**:

```
>> xvals = [8 7 5 4 6];
>> myvar = var(xvals)
yvar =
    2.5000
```

The standard deviation is the square root of the variance:

$$sd = \sqrt{var}$$

The built-in function in MATLAB for the standard deviation is called **std**; the standard deviation can be found either as the **sqrt** of the variance, or using **std**:

```
>> shortx = [2 5 1 4];
>> myvar = var(shortx)
myvar =
    3.3333
```

```
>> sqrt(myvar)
ans =
   1.8257

>> std(shortx)
ans =
   1.8257
```

The less spread out the numbers are, the smaller the standard deviation will be, since it is a way of determining the spread of the data. Likewise, the more spread out the numbers are, the larger the standard deviation will be. For example, here are two data sets that have the same number of values and also the same mean, but the standard deviations are quite different:

```
>> x1 = [9 10 9.4 9.6];
>> mean(x1)
ans =
   9.5000
>> std(x1)
ans =
   0.4163

>> x2 = [2 17 −1.5 20.5];
>> mean(x2)
ans =
   9.5000
>> std(x2)
ans =
   10.8704
```

13.1.3 Mode

The *mode* of a data set is the value that appears most frequently. The built-in function in MATLAB for this is called the **mode**.

```
>> x = [9 10 10 9 8 7 3 10 9 8 5 10];
>> mode(x)
ans =
   10
```

If there is more than one value with the same (highest) frequency, the smaller value is the mode. In the following case, since 3 and 8 appear twice in the vector, the smaller value (3) is the mode:

```
>> x = [3 8 5 3 4 1 8];
>> mode(x)
ans =
   3
```

Therefore, if no value appears more frequently than any other, the smallest value in the vector will be the mode of the vector.

13.1.4 Median

The *median* is defined only for a data set that has been *sorted* first, meaning that the values are in order. The median of a sorted set of n data values is defined as the value in the middle, if n is odd, or the average of the two values in the middle if n is even. For example, for the vector [1 4 5 9 12], the middle value is 5. The function in MATLAB is called **median**:

```
>> median([1 4 5 9 12 ])
ans =
     5
```

For the vector [1 4 5 9 12 33], the median is the average of the 5 and 9 in the middle:

```
>> median([1 4 5 9 12 33])
ans =
     7
```

If the vector is not in sorted order to begin with, the **median** function will still return the correct result (it will sort the vector automatically). For example, scrambling the order of the values in the first example will still result in a median value of 5.

```
>> median([9 4 1 5 12])
ans =
     5
```

PRACTICE 13.1

For the vector [1 1 3 6 9], find the

- Minimum
- Maximum
- Arithmetic mean
- Geometric mean
- Harmonic mean
- Variance
- Standard deviation
- Mode
- Median

13.2 SET OPERATIONS

MATLAB has several built-in functions that perform *set operations* on vectors. These include **union**, **intersect**, **unique**, **setdiff**, and **setxor**. All of these functions can be useful when working with data sets, and all return vectors that are sorted from lowest to highest (*ascending order*). Additionally, there are two "is" functions that work on sets: **ismember** and **issorted**.

For example, given the following vectors:

```
>> v1 = 6:-1:2
     6    5    4    3    2

>> v2 = 1:2:7
v2 =
     1    3    5    7
```

the **union** function returns a vector that contains all of the values from the two input argument vectors, without repeating any.

```
>> union(v1,v2)
ans =
     1    2    3    4    5    6    7
```

The **intersect** function instead returns all of the values that can be found in both of the two input argument vectors.

```
>> intersect(v1,v2)
ans =
     3    5
```

The **setdiff** function receives two vectors as input arguments, and returns a vector consisting of all of the values that are contained in the first vector argument but not the second. Therefore, the order of the two input arguments is important.

```
>> setdiff(v1,v2)
ans =
     2    4    6

>> setdiff(v2,v1)
ans =
     1    7
```

The function **setxor** receives two vectors as input arguments, and returns a vector consisting of all of the values from the two vectors that are not in the intersection of these two vectors. In other words, it is the union of the two vectors obtained using **setdiff** when passing the vectors in different orders as before.

```
>> setxor(v1,v2)
ans =
     1    2    4    6    7

>> union(setdiff(v1,v2), setdiff(v2,v1))
ans =
     1    2    4    6    7
```

The set function **unique** returns all of the unique values from a set argument:

```
>> v3 = [1:5  3:6]
v3 =
     1    2    3    4    5    3    4    5    6

>> unique(v3)
ans =
     1    2    3    4    5    6
```

Many of the set functions return vectors that can be used to index into the original vectors as optional output arguments. For example, the two vectors *v1* and *v2* were defined previously as follows:

```
>> v1
v1 =
     6    5    4    3    2

>> v2
v2 =
     1    3    5    7
```

The **intersect** function returns, in addition to the vector containing the values in the intersection of *v1* and *v2*, an index vector into *v1* and an index vector into *v2* such that *outvec* is the same as *v1(index1)* and also *v2(index2)*.

```
>> [outvec, index1, index2] = intersect(v1,v2)
outvec =
     3    5

index1 =
     4    2

index2 =
     2    3
```

Using these vectors to index into *v1* and *v2* will return the values from the intersection. For example, this expression returns the second and fourth elements of *v1* (it puts them in ascending order):

```
>> v1(index1)
ans =
     3    5
```

This returns the second and third elements of *v2*:

```
>> v2(index2)
ans =
     3    5
```

The function **ismember** receives two vectors as input arguments, and returns a **logical** vector that is the same length as the first argument, containing **logical** 1 for **true** if the element in the first vector is also in the second, or **logical** 0 for **false** if not. The order of the arguments matters for this function.

```
>> v1
v1 =
     6     5     4     3     2

>> v2
v2 =
     1     3     5     7

>> ismember(v1,v2)
ans =
     0     1     0     1     0

>> ismember(v2,v1)
ans =
     0     1     1     0
```

Using the result from the **ismember** function as an index into the first vector argument will return the same values as the **intersect** function (although not necessarily sorted).

```
>> logv = ismember(v1,v2)
logv =
     0     1     0     1     0

>> v1(logv)
ans =
     5     3

>> logv = ismember(v2,v1)
logv =
     0     1     1     0

>> v2(logv)
ans =
     3     5
```

The **issorted** function will return **logical** 1 for **true** if the argument is sorted in *ascending* order (lowest to highest), or **logical** 0 for **false** if not.

```
>> v3 = [1:5  3:6]
v3 =
     1     2     3     4     5     3     4     5     6
```

```
>> issorted(v3)
ans =
     0

>> issorted(v2)
ans =
     1

>> issorted(v1)
ans =
     0
```

In the next section, we will see how to sort a vector.

PRACTICE 13.2

Create two vector variables *vec1* and *vec2* that contain seven random integers, each in the range from 1 to 20. Do each of the following operations by hand first, and then check in MATLAB:

- union
- intersection
- setdiff
- setxor
- unique (for each)

13.3 SORTING

Sorting is the process of putting a list in order—either *descending* (highest to lowest) or *ascending* (lowest to highest) order. For example, here is a list of *n* integers, visualized as a column vector.

```
1    | 85  |
2    | 70  |
3    | 100 |
4    | 95  |
5    | 80  |
6    | 91  |
```

What is desired is to sort this in ascending order in place—by rearranging this vector, not creating another. The following is one basic algorithm:

- Look through the vector to find the smallest number, and then put it in the first element in the vector. How? By exchanging it with the number currently in the first element.

■ Then, scan the rest of the vector (from the second element down) looking for the next smallest (or, the smallest in the rest of the vector). When found, put it in the first element of the rest of the vector (again, by exchanging).

■ Continue doing this for the rest of the vector. Once the next-to-last number has been placed in the correct location in the vector, by default the last number has been as well.

What is important in each pass through the vector is not knowing what the smallest value is, but where it is so that the exchange can be made.

This table shows the progression. The left column shows the original vector. The second column (from the left) shows that the smallest number, 70, is now in the first element in the vector. It was put there by exchanging with what had been in the first element, 85. This continues element by element, until the vector has been sorted.

85	70	70	70	70	70
70	85	80	80	80	80
100	100	100	85	85	85
95	95	95	95	91	91
80	80	85	100	100	95
91	91	91	91	95	100

This is called the *selection sort*; it is one of many different sorting algorithms.

THE PROGRAMMING CONCEPT

The following function implements the **selection sort** to sort a vector:

mysort.m

```
function outv = mysort(vec)
% mysort sorts a vector using the selection sort
% Format: mysort(vector)

% Loop through the elements in the vector to end-1
for i = 1:length(vec)-1
    low = i;    % stores the index of the smallest
    %Select the smallest number in the rest of the vector
    for j = i+1:length(vec)
        if vec(j) < vec(low)
            low = j;
        end
    end
```

```
    % Exchange elements
    temp = vec(i);
    vec(i) = vec(low);
    vec(low) = temp;
  end
  outv = vec;
  end
```

```
>> vec = [85 70 100 95 80 91];
>> vec = mysort(vec)
vec =
    70    80    85    91    95    100
```

THE EFFICIENT METHOD

MATLAB has a built-in function, **sort**, that will sort a vector in ascending order:

```
>> vec = [85 70 100 95 80 91];
>> vec = sort(vec)
vec =
    70    80    85    91    95    100
```

Descending order can also be specified. For example,

```
>> sort(vec,'descend')
ans =
    100    95    91    85    80    70
```

For matrices, the **sort** function will by default sort each column. To sort by rows, the dimension 2 is specified. For example,

```
>> mat
mat =
    4    6    2
    8    3    7
    9    7    1

>> sort(mat)    % sorts by column
ans =
    4    3    1
    8    6    2
    9    7    7
```

```
>> sort(mat,2) % sorts by row
ans =
     2     4     6
     3     7     8
     1     7     9
```

13.3.1 Sorting vectors of structures

When working with a vector of structures, it is common to sort based on a particular field that is within the structures. For example, recall the vector of structures used to store information on different software packages that was created in Chapter 8.

packages

	item_no	cost	price	code
1	123	19.99	39.95	g
2	456	5.99	49.99	l
3	587	11.11	33.33	w

Here is a function that sorts this vector of structures in ascending order based on the *price* field.

mystructsort.m

```
function outv = mystructsort(structarr)
% mystructsort sorts a vector of structs on the price field
% Format: mystructsort(structure vector)

for i = 1:length(structarr)-1
    low = i;
    for j = i+1:length(structarr)
        if structarr(j).price < structarr(low).price
            low = j;
        end
    end
    % Exchange elements
    temp = structarr(i);
    structarr(i) = structarr(low);
    structarr(low) = temp;
end
outv = structarr;
end
```

Note that only the *price* field is compared in the sort algorithm, but the entire structure is exchanged. Consequently, each element in the vector, which is a structure of information about a particular software package, remains intact.

Recall that we created a function *printpackages* also in Chapter 8 that prints the information in a nice table format. Calling the *mystructsort* function and also the function to print will demonstrate this:

```
>> printpackages(packages)

Item #  Cost  Price  Code

   123  19.99  39.95   g
   456   5.99  49.99   l
   587  11.11  33.33   w

>> packByPrice = mystructsort(packages);
>> printpackages(packByPrice)

Item #  Cost  Price  Code

   587  11.11  33.33   w
   123  19.99  39.95   g
   456   5.99  49.99   l
```

This function only sorts the structures based on the *price* field. A more general function is shown in the following, which receives a string that is the name of the field. The function checks first to make sure that the string that is passed is a valid field name for the structure. If it is, it sorts based on that field, and if not, it returns an empty vector.

Strings are created consisting of the name of the vector variable, followed by parentheses containing the element number, the period, and finally the name of the field. The strings are created using square brackets to concatenate the pieces of the string, and the **int2str** function is used to convert the element number to a string. Then, using the **eval** function, the vector elements are compared to determine the lowest.

generalPackSort.m

```
function outv = generalPackSort(inputarg, fname)
% generalPackSort sorts a vector of structs
% based on the field name passed as an input argument

if isfield(inputarg,fname)
   for i = 1:length(inputarg)-1
      low = i;
      for j = i + 1:length(inputarg)
         if eval([ 'inputarg(' int2str(j) ').' fname] ) < ...
            eval([ 'inputarg(' int2str(low)  ').' fname] )
            low = j;
         end
      end
      % Exchange elements
```

```
      temp = inputarg(i);
      inputarg(i) = inputarg(low);
      inputarg(low) = temp;
   end
   outv = inputarg;
else
   outv =[];
end
end
```

The following are examples of calling the function:

```
>> packByPrice = generalPackSort(packages,'price');
>> printpackages(packByPrice)

Item #   Cost   Price   Code

   587   11.11   33.33     w
   123   19.99   39.95     g
   456    5.99   49.99     l

>> packByCost = generalPackSort(packages,'cost');
>> printpackages(packByCost)

Item #   Cost   Price   Code

   456    5.99   49.99     l
   587   11.11   33.33     w
   123   19.99   39.95     g

>> packByProfit = generalPackSort(packages,'profit')
packByProfit =
   []
```

QUICK QUESTION!

Is this *generalPackSort* function truly general? Would it work for any vector of structures, not just one configured like *packages*? **Answer:** It is fairly general. It will work for any vector of structures. However, the comparison will only work for numerical or character fields. Thus, as long as the field is a number or character, this function will work for any vector of structures. If the field is a vector itself (including a string), it will not work.

13.3.2 Sorting strings

For a matrix of strings, the **sort** function works exactly as shown previously for numbers. For example,

```
>> words = char('Hello', 'Howdy', 'Hi', 'Goodbye', 'Ciao')
words =
Hello
Howdy
Hi
Goodbye
Ciao
```

The following sorts column by column using the ASCII equivalents of the characters. It can be seen from the results that the space character comes before the letters of the alphabet in the character encoding:

```
>> sort(words)
ans =
Ce
Giad
Hildb
Hoolo
Howoyye
```

To sort on the rows instead, the second dimension must be specified.

```
>> sort(words,2)
ans =
   Hello
   Hdowy
      Hi
 Gbdeooy
   Caio
```

It can be seen that the blank space comes before the letters of the alphabet in the character encoding, and also that the uppercase letters come before the lowercase letters.

How could the strings be sorted alphabetically? MATLAB has a function **sortrows** that will do this. The way it works is that it examines the strings column by column starting from the left. If it can determine which letter comes first, it picks up the entire string and puts it in the first row. In this example, the first two strings are placed based on the first character, C and G. For the other three strings, they all begin with H so the next column is examined. In this case the strings are placed based on the second character, e, i, o.

```
>> sortrows(words)
ans =
Ciao
Goodbye
Hello
Hi
Howdy
```

The **sortrows** function sorts each row as a block, or group, and it will also work on numbers. In this example the rows beginning with 3 and 4 are placed first. Then, for the rows beginning with 5, the values in the second column (6 and 7) determine the order.

```
>> mat = [5 7 2; 4 6 7; 3 4 1; 5 6 2]
mat =
     5     7     2
     4     6     7
     3     4     1
     5     6     2

>> sortrows(mat)
ans =
     3     4     1
     4     6     7
     5     6     2
     5     7     2
```

In order to sort a cell array of strings, the **sort** function can be used. For example,

```
>> engcellnames = {'Chemical','Mechanical',...
     'Biomedical','Electrical', 'Industrial'};
>> sort(engcellnames')
ans =
    'Biomedical'
    'Chemical'
    'Electrical'
    'Industrial'
    'Mechanical'
```

13.4 INDEX VECTORS

Using index vectors is an alternative to sorting a vector. With indexing, the vector is left in its original order. An index vector is used to "point" to the values in the original vector in the desired order.

For example, here is a vector of exam grades:

```
          grades
     1   2   3   4   5   6
   ┌────┬────┬─────┬────┬────┬────┐
   │ 85 │ 70 │ 100 │ 95 │ 80 │ 91 │
   └────┴────┴─────┴────┴────┴────┘
```

In ascending order, the lowest grade is in element 2, the next lowest grade is in element 5, and so on. The index vector *grade_index* gives the order that follows.

```
                        grade_index
                  1    2    3    4    5    6
                ┌────┬────┬────┬────┬────┬────┐
                │ 2  │ 5  │ 1  │ 6  │ 4  │ 3  │
                └────┴────┴────┴────┴────┴────┘
```

Note

This is a particular type of index vector in which all of the indices of the original vector appear in the desired order.

The index vector is then used as the indices for the original vector. To get the *grades* vector in ascending order, the indices used would be *grades(2)*, *grades(5)*, and so on. Using the index vector to accomplish this, `grades(grade_index(1))` would be the lowest grade, 70, and `grades(grade_index(2))` would be the second-lowest grade. In general, `grades(grade_index(i))` would be the *i*th lowest grade.

To create these in MATLAB:

```
>> grades = [85 70 100 95 80 91];
>> grade_index = [2 5 1 6 4 3];
>> grades(grade_index)
ans =
    70    80    85    91    95    100
```

In general, instead of creating the index vector manually as shown here, the procedure to initialize the index vector is to use a sort function. The following is the algorithm:

- Initialize the values in the index vector to be the indices 1, 2, 3, ... to the length of the original vector.
- Use any sort algorithm, but compare the elements in the original vector using the index vector to index into it (e.g., using *grades(grade_index(i))* as previously shown).
- When the sort algorithm calls for exchanging values, exchange the elements in the index vector, not in the original vector.

Here is a function that implements this algorithm:

createind.m

```
function indvec = createind(vec)
% createind returns an index vector for the
% input vector in ascending order
% Format: createind(inputVector)

% Initialize the index vector
len = length(vec);
indvec = 1:len;

for i = 1:len-1
    low = i;
    for j = i+1:len
```

```
        % Compare values in the original vector
        if vec(indvec(j)) < vec(indvec(low))
            low = j;
        end
    end
    % Exchange elements in the index vector
    temp = indvec(i);
    indvec(i) = indvec(low);
    indvec(low) = temp;
end
end
```

For example, for the grades vector just given:

```
>> clear grade_index

>> grade_index = createind(grades)
grade_index =
     2     5     1     6     4     3

>> grades(grade_index)
ans =
    70    80    85    91    95   100
```

13.4.1 Indexing into vectors of structures

Often, when the data structure is a vector of structures, it is necessary to iterate through the vector in order by different fields. For example, for the *packages* vector defined previously, it may be necessary to iterate in order by the *cost*, or by the *price* fields.

Rather than sorting the entire vector of structures based on these fields, it may be more efficient to index into the vector based on these fields; so, for example, to have an index vector based on *cost* and another based on *price*.

packages

	item_no	cost	price	code
1	123	19.99	39.95	g
2	456	5.99	49.99	1
3	587	11.11	33.33	w

cost_ind

1	2
2	3
3	1

price_ind

1	3
2	1
3	2

These index vectors would be created as before, comparing the fields but exchanging the entire structures. Once the index vectors have been created, they can be used to iterate through the *packages* vector in the desired order. For example, the function to print the information from *packages* has been modified so that in addition to the vector of structures, the index vector is also passed, and the function iterates using that index vector.

printpackind.m

```
function printpackind(packstruct, indvec)
% printpackind prints a table showing all
% values from a vector of packages structures
% using an index vector for the order
% Format: printpackind(vector of packages, index vector)

fprintf('Item #  Cost  Price  Code\n')
no_packs = length(packstruct);
for i = 1:no_packs
    fprintf('%6d %6.2f %6.2f %3c\n', ...
        packstruct(indvec(i)).item_no, ...
        packstruct(indvec(i)).cost, ...
        packstruct(indvec(i)).price, ...
        packstruct(indvec(i)).code)
end
end
```

```
>> printpackind(packages,cost_ind)
Item #  Cost  Price  Code
  456    5.99  49.99   l
  587   11.11  33.33   w
  123   19.99  39.95   g

>> printpackind(packages,price_ind)
Item #  Cost  Price  Code
  587   11.11  33.33   w
  123   19.99  39.95   g
  456    5.99  49.99   l
```

PRACTICE 13.3

Modify the function *createind* to create the *cost_ind* index vector.

13.5 SEARCHING

Searching means looking for a value (a *key*) in a list or in a vector. We have already seen that MATLAB has a function, **find**, which will return the indices in an array that meet a criterion. To examine the programming methodologies, we will in this section examine two search algorithms:

- sequential search
- binary search

13.5.1 Sequential search

A *sequential search* is accomplished by looping through the vector element by element starting from the beginning, looking for the key. Normally the index of the element in which the key is found is what is returned. For example, here is a function that will search a vector for a key and return the index or the value 0 if the key is not found:

seqsearch.m

```
function index = seqsearch(vec, key)
% seqsearch performs an inefficient sequential search
% through a vector looking for a key; returns the index
% Format: seqsearch(vector, key)

len = length(vec);
index = 0;

for i = 1:len
    if vec(i) == key
        index = i;
    end
end
end
```

Here are two examples of calling this function:

```
>> values = [85 70 100 95 80 91];
>> key = 95;
>> seqsearch(values, key)
ans =
    4

>> seqsearch(values, 77)
ans =
    0
```

This example assumes that the key is found only in one element in the vector. Also, although it works, it is not a very efficient algorithm. If the vector is large, and the key is found in the beginning, this still loops through the rest of the vector. An improved version would loop until the key is found or the entire vector has been searched. In other words, a **while** loop is used rather than a **for** loop; there are two parts to the condition.

smartseqsearch.m

```
function index = smartseqsearch(vec, key)
% Smarter sequential search; searches a vector
% for a key but ends when it is found
% Format: smartseqsearch(vector, key)
```

Continued

```
len = length(vec);
index = 0;
i = 1;

while i < len && vec(i) ~= key
    i = i + 1;
end

if vec(i) == key
    index = i;
end
end
```

13.5.2 Binary search

The *binary search* assumes that the vector has been sorted first. The algorithm is similar to the way it works when looking for a name in a phone directory (which is sorted alphabetically). To find the value of a key:

- Look at the element in the middle:
  - If that is the key, the index has been found.
  - If it is not the key, decide whether to search the elements before or after this location and adjust the range of values in which the search is taking place and start this process again.

To implement this, we will use variables *low* and *high* to specify the range of values in which to search. To begin, the value of *low* will be 1, and the value of *high* will be the length of the vector. The variable *mid* will be the index of the element in the middle of the range from *low* to *high*. If the key is not found at *mid*, there are two possible ways to adjust the range. If the key is less than the value at *mid*, we change *high* to *mid* – 1. If the key is greater than the value at *mid*, we change *low* to *mid* + 1.

An example is to search for the key of 91 in the vector

	1	2	3	4	5	6
	70	80	85	91	95	100

The following table shows what will happen in each iteration of this search algorithm.

Iteration	Low	High	Mid	Found?	Action
1	1	6	3	No	Move low to mid + 1
2	4	6	5	No	Move high to mid – 1
3	4	4	4	Yes	Done! Index is mid

The key was found in the fourth element of the vector.

Another example: Search for the key of 82.

Iteration	Low	High	Mid	Found?	Action
1	1	6	3	No	Move high to mid −1
2	1	2	1	No	Move low to mid + 1
3	2	2	2	No	Move low to mid + 1
4	3	2	This ends it!		

The value of *low* cannot be greater than *high*; this means that the key is not in the vector. So, the algorithm repeats until either the key is found or until *low* > *high*, which means that the key is not there.

The following function implements this binary search algorithm. The function receives two arguments: the sorted vector and a key (alternatively, the function could sort the vector). The values of *low* and *high* are initialized to the first and last indices in the vector. The output argument *outind* is initialized to 0, which is the value that the function will return if the key is not found. The function loops until either *low* is greater than *high*, or until the key is found.

binsearch.m
```
function outind = binsearch(vec, key)
%   binsearch searches through a sorted vector
%   looking for a key using a binary search
%   Format: binsearch(sorted vector, key)

low = 1;
high = length(vec);
outind = 0;

while low <= high && outind == 0
   mid = floor((low + high)/2);
   if vec(mid) == key
       outind = mid;
   elseif key < vec(mid)
       high = mid - 1;
   else
       low = mid + 1;
   end
end
end
```

The following are examples of calling this function:
```
>> vec = round(rand(1,7)*29+1)
vec =
     2    11    25     1     5     7     6
```

```
>> svec = sort(vec)
svec =
     1    2    5    6    7   11   25

>> binsearch(svec, 4)
ans =
     0

>> binsearch(svec, 25)
ans =
     7

>> binsearch(svec, 5)
ans =
     3
```

The binary search can also be implemented as a recursive function. The following recursive function implements this binary search algorithm. The function receives four arguments: a sorted vector, a key to search for, and the values of *low* and *high* (which to begin with will be 1 and the length of the vector). It will return 0 if the key is not in the vector or the index of the element in which it is found. The base cases in the algorithm are when *low > high*, which means that the key is not in the vector, or when it is found. Otherwise, the general case is to adjust the range and call the binary search function again.

recbinsearch.m

```
function outind = recbinsearch(vec, key, low, high)
%   recbinsearch recursively searches through a vector
%   for a key; uses a binary search function
%   The min and max of the range are also passed
%   Format: recbinsearch(vector, key, rangemin, rangemax)

mid = floor((low + high)/2);

if low > high
    outind = 0;
elseif vec(mid) == key
    outind = mid;
elseif key < vec(mid)
    outind = recbinsearch(vec,key,low,mid-1);
else
    outind = recbinsearch(vec,key,mid+1,high);
end
end
```

Examples of calling this function follow:

```
>> recbinsearch(svec, 5,1,length(svec))
ans =
     3
```

```
>> recbinsearch(svec, 25,1,length(svec))
ans =
     7

>> recbinsearch(svec, 4,1,length(svec))
ans =
     0
```

SUMMARY

Common Pitfalls

- Forgetting that **max** and **min** return the index of only the first occurrence of the maximum or minimum value.
- Not realizing that a data set has outliers that can drastically alter the results obtained from the statistical functions.
- When sorting a vector of structures on a field, forgetting that although only the field in question is compared in the sort algorithm, entire structures must be interchanged.
- Forgetting that a data set must be sorted before using a binary search.

Programming Style Guidelines

- Remove the largest and smallest numbers from a large data set before performing statistical analyses, to handle the problem of outliers.
- Use **sortrows** to sort strings stored in a matrix alphabetically; for cell arrays, **sort** can be used.
- When it is necessary to iterate through a vector of structures in order based on several different fields, it may be more efficient to create index vectors based on these fields rather than sorting the vector of structures multiple times.

MATLAB Functions and Commands		
mean	union	ismember
var	intersect	issorted
std	unique	sort
mode	setdiff	sortrows
median	setxor	

Exercises

1. Experimental data values are stored in a file. Create a file in a matrix form with random values for testing. Write a script that will **load** the data, and then determine the difference between the largest and smallest numbers in the file.

2. The *range* of a data set is the difference between the largest value and the smallest. A data file called *tensile.dat* stores the tensile strength of some aluminum samples. Create a test data file; read in the tensile strengths and print the minimum, maximum, and the range.

Note

The function is not receiving a vector; rather, all of the values are separate arguments.

3. Write a function *mymin* that will receive any number of arguments and will return the minimum. For example,

```
>> mymin(3, 6, 77, 2, 99)
ans =
2
```

4. In a marble manufacturing plant, a quality control engineer randomly selects eight marbles from each of the two production lines and measures the diameter of each marble in millimeters. For each data set here, determine the mean, median, mode, and standard deviation using built-in functions.

```
Prod. line A:15.94 15.98 15.94 16.16 15.86 15.86 15.90 15.88
Prod. line B:15.96 15.94 16.02 16.10 15.92 16.00 15.96  16.02
```

Suppose that the desired diameter of the marbles is 16 mm. Based on the results you have, which production line is better in terms of meeting the specification? (*Hint:* Think in terms of the mean and the standard deviation.)

5. The chemical balance of a swimming pool is important for the safety of the swimmers. The pH level of a pool has been measured every day and the results are stored in a file. Create a data file to simulate these measurements; the values should be random numbers in the range from 7 to 8. Read the pH values from the file and calculate the mean and standard deviation of the pH values.

6. A batch of 500-ohm resistors is being tested by a quality engineer. A file called *testresist.dat* stores the resistance of some resistors that have been measured. The resistances have been stored one per line in the file. Create a data file in this format. Then, load the information and calculate and print the mean, median, mode, and standard deviation of the resistances. Also, calculate how many of the resistors are within 1% of 500 ohms.

7. Write a function *calcvals* that will calculate the maximum, minimum, and mean value of a vector based on how many output arguments are used to call the function. The following are examples of function calls:

```
>> vec=[4 9 5 6 2 7 16 0];
>> [mmax, mmin, mmean]= calcvals(vec)
mmax=
        16
mmin=
        0
mmean=
      6
>> [mmax, mmin]= calcvals(vec)
```

```
mmax=
    16
mmin=
    0
>> mmax= calcvals(vec)
mmax=
    16
```

8. Write a script that will do the following. Create two vectors with 20 random integers in each; in one, the integers should range from 1 to 5, and in the other, from 1 to 500. For each vector, would you expect the mean and median to be approximately the same? Would you expect the standard deviation of the two vectors to be approximately the same? Answer these questions, and then use the built-in functions to find the minimum, maximum, mean, median, standard deviation, and mode of each. Do a histogram for each in a subplot. Run the script a few times to see the variations.

9. Write a function that will return the mean of the values in a vector, not including the minimum and maximum values. Assume that the values in the vector are unique. It is okay to use the built-in **mean** function. To test this, create a vector of 10 random integers, each in the range from 0 to 50, and pass this vector to the function.

10. A moving average of a data set $x = \{ x_1, x_2, x_3, x_4, \ldots, x_n \}$ is defined as a set of averages of subsets of the original data set. For example, a moving average of every two terms would be $1/2 * \{ x_1 + x_2, x_2 + x_3, x_3 + x_4, \ldots, x_{n-1} + x_n \}$. Write a function that will receive a vector as an input argument, and will calculate and return the moving average of every two elements.

11. Write a function *mymedian* that will receive a vector as an input argument, and will sort the vector and return the median. Any built-in functions may be used, except the **median** function. Loops may not be used.

12. In statistical analyses, quartiles are points that divide an ordered data set into four groups. The second quartile, Q2, is the median of the data set. It cuts the data set in half. The first quartile, Q1, cuts the lower half of the data set in half. Q3 cuts the upper half of the data set in half. The interquartile range is defined as Q3 − Q1. Write a function that will receive a data set as a vector and will return the interquartile range.

Eliminating or reducing noise is an important aspect of any signal processing. For example, in image processing noise can blur an image. One method of handling this is called median filtering.

13. A median filter on a vector has a size; for example, a size of 3 means calculating the median of every three values in the vector. The first and last elements are left alone. Starting from the second element to the next-to-last element, every element of a vector vec(i) is replaced by the median of [vec(i − 1) vec(i) vec(i + 1)]. For example, if the signal vector is

```
signal =[ 5   11   4   2   6   8   5   9]
```

the median filter with a size of 3 is

```
medianFilter3 = [ 5  5  4  4  6  6  8  9]
```

Write a function to receive the original signal vector and return the median filtered vector.

14. Modify the function from Exercise 13 so that the size of the filter is also passed as an input argument.

15. What is the difference between the mean and the median of a data set if there are only two values in it?

16. A student missed one of four exams in a course and the professor decided to use the "average" of the other three grades for the missed exam grade. Which would be better for the student: the mean or the median if the three recorded grades were 99, 88, and 95? What if the grades were 99, 70, and 77?

17. A weighted mean is used when there are varying weights for the data values. For a data set given by $x = \{ x_1, x_2, x_3, x_4, \ldots, x_n\}$ and corresponding weights for each x_i, $w = \{w_1, w_2, w_3, w_4, \ldots, w_n\}$, the weighted mean is

$$\frac{\sum\limits_{i=1}^{n} x_i w_i}{\sum\limits_{i=1}^{n} w_i}$$

For example, assume that in an economics course there are three quizzes and two exams, and the exams are weighted twice as much as the quizzes. If the quiz scores are 95, 70, and 80 and the exam scores are 85 and 90, the weighted mean would be:

$$\frac{95 * 1 + 70 * 1 + 80 * 1 + 85 * 2 + 90 * 2}{1 + 1 + 1 + 2 + 2} = \frac{595}{7} = 85$$

Write a function that will receive two vectors as input arguments: one for the data values and one for the weights, and will return the weighted mean.

18. A production facility is producing some nails that are supposed to have a diameter of 0.15 inch. At five different times, 10 sample nails were measured; their diameters were stored in a file that has five lines and 10 diameters in each. First, create a data file to simulate these data. Then, write a script to print the mean and standard deviation for each of the five sets of sample nails.

19. The *coefficient of variation* is useful when comparing data sets that have quite different means. The formula is CV = (standard deviation/mean) * 100%. A history course has two different sections; their final exam scores are stored in two separate rows in a file. For example,

```
99  100  95  92  98  89  72  95  100  100
83   85  77  62  68  84  91  59   60
```

Create the data file, read the data into vectors, and then use the CV to compare the two sections of this course.

20. Write a function *allparts* that will read in lists of part numbers for parts produced by two factories. These are contained in data files called *xyparts.dat* and *qzparts.dat*. The function will return a vector of all parts produced, in sorted order (with no repeats). For example, if the file *xyparts.dat* contains

```
123  145  111  333  456  102
```

and the file *qzparts.dat* contains

```
876  333  102  456  903  111
```

calling the function would return the following:

```
>> partslist = allparts
   partslist =
      102   111   123   145   333   456   876   903
```

21. The set functions can be used with cell arrays of strings. Create two cell arrays to store (as strings) course numbers taken by two students. For example,

```
s1 = {'EC 101', 'CH 100', 'MA 115'};
s2 = {'CH 100', 'MA 112', 'BI 101'};
```

Use a set function to determine which courses the students have in common.

22. A vector *v* is supposed to store unique random numbers. Use set functions to determine whether or not this is true.

23. A function *generatevec* generates a vector of n random integers (where n is a positive integer), each in the range from 1 to 100, but all of the numbers in the vector must be different from each other (no repeats). So, it uses **rand** to generate the vector and then uses another function *alldiff* that will return **logical** 1 for **true** if all of the numbers in the vector are different, or **logical** 0 for **false** if not in order to check. The *generatevec* function keeps looping until it does generate a vector with n non-repeating integers. It also counts how many times it has to generate a vector until one is generated with n nonrepeating integers and returns the vector and the count. Write the *alldiff* function.

generatevec.m

```
function [ outvec, count] = generatevec(n)
trialvec = round(rand*99)+1;
count = 1;
while ~alldiff(trialvec)
    trialvec = round(rand*99)+1;
    count = count + 1;
end
outvec = trialvec;
end
```

24. Write a function *mydsort* that sorts a vector in descending order (using a loop, not the built-in sort function).

25. In product design, it is useful to gauge how important different features of the product would be to potential customers. One method of determining which features are most important is a survey in which people are asked "Is this feature important to you?" when shown a number of features. The number of potential customers who responded "Yes" is then tallied. For example, a company conducted such a survey for 10 different features; 200 people took part in the survey. The data were collected into a file that might look like this:

1	2	3	4	5	6	7	8	9	10
30	83	167	21	45	56	55	129	69	55

A *Pareto chart* is a bar chart in which the bars are arranged in decreasing values. The bars on the left in a Pareto chart indicate the most important features. Create a data file, and then a **subplot** to display the data with a bar chart organized by question on the left and a Pareto chart on the right.

26. DNA is a double-stranded helical polymer that contains basic genetic information in the form of patterns of nucleotide bases. The patterns of the base molecules A, T, C, and G encode the genetic information. Construct a cell array to store some DNA sequences as strings, such as

```
TACGGCAT
ACCGTAC
```

and then sort these alphabetically. Next, construct a matrix to store some DNA sequences of the same length and then sort them alphabetically.

27. Write a function *matsort* to sort all of the values in a matrix (decide whether the sorted values are stored by row or by column). It will receive one matrix argument and return a sorted matrix. Do this without loops, using the built-in functions **sort** and **reshape**. For example:

```
>> mat
mat =
     4     5     2
     1     3     6
     7     8     4
     9     1     5
>> matsort(mat)
ans =
     1     4     6
     1     4     7
     2     5     8
     3     5     9
```

28. Write a function that will receive two arguments: a vector and a character (either 'a' or 'd') and will sort the vector in the order specified by the character (ascending or descending).

29. Write a function that will receive a vector and will return two index vectors: one for ascending order and one for descending order. Check the function by writing a script that will call the function and then use the index vectors to print the original vector in ascending and descending order.

30. Write a function *myfind* that will search for a key in a vector and return the indices of all occurrences of the key, like the built-in **find** function. It will receive two arguments—the vector and the key—and will return a vector of indices (or the empty vector [] if the key is not found).

Sights and Sounds

sound signal	pixels	Graphical user
sampling	true color	interfaces
frequency	RGB	event
audio file formats	colormap	callback function

CONTENTS

The MATLAB® product has functions that manipulate audio or sound files and also images. This chapter will start with a brief introduction to some of the sound processing functions. Image processing functions will be introduced, and the two basic methods for representing color in images will be explained. Finally, this chapter will introduce the topic of graphical user interfaces from a programming standpoint.

14.1 SOUND FILES

A *sound signal* is an example of a continuous signal that is sampled to result in a discrete signal. In this case, sound waves traveling through the air are recorded as a set of measurements that can then be used to reconstruct the original sound signal, as closely as possible. The sampling rate or *sampling frequency* is the number of samples taken per time unit, for example, per second. Sound signals are usually measured in Hertz (Hz).

In MATLAB, the discrete sound signal is represented by a vector, and the frequency is measured in Hertz. MATLAB has several MAT-files that store for various sounds the signal vector in a variable y and the frequency in a variable *Fs*. These MAT-files include **chirp**, **gong**, **laughter**, **splat**, **train**, and **handel**. There is a built-in function, **sound**, that will send a sound signal to an output device such as speakers.

The following function call:

```
>> sound(y,Fs)
```

will play the sound represented by the vector *y* at the frequency *Fs*. For example, to hear a gong, load the variables from the MAT-file and then play the sound using the **sound** function:

```
>> load gong
>> sound(y,Fs)
```

Sound is a wave; the amplitudes are what are stored in the sound signal variable *y*. These are supposed to be in the range from -1 to 1. The **plot** function can be used to display the data. For example, the following script creates a **subplot** that displays the signals from **chirp** and from **train**, as shown in Figure 14.1.

chirptrain.m

```
% Display the sound signals from chirp and train
load chirp
subplot(2,1,1)
plot(y)
ylabel('Amplitude')
title('Chirp')
subplot(2,1,2)
ylabel('Amplitude')
title('Train')
```

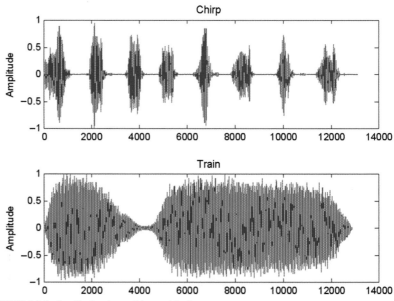

FIGURE 14.1 Amplitudes from **chirp** and **train**

The first argument to the **sound** function can be an $n \times 2$ matrix for stereo sound. Also, the second argument can be omitted when calling the **sound** function, in which case the default sample frequency of 8192 Hz is used. This is the frequency stored in the built-in sound MAT-files.

```
>> load train
Fs
Fs =
        8192
```

PRACTICE 14.1

If you have speakers, try loading one of the sound MAT-files, and use the **sound** function to play the sound. Then, change the frequency—for instance, multiply the variable *Fs* by 2 and by 0.5—and play the sounds again.

```
>> load train
>> sound(y, Fs)
>> sound(y, Fs * 2)
>> sound(y, Fs * .5)
```

MATLAB has several other functions that read sound or audio files and play them. In audio files, sampled data for each audio channel are stored. Several *audio file formats* are used in industry on different computer platforms. Audio files with the extension *.au* were developed by Sun Microsystems, and typically they are used with Java and Unix, whereas Windows PCs typically use *.wav* files that were developed by Microsoft.

MATLAB has functions **wavread** that will read a *.wav* file, **wavrecord** that will record, **wavwrite** that will write a sound file, and **wavplay** that will play one. The default frequency for these functions is 11,025 Hz.

For *.au* files, there are functions **auread** to read and **auwrite** to write in this format.

14.2 IMAGE PROCESSING

Images are represented as grids, or matrices, of picture elements (called *pixels*). In MATLAB an image is typically represented as a matrix in which each element corresponds to a pixel in the image. Each element that represents a particular pixel stores the color for that pixel. There are two basic ways that the color can be represented:

- *True color*, or *RGB*, in which the three color components are stored (red, green, and blue, in that order)
- Index into a *colormap*, in which the value stored is an integer that refers to a row in a matrix called a colormap. The colormap stores the red, green, and blue components in three separate columns.

For an image that has $m \times n$ pixels, the true color matrix would be a three-dimensional (3D) matrix with the size $m \times n \times 3$. The first two dimensions represent the coordinates of the pixel. The third index is the color component: `(:,:,1)` is the red, `(:,:,2)` is the green, and `(:,:,3)` is the blue.

The indexed representation instead would be an $m \times n$ matrix of integers, each of which is an index into a colormap matrix, which is size $p \times 3$ (where p is the number of colors available in that particular colormap). Each row in the colormap has three numbers representing one color: first the red, then the green, and then the blue component.

14.2.1 Colormaps

When an image is represented using a colormap, there are two matrices:

- The colormap matrix, which has dimensions $p \times 3$ where p is the number of available colors. Every row stores three real numbers in the range from 0 to 1, representing the red, green, and blue components of the color.
- The image matrix, with dimensions $m \times n$. Every element is an index into the colormap, which means that it is an integer in the range 1 to p.

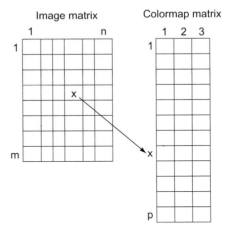

MATLAB has several built-in colormaps that are named; these can be seen and can be set using the built-in function **colormap**. The reference page on **colormap** displays them. Calling the function **colormap** without passing any arguments will return the current colormap, which by default is called **jet**.

The following stores the current colormap in a variable *map*, gets the size of the matrix (which will be the number of rows in this matrix, or in other words, the number of colors, by three columns), and displays the first five rows in this colormap. If the current colormap is the default **jet**, the following will be the result:

```
>> map = colormap;
>> [r c] = size(map)
r =
    64
c =
    3
>> map(1:5,:)
ans =
         0         0    0.5625
         0         0    0.6250
         0         0    0.6875
         0         0    0.7500
         0         0    0.8125
```

This shows that there are 64 rows, or in other words, 64 colors, in this particular colormap. It also shows that the first five colors are shades of blue.

The format for calling the **image** function is

```
image(mat)
```

where the matrix *mat* represents the colors in an $m \times n$ image ($m \times n$ pixels in the image). If the matrix has the size $m \times n$, then each element is an index into the current colormap.

One way to display the colors in the **jet** colormap (which has 64 colors) is to create a matrix that stores the values 1 through 64, and pass that to the **image** function, as shown in Figure 14.2. When the matrix is passed to the **image** function, the value in each element in the matrix is used as an index into the colormap.

For example, the value in cmap(1,2) is 9, so the color displayed in location (1,2) in the image will be the color represented by the 9th row in the colormap. By using the numbers 1 through 64, we can see all colors in this colormap. The figure shows that the first colors are shades of blue, the last colors are shades of red, and in between are shades of aqua, green, yellow, and orange.

```
>> cmap = reshape(1:64, 8,8)
cmap =

     1     9    17    25    33    41    49    57
     2    10    18    26    34    42    50    58
     3    11    19    27    35    43    51    59
     4    12    20    28    36    44    52    60
     5    13    21    29    37    45    53    61
     6    14    22    30    38    46    54    62
     7    15    23    31    39    47    55    63
     8    16    24    32    40    48    56    64

>> image(cmap)
```

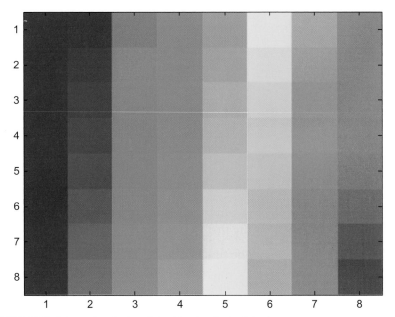

FIGURE 14.2 Columnwise display of the 64 colors in the **jet** colormap

Another example creates a 5×5 matrix of random integers in the range from 1 to the number of colors (stored in a variable r); the resulting image appears in Figure 14.3.

```
>> mat = round(rand(5)*(r - 1) + 1)
    54    33    13    45    32
     2    46    44    25    58
    44    28    20    56    53
    25    20    35    55    42
    54    13    10    38    53

>> image(mat)
```

Of course, these "images" are rather crude; the elements representing the pixel colors are quite large blocks. A larger matrix would result in something more closely resembling an image, as shown in Figure 14.4.

```
>> mat = round(rand(500)*(r - 1) + 1);
>> image(mat)
```

Although MATLAB has built-in colormaps, it is also possible to create others using any color combinations. For example, the following creates a customized colormap with just three colors: black, white, and red. This is then set to be the

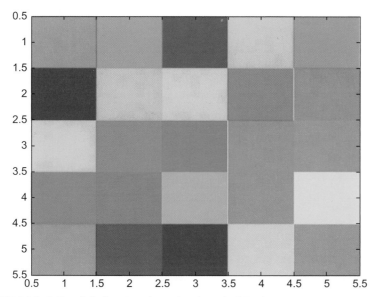

FIGURE 14.3 A *5 × 5* display of random colors from the **jet** colormap

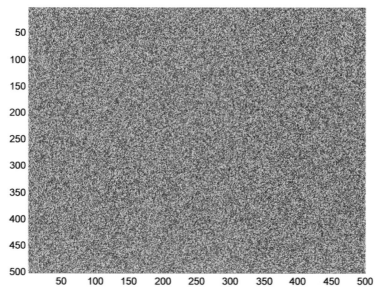

FIGURE 14.4 A *500 × 500* display of random colors

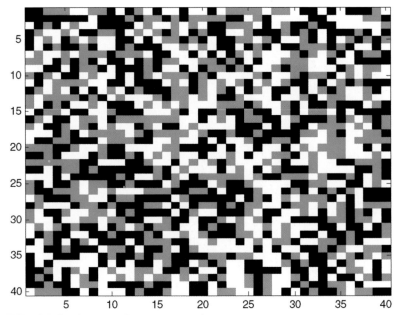

FIGURE 14.5 Random colors from a custom colormap

current colormap by passing the colormap matrix to the **colormap** function. Then, a *40 × 40* matrix of random integers in the range 1 to 3 (since there are just three colors) is created, and that is passed to the **image** function; the results are shown in Figure 14.5.

```
>> mycolormap = [0  0  0;  1  1  1;  1  0  0]
mycolormap =
      0      0      0
      1      1      1
      1      0      0

>> colormap(mycolormap)
>> mat = round(rand(40)*(3-1)+1);
>> image(mat)
```

The numbers in the colormap do not have to be integers; real numbers represent different shades as seen with the default colormap **jet**. For example, the following colormap gives us a way to visualize different shades of red as shown in Figure 14.6.

```
>> colors = [0  0  0;  0.2  0  0;  0.4  0  0;...
             0.6  0  0;  0.8  0  0;  1  0  0];
>> colormap(colors)
>> vec = 1:length(colors);
>> image(vec)
```

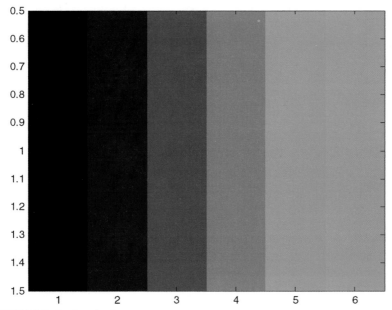

FIGURE 14.6 Shades of red

PRACTICE 14.2

Given the following colormap, "draw" the scene shown in Figure 14.7. (*Hint:* Preallocate the image matrix. The fact that the first color in the colormap is white makes this easier.)

```
>> mycolors = [ 1 1 1; 0 1 0; 0 0.5 0; ...
               0 0 1; 0 0 0.5; 0.3 0 0];
```

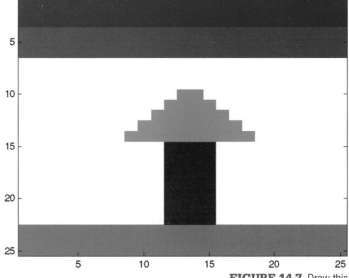

FIGURE 14.7 Draw this tree with grass and sky

14.2.2 True color matrices

True color matrices are another way to represent images. True color matrices are 3D matrices. The first two coordinates are the coordinates of the pixel. The third index is the color component; (:,:,1) is the red, (:,:,2) is the green, and (:,:,3) is the blue component. Each element in the matrix is of the type **uint8**, which is an unsigned integer type storing values in the range from 0 to 255. The minimum value, 0, represents the darkest hue available, so all 0s results in a black pixel. The maximum value, 255, represents the

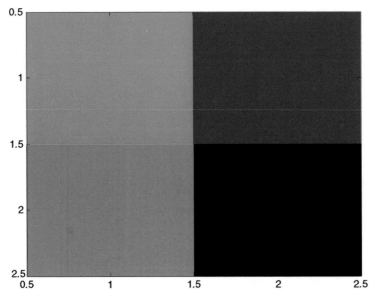

FIGURE 14.8 Image from
a true color matrix

brightest hue. For example, if the values for given pixel coordinates px and py are: (px, py,1) is 255, (px,py,2) is 0, and (px,py,3) is 0, then that pixel will be bright red. All 255s results in a white pixel.

The **image** function displays the information in the 3D matrix as an image. For example, the following creates a 2 × 2 image as shown in Figure 14.8. The matrix is 2 × 2 × 3 where the third dimension is the color. The pixel in location (1,1) is red, the pixel in location (1,2) is blue, the pixel in location (2,1) is green, and the pixel in location (2,2) is black. It is necessary to cast the matrix to the type **uint8**.

```
>> mat(1,1,1) = 255;
>> mat(1,1,2) = 0;
>> mat(1,1,3) = 0;
>> mat(1,2,1) = 0;
>> mat(1,2,2) = 0;
>> mat(1,2,3) = 255;
>> mat(2,1,1) = 0;
>> mat(2,1,2) = 255;
>> mat(2,1,3) = 0;
>> mat(2,2,1) = 0;
>> mat(2,2,2) = 0;
>> mat(2,2,3) = 0;
>> mat = uint8(mat);
>> image(mat)
```

The function **imread** can read in an image file, for example, a JPEG (.jpg) file. The function reads color images into a 3D matrix.

```
>> myimage1 = imread('Fishing_1.JPG');
>> size(myimage1)
ans =
        1536      2048       3
```

In this case, the image is represented as a true color matrix. This indicates that the image has 1536 × 2048 pixels. The **image** function displays the information in this 3D matrix as an image, as shown in Figure 14.9.

```
>> image(myimage1)
```

FIGURE 14.9 Image from a JPEG file displayed using **image**

The image can be changed by manipulating the numbers in the matrix. For example, multiplying every number by 0.75 will result in a range of values from 0 to 191 instead of 0 to 255. Since the larger numbers are brighter, this has the effect of dimming the hues in the pixels, as shown in Figure 14.10.

```
>> dimmer = 0.75*myimage1;
>> image(dimmer)
```

FIGURE 14.10 Image dimmed by manipulating the matrix

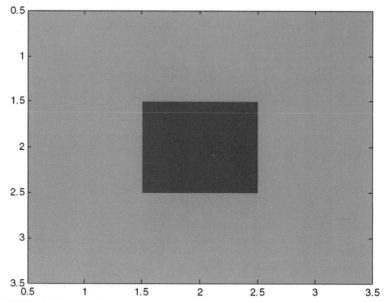

FIGURE 14.11 Create this true color matrix

PRACTICE 14.3

Create the $3 \times 3 (\times 3)$ true color matrix shown in Figure 14.11 (the axes are defaults).

14.3 INTRODUCTION TO GRAPHICAL USER INTERFACES

Graphical user interfaces, or GUIs, are essentially objects that allow users to have input using graphical interfaces such as pushbuttons, sliders, radio buttons, toggle buttons, pop-up menus, and so forth. GUIs are an example of object-oriented programming in which there is a hierarchy. For example, the parent may be a Figure Window and its children would be graphics objects such as pushbuttons and text boxes.

The parent user interface object can be a **figure**, **uipanel**, or **uibuttongroup**. A **figure** is a Figure Window created by the **figure** function. A **uipanel** is a means of grouping together user interface objects (the "ui" stands for user interface). A **uibuttongroup** is a means of grouping together buttons (both radio buttons and toggle buttons).

In MATLAB there are two basic methods for creating GUIs: writing the GUI program from scratch, or using the built-in Graphical User Interface Development Environment (GUIDE). GUIDE allows the user to graphically lay out the GUI and MATLAB generates the code for it automatically. However, to be able to understand and modify this code, it is important to understand the underlying programming concepts. Therefore, this section will concentrate on the programming methodology.

A Figure Window is the parent of any GUI. Just calling the **figure** function will bring up a blank Figure Window. Assigning the handle of this Figure Window to a variable and then using the **get** function will show the default properties. These properties, such as the color of the window, its position on the screen, and so forth can be changed using the **set** function or when calling the **figure** function to begin with. For example,

```
>> f = figure;
```

brings up a grey figure box near the top of the screen as shown in Figure 14.12.

Some of its properties are excerpted here:

```
>> get(f)
    Color = [ 0.8 0.8 0.8]
    Colormap = [  (64 by 3) double array]
    Position = [360 502 560 420]
    Units = pixels
    Children = [ ]
    Visible = on
```

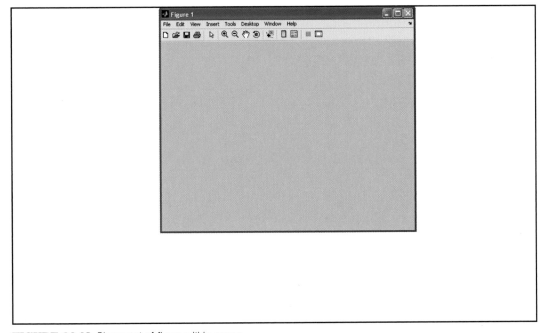

FIGURE 14.12 Placement of figure within screen

The position vector specifies `[ left bottom width height]`. The first two numbers, the left and bottom, are the distance that the lower left corner of the figure box is from the lower left of the monitor screen (first from the left and then from the bottom). The last two are the width and height of the figure box itself. All of these are in the default units of pixels.

The 'Visible' property "on" means that the Figure Window can be seen. When creating a GUI, however, the normal procedure is to create the parent Figure Window but make it invisible. Then, all user interface objects are added to it, and properties are set. When everything has been completed, the GUI is made visible.

Most user interface objects are created using the **uicontrol** function. The 'Style' property defines the type of object, as a string. For example, 'text' is the Style of a static text box, which is normally used as a label for other objects in the GUI, or for instructions.

The following example creates a GUI that just consists of a static text box in a Figure Window. The figure is first created but made invisible. The color is white, and it is given a position. Storing the handle of this figure in a variable allows the script to refer to it later on, to set properties, for example. The **uicontrol** function is used to create a text box, position it (the vector specifies the [left bottom width height] within the Figure Window itself), and put a string in it.

Note that the position is within the Figure Window, not within the screen. A name is put on the top of the figure. The **movegui** function moves the GUI (the figure) to the center of the screen. Finally, when everything has been completed, the GUI is made visible.

```
simpleGui.m
```

```
function simpleGui
% simpleGui creates a simple GUI with just a static text box
% Format: simpleGui or simpleGui()

% Create the GUI but make it invisible for now while
%   it is being initialized
f = figure('Visible', 'off','Color','white','Position',...
    [300, 400, 450,250] );
htext = uicontrol('Style','text','Position', ...
  [200,50, 100, 25] , 'String','My First GUI string');

% Put a name on it and move to the center of the screen
set(f,'Name','Simple GUI')
movegui(f,'center')

% Now the GUI is made visible
set(f,'Visible','on');
end
```

The Figure Window shown in Figure 14.13 will appear in the middle of the screen. The static text box requires no interaction with the user.

In the next example, we will allow the user to enter a string in an editable text box, and then the GUI will print the user's string in red. In this example, there will be user interaction. First the user must type in a string, and once this happens the user's entry in the editable text box will no longer be shown but instead the string that the user typed will be displayed in a larger red font in a static text box. When the user's action (which is called an *event*) causes a response, what happens is that a ***callback function*** is called, or invoked. This is a ***nested function*** within the GUI function. The algorithm for this example is:

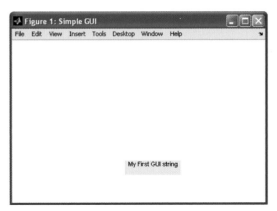

FIGURE 14.13 Simple GUI with a static text box

- Create the Figure Window, but make it invisible.
- Make the color of the figure white, put a title on it, and move it to the center.
- Create a static text box with an instruction to enter a string.
- Create an editable text box.
   - The Style of this is 'edit'.
   - The callback function must be specified since the user's entry of a string necessitates a response (the function handle of the nested function is used).
- Make the GUI visible so that the user can see the instruction and type in a string.
- When the string is entered, the callback function *callbackfn* is called. Note that in the function header, there are two input arguments, *source* and *eventdata*. The input argument *source* refers to the handle of the **uicontrol** object that called it; *eventdata* can store in a structure information about actions performed by the user (e.g., pressing keys).
- The algorithm for the nested function *callbackfn* is:
   - Make the previous GUI objects invisible.
   - Get the string that the user typed. (*Note:* Either *source* or the function handle name *huitext* can be used to refer to the object in which the string was entered.)
   - Create a static text box to print the string in red with a larger font.
   - Make this new object visible.

guiWithEditbox.m

```
function guiWithEditbox
% guiWithEditbox has an editable text box
%   and a callback function that prints the user's
%   string in red
```

```
% Format: guiWithEditbox or guiWithEditbox()

% Create the GUI but make it invisible for now
f = figure('Visible', 'off','Color','white','Position',...
    [360, 500, 800,600]);
% Put a name on it and move it to the center of the screen
set(f,'Name','GUI with editable text')
movegui(f,'center')
% Create two objects: a box where the user can type and
% edit a string and also a text title for the edit box
hsttext = uicontrol('Style','text',...
    'BackgroundColor','white',...
    'Position',[100,425,400, 55],...
    'String','Enter your string here');
huitext = uicontrol('Style','edit',...
    'Position',[100,400,400,40] , ...
    'Callback',@callbackfn);

% Now the GUI is made visible
set(f,'Visible','on');

    % Call back function
    function callbackfn(source,eventdata)
       % callbackfn is called by the 'Callback' property
       % in the editable text box
       set([hsttext huitext] ,'Visible','off');
       % Get the string that the user entered and print
       %    it in big red letters
       printstr = get(huitext,'String');
       hstr = uicontrol('Style','text',...
           'BackgroundColor','white',...
           'Position',[100,400,400,55] ,...
           'String',printstr,...
           'ForegroundColor','Red','FontSize',30);
       set(hstr,'Visible','on')
    end
end
```

When the Figure Window is first made visible, the static text and the editable text box are shown. In this case, the user entered "hi and how are you?" Note that to enter the string, the user must first click the mouse in the editable text box. The string that was entered by the user is shown in Figure 14.14.

After the user enters the string and hits the Enter key, the callback function is executed; the results are shown in Figure 14.15.

Now, we'll add a pushbutton to the GUI. This time, the user will enter a string but the callback function will be invoked when the pushbutton is pushed.

FIGURE 14.14 String entered by user in editable text box

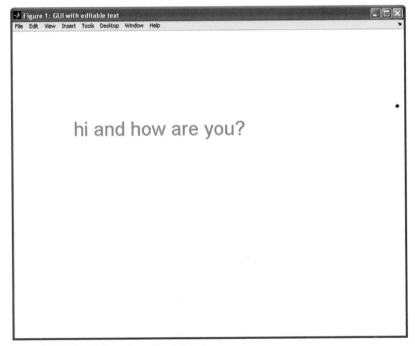

FIGURE 14.15 Result from the callback function

guiWithPushbutton.m

```
function guiWithPushbutton
% guiWithPushbutton has an editable text box and a pushbutton
% Format: guiWithPushbutton or guiWithPushbutton()

% Create the GUI but make it invisible for now while
%    it is being initialized
f = figure('Visible', 'off','Color','white','Position',...
    [360, 500, 800,600] );
hsttext = uicontrol('Style','text','BackgroundColor', 'white',...
    'Position',[100,425,400, 55] ,...
    'String','Enter your string here');
huitext = uicontrol('Style','edit','Position',[100,400, 400,40] );
set(f,'Name','GUI with pushbutton')
movegui(f,'center')

% Create a pushbutton that says "Push me!!"
hbutton = uicontrol('Style','pushbutton','String',...
    'Push me!!', 'Position',[600,50,150,50] , ...
    'Callback',@callbackfn);

% Now the GUI is made visible
set(f,'Visible','on');

% Call back function
    function callbackfn(source,eventdata)
        % callbackfn is called by the 'Callback' property
        % in the pushbutton
        set([hsttext huitext hbutton] ,'Visible','off');
        printstr = get(huitext,'String');
        hstr = uicontrol('Style','text','BackgroundColor',...
            'white', 'Position',[100,400,400,55] ,...
            'String',printstr, ...
            'ForegroundColor','Red','FontSize',30);
        set(hstr,'Visible','on')
    end
end
```

In this case, the user types the string into the edit box. Hitting Enter, however, does not cause the callback function to be called; instead, the user must push the button with the mouse. The callback function is associated with the pushbutton object. So, pushing the button will bring up the string in a larger red font. The initial configuration with the pushbutton is shown in Figure 14.16.

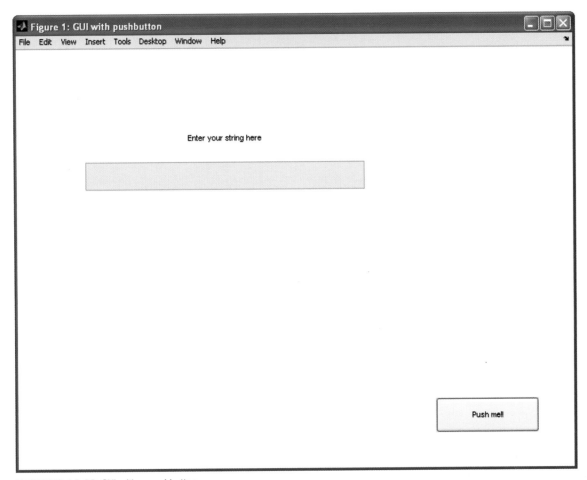

FIGURE 14.16 GUI with a pushbutton

PRACTICE 14.4

Create a GUI that will convert a length from inches to centimeters. The GUI should have an editable text box in which the user enters a length in inches, and a pushbutton that says "Convert me!" Pushing the button causes the GUI to calculate the length in centimeters and display that. The callback function that accomplishes this should leave all objects visible. That means that the user can continue converting lengths until the Figure Window is closed. The GUI should display a default length to begin with (e.g., 1 in.). For example, calling the function might bring up the Figure Window shown in Figure 14.17.

Then, when the user enters a length (e.g., 5.2 in.) and pushes the button, the Figure Window will show the new calculated length in centimeters (as seen in Figure 14.18).

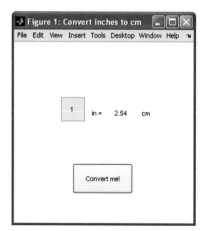

FIGURE 14.17 Length conversion GUI with pushbutton

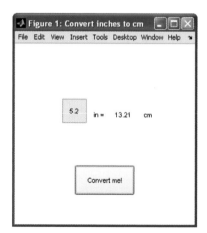

FIGURE 14.18 Result from conversion GUI

Another GUI object that can be created is a slider. The slider object has a numerical value, and can be controlled by either clicking on the arrows to move the value up or down, or by sliding the bar with the mouse. By default the numerical value ranges from 0 to 1, but these values can be modified using the 'Min' and 'Max' properties.

The function *guiSlider* creates in a Figure Window a slider that has a minimum value of 0 and a maximum value of 5. It uses text boxes to show the minimum and maximum values, and also the current value of the slider.

guiSlider.m

```
function guiSlider
% guiSlider is a GUI with a slider

f = figure('Visible', 'off','Color','white','Position',...
    [360, 500, 300,300]);

% Minimum and maximum values for slider
minval = 0;
maxval = 5;

% Create the slider object
slhan = uicontrol('Style','slider','Position',[80,170,100, 50] , ...
    'Min', minval, 'Max', maxval,'Callback', @callbackfn);
% Text boxes to show the minimum and maximum values
hmintext = uicontrol('Style','text','BackgroundColor ','white', ...
    'Position', [40, 175, 30,30] , 'String', num2str(minval));
hmaxtext = uicontrol('Style', 'text', 'BackgroundColor', 'white',...
    'Position', [190, 175, 30,30] , 'String', num2str(maxval));
% Text box to show the current value (off for now)
```

```
hsttext = uicontrol('Style','text','BackgroundColor', 'white',...
    'Position',[120,100,40,40],'Visible', 'off');

set(f,'Name','Slider Example')
movegui(f,'center')
set(f,'Visible','on');

% Call back function displays the current slider value
  function callbackfn(source,eventdata)
      % callbackfn is called by the 'Callback' property
      % in the slider

      num=get(slhan, 'Value');
      set(hsttext,'Visible','on','String',num2str(num))
  end
end
```

Calling the function brings up the initial configuration shown in Figure 14.19.

Then, when the user interacts by sliding the bar or clicking on an arrow, the current value of the slider is shown under it, as shown in Figure 14.20.

PRACTICE 14.5

Use the Help browser to find the property that controls the increment value on the slider, and modify the *guiSlider* function to move in increments of 0.5, regardless of whether an arrow or slider is used.

It is possible to have a callback function invoked, or called, by multiple objects. For example, the function *guiMultiplierIf* has two editable text boxes for numbers to be multiplied together, as well as a pushbutton that says "Multiply me!" as shown in Figure 14.21. Three static text boxes show the 'x', '=', and the result

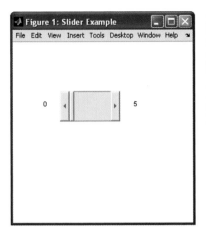

FIGURE 14.19 GUI with slider

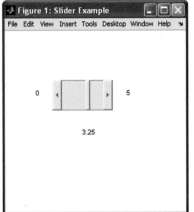

FIGURE 14.20 GUI with slider result shown

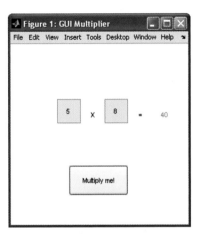

FIGURE 14.21 Multiplier GUI

of the multiplication. The callback function is associated with both the pushbutton and the second editable text box. The callback function uses the input argument *source* to determine which object called it; it displays the result of the multiplication in red if called by the editable text box, or it displays the result in green if called by the pushbutton.

guiMultiplierIf.m

```
function guiMultiplierIf
% guiMultiplierIf has 2 edit boxes for numbers and
%  multiplies them
% Format: guiMultiplierIf or guiMultiplierIf()

f = figure('Visible', 'off','Color','white','Position',...
   [360, 500, 300,300] );
firstnum = 0;
secondnum = 0;
product = 0;

hsttext = uicontrol('Style','text','BackgroundColor',' white',...
   'Position',[120,150,40,40] ,'String','X');
hsttext2 = uicontrol('Style','text','BackgroundColor', 'white',...
   'Position',[200,150,40,40] ,'String','=');
hsttext3 = uicontrol('Style','text','BackgroundColor', 'white',...
   'Position',[240,150,40,40] ,'Visible','off');
huitext = uicontrol('Style','edit','Position',[80,170,40, 40] ,...
   'String',num2str(firstnum));
huitext2 = uicontrol('Style','edit','Position',[160,170,40, 40] ,...
   'String',num2str(secondnum),...
   'Callback',@callbackfn);

set(f,'Name','GUI Multiplier')
movegui(f,'center')

hbutton = uicontrol('Style','pushbutton',...
   'String','Multiply me!',...
   'Position',[100,50,100,50] , 'Callback',@callbackfn);


set(f,'Visible','on');

  function callbackfn(source,eventdata)
      % callbackfn is called by the 'Callback' property
      % in either the second edit box or the pushbutton

     firstnum=str2num(get(huitext,'String'));
```

```
        secondnum=str2num(get(huitext2,'String'));
        set(hsttext3,'Visible','on',...
              'String',num2str(firstnum*secondnum))
        if source == hbutton
            set(hsttext3,'ForegroundColor','g')
        else
            set(hsttext3,'ForegroundColor','r')
        end
    end
end
```

GUI functions can also have multiple callback functions. In the example *gui-WithTwoPushbuttons*, there are two buttons that could be pushed (see Figure 14.22). Each of them has a unique callback function associated with it. If the top button is pushed, its callback function prints red exclamation points (as shown in Figure 14.23). If the bottom button is instead pushed, its callback function prints blue asterisks.

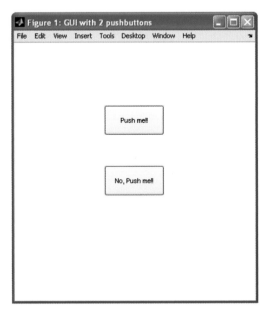

FIGURE 14.22 GUI with two pushbuttons and two callback functions

FIGURE 14.23 The result from the first callback function

function guiWithTwoPushbuttons.m

```
function guiWithTwoPushbuttons
% guiWithTwoPushbuttons has two pushbuttons, each
%    of which has a separate callback function
% Format: guiWithTwoPushbuttons

% Create the GUI but make it invisible for now while
%    it is being initialized
f = figure('Visible', 'off','Color','white',...
    'Position', [360, 500, 400,400]);
set(f,'Name','GUI with 2 pushbuttons')
movegui(f,'center')

% Create a pushbutton that says "Push me!!"
hbutton1 = uicontrol('Style','pushbutton','String',...
    'Push me!!', 'Position',[150,275,100,50], ...
    'Callback',@callbackfn1);

% Create a pushbutton that says "No, Push me!!"
hbutton2 = uicontrol('Style','pushbutton','String',...
    'No, Push me!!', 'Position',[150,175,100,50], ...
    'Callback',@callbackfn2);
% Now the GUI is made visible
set(f,'Visible','on');

  % Call back function for first button
  function callbackfn1(source,eventdata)
     % callbackfn is called by the 'Callback' property
     % in the first pushbutton

     set([hbutton1 hbutton2],'Visible','off');
     hstr = uicontrol('Style','text',...
         'BackgroundColor', 'white', 'Position',...
         [150,200,100,100], 'String','!!!!!', ...
         'ForegroundColor','Red','FontSize',30);
     set(hstr,'Visible','on')
  end

  % Call back function for second button
  function callbackfn2(source,eventdata)
     % callbackfn is called by the 'Callback' property
     % in the second pushbutton

     set([hbutton1 hbutton2],'Visible','off');
     hstr = uicontrol('Style','text',...
         'BackgroundColor','white', ...
         'Position',[150,200,100,100] ,...
         'String','*****', ...
```

```
              'ForegroundColor','Blue','FontSize',30);
        set(hstr,'Visible','on')
    end

end
```

If the first button is pushed, the first callback function is called, which would produce the image in Figure 14.23.

Plots and images can be imbedded in a GUI. In the next example, *guiSliderPlot* shows a plot of **sin(x)** from 0 to the value of a slider bar. The axes are positioned within the Figure Window, and then when the slider is moved the callback function plots. Note the use of the 'Units' property: when set to 'normalized', the Figure Window can be resized and all of the objects will resize accordingly.

```
function guiSliderPlot.m
```

```
function guiSliderPlot
% guiSliderPlot has a slider
% It plots sin(x) from 0 to the value of the slider
% Format: guisliderPlot

f = figure('Visible', 'off','Position',...
    [360, 500, 400,400]);

% Minimum and maximum values for slider
minval = 0;
maxval = 4*pi;
% Create the slider object
slhan = uicontrol('Style','slider','Position',[140,280,100, 50],...
        'Min', minval, 'Max', maxval, 'Callback', @callbackfn);
% Text boxes to show the min and max values and slider value
hmintext = uicontrol('Style','text','BackgroundColor', 'white',...
        'Position', [90, 285, 40,15], 'String', num2str(minval));
hmaxtext = uicontrol('Style','text', 'BackgroundColor', 'white',...
        'Position', [250, 285, 40,15], 'String', num2str(maxval));
hsttext = uicontrol('Style','text','BackgroundColor', 'white',...
        'Position', [170,340,40,15], 'Visible','off');
% Create axes handle for plot
axhan = axes('Units','Pixels','Position', [100,50,200,200]);

set(f,'Name','Slider Example with sin plot')
movegui(f,'center')
set([slhan,hmintext,hmaxtext,hsttext,axhan], 'Units','normalized')
set(f,'Visible','on');

% Call back function displays the current slider value & plots sin
    function callbackfn(source,eventdata)
```

Continued

```
      % callbackfn is called by the 'Callback' property
      % in the slider
      num=get(slhan, 'Value');
      set(hsttext,'Visible','on','String',num2str(num))
      x = 0:num/50:num;
      y = sin(x);
      plot(x,y);
   end
end
```

Figure 14.24 shows the initial configuration of the window, with the slider bar, static text boxes to the left and right showing the minimum and maximum values, and the axes positioned underneath.

After the slider bar is moved, the callback function plots **sin(x)** from 0 to the position of the slider bar, as shown in Figure 14.25.

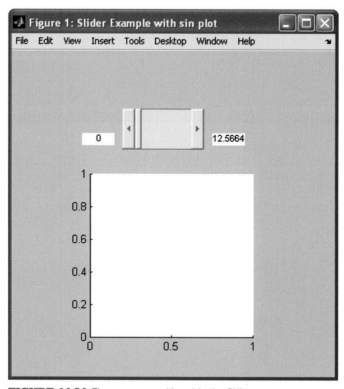

FIGURE 14.24 The axes are positioned in the GUI

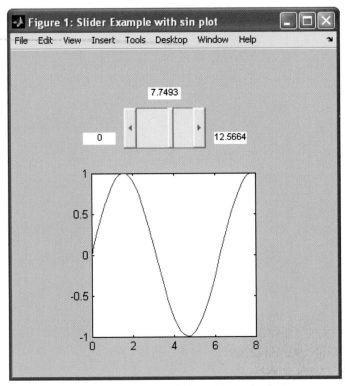

FIGURE 14.25 Plot shown in a GUI Figure Window

The next example illustrates several features: radio buttons, grouping objects together (in this case in a button group), and the use of normalized units when setting the positions. The GUI presents the user with a choice of colors using two radio buttons, only one of which can be pressed at any given time. The GUI prints a string to the right of the radio buttons, in the chosen color.

The function **uibuttongroup** creates a mechanism for grouping together the buttons. Since only one button can be chosen at a time, there is a type of callback function called **SelectionChangeFcn** that is called when a button is chosen.

This function gets from the button group which button is chosen with the 'SelectedObject' property. It then chooses the color based on this. This property is initially set to the empty vector, so that neither button is selected; the default is that the first button would be selected.

The 'Units' property of the objects is set to 'Normalized', which means that rather than specifying in pixels the position, it is done as a percentage of the Figure Window. This allows the Figure Window to be resized.

```matlab
function guiWithButtongroup
% guiWithButtongroup has a button group with 2 radio buttons
% Format: guiWithButtongroup

% Create the GUI but make it invisible for now while
% it is being initialized
f = figure('Visible', 'off','Color','white','Position',...
    [360, 500, 400,400] );

% Create a button group
grouph = uibuttongroup('Parent',f,'Units','Normalized',...
    'Position',[.2 .5 .4 .4], 'Title','Choose Color',...
    'SelectionChangeFcn',@whattodo);

% Put two radio buttons in the group
toph = uicontrol(grouph,'Style','radiobutton',...
    'String','Blue','Units','Normalized',...
    'Position', [.2 .7 .4 .2] );

both = uicontrol(grouph, 'Style','radiobutton',...
    'String','Green','Units','Normalized',...
    'Position',[.2 .4 .4 .2] );

% Put a static text box to the right
texth = uicontrol('Style','text','Units','Normalized',...
    'Position',[.6 .5 .3 .3] ,'String','Hello',...
    'Visible','off','BackgroundColor','white');

set(grouph,'SelectedObject',[ ] ) % No button selected yet

set(f,'Name','GUI with button group')
movegui(f,'center')
% Now the GUI is made visible
set(f,'Visible','on');

  function whattodo(source, eventdata)
    % whattodo is called by the 'SelectionChangeFcn' property
    % in the button group

  which = get(grouph,'SelectedObject');

  if which == toph
      set(texth,'ForegroundColor','blue')
    else
      set(texth,'ForegroundColor','green')
  end

  set(texth,'Visible','on')
  end

end
```

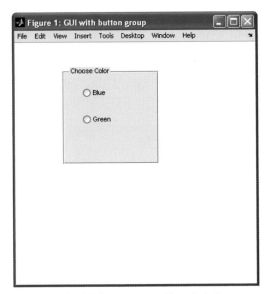

FIGURE 14.26 Button group with radio buttons

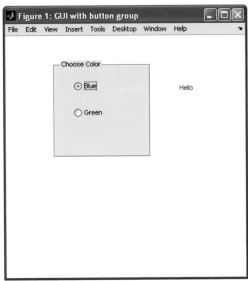

FIGURE 14.27 Button group: choice of color for string

Figure 14.26 shows the initial configuration of the GUI: the button group is in place, as are the buttons (but neither is chosen).

Once a radio button has been chosen, the **SelectionChosenFcn** chooses the color for the string, which is printed in a static text box on the right as shown in Figure 14.27.

The **uibuttongroup** function is used specifically to group together buttons; other objects can be grouped together similarly using the **uipanel** function.

SUMMARY

Common Pitfalls
- Confusing true color and colormap images.
- Forgetting that **uicontrol** object positions are within the Figure Window, not within the screen.

Programming Style Guidelines
- Make a GUI invisible while it is being created, so that everything becomes visible at once.

MATLAB Functions and Commands		
chirp	laughter	train
gong	splat	handel

Continued

MATLAB Functions and Commands—cont'd		
sound	colormap	uibuttongroup
wavread	jet	uicontrol
wavrecord	image	movegui
wavwrite	imread	
wavplay	uipanel	

Exercises

1. Load two of the built-in MAT-file sound files (e.g., **gong** and **chirp**). Store the sound vectors in two separate variables. Determine how to concatenate these so that the **sound** function will play one immediately followed by the other; fill in the blank here:

   ```
   sound(     , 8192)
   ```

2. The following function *playsound* plays one of the built-in sounds. The function has a cell array that stores the names. When the function is called, an integer is passed, which is an index into this cell array indicating the sound to be played. The default is 'train', so if the user passes an invalid index, the default is used. The appropriate MAT-file is loaded. If the user passes a second argument, it is the frequency at which the sound should be played (otherwise, the default frequency is used). The function prints what sound is about to be played and at which frequency, and then actually plays this sound. You are to fill in the rest of the following function. Here are examples of calling it (you can't hear it here, but the sound will be played!).

   ```
   >> playsound(-4)
   You are about to hear train at frequency 8192.0
   >> playsound(2)
   You are about to hear gong at frequency 8192.0
   >> playsound(3,8000)
   You are about to hear laughter at frequency 8000.0
   ```

   ```
   function playsound(caind, varargin)
   %This function plays a sound

   soundarray ={ 'chirp','gong','laughter','splat','train'};
   if caind < 1 || caind > length(soundarray)
       caind = length(soundarray);
   end
   mysound = soundarray{ caind};
   eval(['load ' mysound] )

   % Fill in the rest
   ```

3. Write a script that will create the image shown in Figure 14.28 using a colormap.

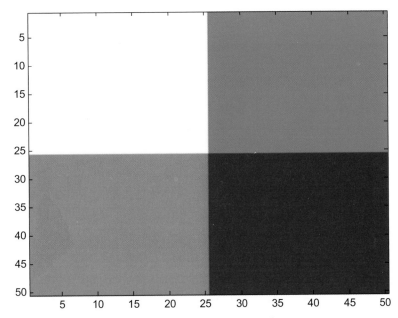

FIGURE 14.28 Image displaying four colors using a custom colormap

4. Write a script that will create the same image as in Exercise 3, using a 3D true color matrix.

5. Write a script that will generate a 50 × 50 image of pixels. The lower triangular part (including the diagonal) will be all white. The upper triangular part will be randomly either red or green for each element, as shown in Figure 14.29.

6. A script *rancolors* displays random colors in the Figure Window as shown in Figure 14.30. It starts with a variable *nColors*, which is the number of random colors to display (e.g., below this is 10). It then creates a colormap variable *mycolormap*, which has that many random colors, meaning that all three of the color components (red, green, and blue) are random real numbers in the range from 0 to 1. For example, if the output is not suppressed, the variable might store the following values:

```
>> rancolors
mycolormap =
      0.3804    0.0119    0.6892
      0.5678    0.3371    0.7482
      0.0759    0.1622    0.4505
      0.0540    0.7943    0.0838
      0.5308    0.3112    0.2290
      0.7792    0.5285    0.9133
      0.9340    0.1656    0.1524
      0.1299    0.6020    0.8258
      0.5688    0.2630    0.5383
      0.4694    0.6541    0.9961
```

The script then displays these colors in an image in the Figure Window.

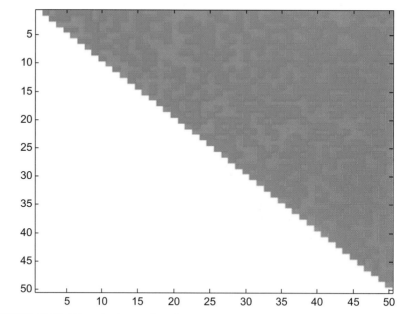

FIGURE 14.29 Triangular image of random red and green

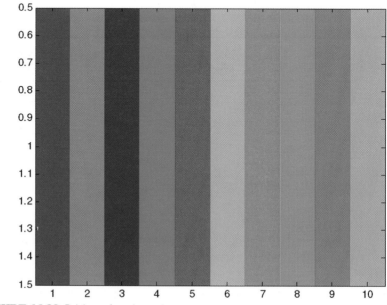

Note
Ignore the numbers on the y-axis in Figure 14.30 (they are defaults).

FIGURE 14.30 Rainbow of random colors

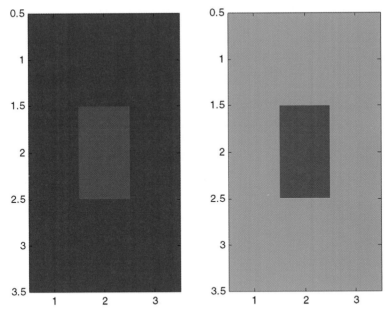

FIGURE 14.31 Depiction of brightness perception

7. It is sometimes difficult for the human eye to perceive the brightness of an object correctly. For example, in Figure 14.31, the middle of both images is the same color, and yet, because of the surrounding colors, the one on the left looks lighter than the one on the right.

   Write a script to generate a Figure Window similar to this one. Two 3×3 matrices were created. Using the default colormap, the middle elements in both were given a value of 12. For the image on the left, all other elements were given a value of 1, and for the image on the right, all other elements were given the value 32. Use **subplot** to display both images side by side (the axes shown here are the defaults).

8. Write a script that will produce the output shown in Figure 14.32. Use **eye** and **repmat** to generate the required matrix efficiently. Also, use **axis image** to correct the aspect ratio.

FIGURE 14.32 Checkerboard

   In a random walk, every time a "step" is taken, a direction is randomly chosen. Watching a random walk as it evolves, by viewing it as an image, can be very entertaining. However, there are actually very practical applications of random walks; they can be used to simulate diverse events such as the spread of a forest fire or the growth of a dendritic crystal.

9. The following function simulates a "random walk," using a matrix to store the random walk as it progresses. To begin with, all elements are initialized to 1. Then,

the "middle" element is chosen to be the starting point for the random walk; a 2 is placed in that element. (*Note:* These numbers will eventually represent colors.) Then, from this starting point another element next to the current one is chosen randomly and the *color* stored in that element is incremented; this repeats until one of the edges of the matrix is reached. Every time an element is chosen for the next element, it is done randomly by either adding or subtracting one to/from each coordinate (*x* and *y*), or leaving it alone. The resulting matrix that is returned is an $n \times n$ matrix.

```
function walkmat = ranwalk(n)
walkmat = ones(n);
x = floor(n/2);
y = floor(n/2);
color = 2;
walkmat(x,y) = color;
while x ~= 1 && x ~= n && y ~= 1 && y ~= n
    x = x + randint(1,1,[ -1 1] );
    y = y + randint(1,1,[ -1 1] );
    color = color + 1;
    walkmat(x,y) = mod(color,65);
end
```

You are to write a script that will call this function twice (once passing 8 and once passing 100) and display the resulting matrices as images side by side. Your script must create a custom colormap that has 65 colors; the first is white and the rest are from the colormap *jet*. For example, the result may look like Figure 14.33. (Note that with the 8×8 matrix, the colors are not likely to get out of the blue

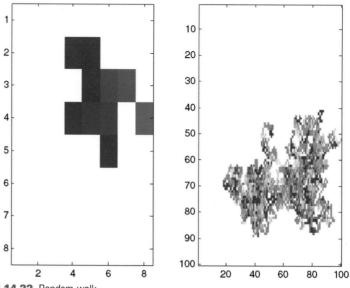

FIGURE 14.33 Random walk

range, but with *100 × 100*, it cycles through all colors multiple times until an edge is reached.)

10. A script *colorguess* plays a guessing game. It creates an $n \times n$ matrix, and randomly picks one element in the matrix. It prompts the user to guess the element (meaning the row index and column index). Every time the user guesses, that element is displayed as red. When the user correctly guesses the randomly picked element, that element is displayed in blue and the script ends. Here is an example of running the script (the randomly picked element in this case is (8,4)). Only the last version of the Figure Window is shown in Figure 14.34.

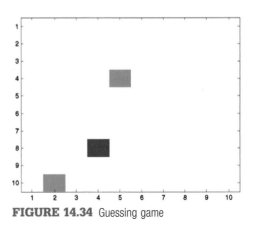

FIGURE 14.34 Guessing game

```
>> colorguess
Enter the row #: 4
Enter the col #: 5
Enter the row #: 10
Enter the col #: 2
Enter the row #: 8
Enter the col #: 4
```

11. Write a script that will create a colormap that has nine colors: three shades each of red, green, and blue. It then creates a 25×25 image matrix in which every element is a random integer in the range from 1 to 9. Next, it creates a new image matrix in which any pixel that is a shade of a particular color is replaced by that color. The images are to be shown side by side, as seen in Figure 14.35.

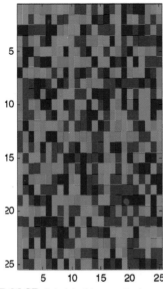 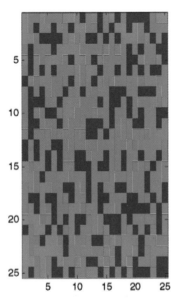

FIGURE 14.35 Subplot of image shades

12. Put a JPEG file in your Current Folder and use **imread** to load it into a matrix. Calculate and print the **mean** separately of the red, green, and blue components in the matrix and also the standard deviation for each.

13. Some image acquisition systems are not very accurate, and the result is *noisy* images. To see this effect, put a JPEG file in your Current Folder and use **imread** to load it. Then, create a new image matrix by randomly adding or subtracting a value n to every element in this matrix. Experiment with different values of n. Create a script that will use **subplot** to display both images side by side, using **imshow** instead of **image**.

14. The *dynamic range* of an image is the range of colors in the image (the minimum value to the maximum value). Put a JPEG file into your Current Folder. Read the image into a matrix. Use the built-in functions **min** and **max** to determine the dynamic range, and print the range. Note that if the image is a true color image, the matrix will be 3D; thus, it will be necessary to nest the functions three times to get the overall minimum and maximum values.

15. A part of an image is represented by an $n \times n$ matrix. After performing data compression and then data reconstruction techniques, the resulting matrix has values that are close to but not exactly equal to the original matrix. For example, the following 4×4 matrix variable *orig_im* represents a small part of a true color image, and *fin_im* represents the matrix after it has undergone data compression and then reconstruction.

```
orig_im =
     156      44     129      87
      18     158     118     102
      80      62     138      78
     155     150     241     105
fin_im =
     153      43     130      92
      16     152     118     102
      73      66     143      75
     152     155     247     114
```

Write a script that will simulate this by creating a square matrix of random integers, each in the range from 0 to 255. It will then modify this to create the new matrix by randomly adding or subtracting a random number (in a relatively small range, say 0 to 10) from every element in the original matrix. Then, calculate the average difference between the two matrices.

16. Put a JPEG file into your Current Folder. Type in the following script, using your own JPEG filename.

```
I1 = imread('xxx.jpg');
[r c h] = size(I1);
Inew(:,:,:) = I1(:,c:-1:1,:);
figure(1)
subplot(2,1,1)
image(I1);
subplot(2,1,2)
image(Inew);
```

Determine what the script does. Put comments into the script to explain it step by step. Also, try it using **imshow** instead of **image**.

17. Put two different JPEG files into your Current Folder. Read both into matrix variables. To superimpose the images, if the matrices are the same size, the elements can simply be added element by element. However, if they are not the same size, one method of handling this is to crop the larger matrix to be the same size as the smaller, and then add them. Write a script to do this.

18. Write a function that will create a simple GUI with one static text box near the middle of the Figure Window. Put your name in the string, and make the background color of the text box white.

19. Write a function that will create a GUI with one editable text box near the middle of the Figure Window. Put your name in the string. The GUI should have a callback function that prints the user's string twice, one under the other.

20. Write a function that creates a GUI to calculate the area of a rectangle. It should have edit text boxes for the length and width, and a pushbutton that causes the area to be calculated and printed in a static text box.

21. Write a function that creates a simple calculator with a GUI. The GUI should have two editable text boxes in which the user enters numbers. There should be four pushbuttons to show the four operations ($+$, $-$, $*$, $/$). When one of the four pushbuttons is pressed the type of operation should be shown in a static text box between the two editable text boxes and the result of the operation should be displayed in a static text box. If the user tries to divide by zero display an error message in a static text box.

22. Write a function that will create a GUI in which there is a plot of $\cos(x)$. There should be two editable text boxes in which the user can enter the range for x.

23. Write a function that will create a GUI in which there is a plot. Use a button group to allow the user to choose among several functions to plot.

24. Modify any example GUI from the chapter to use normalized units instead of pixels.

25. Modify any example GUI to use the 'HorizontalAlignment' property to left-justify text within an edit text box.

26. Modify the *gui_slider* example in the text to include a **persistent** count variable in the callback function that counts how many times the slider is moved. This count should be displayed in a static text box in the upper right corner, as shown in Figure 14.36.

FIGURE 14.36 Slider with count

27. The wind chill factor (WCF) measures how cold it feels with a given air temperature (T, in degrees Fahrenheit) and wind speed (V, in miles per hour). The formula is approximately

$$WCF = 35.7 + 0.6\,T - 35.7(V^{0.16}) + 0.43\,T(V^{0.16})$$

Write a GUI function that will display sliders for the temperature and wind speed. The GUI will calculate the WCF for the given values, and display the result in a text box. Choose appropriate minimum and maximum values for the two sliders.

Advanced Mathematics

KEY TERMS

curve fitting

best fit

symbolic mathematics

polynomials

degree

order

discrete

continuous

data sampling

interpolation

extrapolation

least squares

regression

complex number

real part

imaginary part

purely imaginary

complex conjugate

magnitude

complex plane

In this chapter, selected advanced mathematics and related built-in functions in the MATLAB® software are introduced. In many applications data are sampled, which results in discrete data points. Fitting a curve to the data is often desired. *Curve fitting* is finding the curve that *best fits* the data.

This chapter first explores fitting curves that are simple polynomials to data. Other topics include complex numbers and a brief introduction to differentiation and integration in calculus. *Symbolic mathematics* means doing mathematics on symbols. Some of the symbolic math functions, all of which are in the Symbolic Math Toolbox in MATLAB, are also introduced, including the **solve** function for solving equations. (Note that this is a Toolbox, and as a result may not be universally available.)

15.1 FITTING CURVES TO DATA

MATLAB has several curve-fitting functions; the Curve Fitting Toolbox has many more of these functions. Some of the simplest curves are polynomials of different degrees, which are described next.

473

15.1.1 Polynomials

Simple curves are *polynomials* of different *degrees* or *orders*. The degree is the integer of the highest exponent in the expression. For example:

- A straight line is a first-order (or degree 1) polynomial of the form $ax + b$, or more explicitly $ax^1 + b$.
- A quadratic is a second-order (or degree 2) polynomial of the form $ax^2 + bx + c$.
- A cubic (degree 3) is of the form $ax^3 + bx^2 + cx + d$.

MATLAB represents a polynomial as a row vector of coefficients. For example, the polynomial $x^3 + 2x^2 - 4x + 3$ would be represented by the vector $[1\ 2\ -4\ 3]$. The polynomial $2x^4 - x^2 + 5$ would be represented by $[2\ 0\ -1\ 0\ 5]$; note the zero terms for x^3 and x^1.

The built-in functions **sym2poly** and **poly2sym** convert from symbolic expressions to polynomial vectors and vice versa. For example:

```
>> myp = [1 2 −4 3];
>> poly2sym(myp)
ans =
x^3 + 2 * x^2 - 4 * x + 3
>> mypoly = [2 0 −1 0 5];
>> poly2sym(mypoly)
ans =
2 * x^4 - x^2 + 5

>> sym2poly(ans)
ans =
     2     0    -1     0     5
```

Note

This is a mathematical expression, not MATLAB!

The **roots** function in MATLAB can be used to find the roots of an equation represented by a polynomial. For example, for the mathematical function

$$f(x) = 4x^3 - 2x^2 - 8x + 3$$

to solve the equation $f(x) = 0$:

```
>> roots([4 −2 −8 3])
ans =
   -1.3660
    1.5000
    0.3660
```

The function **polyval** will evaluate a polynomial p at x; the form is **polyval(p,x)**. For example, the polynomial $-2x^2 + x + 4$ is evaluated at $x = 3$, which yields $-2 * 9 + 3 + 4$, or -11:

```
>> p = [-2 1 4];
>> polyval(p,3)
ans =
   -11
```

The argument *x* can be a vector:

```
>> polyval(p,1:3)
ans =
     3    -2   -11

>> polyval(p, [5 7])
ans =
  -41    -87
```

15.1.2 Curve fitting

Data can be either *discrete* (e.g., a set of object weights) or *continuous*. In many applications, continuous properties are *sampled*, such as:

- The temperature recorded every hour
- The speed of a car recorded every one-tenth of a mile
- The mass of a radioactive material recorded every second as it decays
- Audio from a sound wave as it is converted to a digital audio file

Sampling provides data in the form of (x,y) points, which can then be plotted. For example, let's say the temperature was recorded every hour one afternoon from 2 to 6 pm; the vectors might be:

```
>> x = 2:6;
>> y = [65 67 72 71 63];
```

and then the plot might look like Figure 15.1.

15.1.3 Interpolation and extrapolation

In many cases, estimating values other than at the sampled data points is desired. For example, we might want to estimate the temperature at 2:30 pm or at 1 pm. *Interpolation* means estimating the values in between recorded data points. *Extrapolation* is estimating outside of the bounds of the recorded data. One way to do this is to fit a curve to the data, and use this for the estimations. *Curve fitting* is finding the curve that "best fits" the data.

Simple curves are polynomials of different degrees as described previously. Thus, curve fitting involves finding the best polynomials to fit the data—for example, for a quadratic

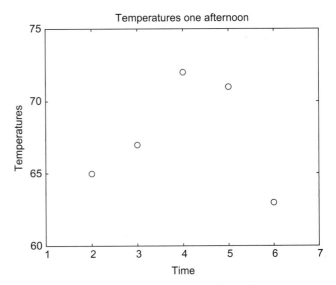

FIGURE 15.1 Plot of temperatures sampled every hour

polynomial in the form $ax^2 + bx + c$, it means finding the values of a, b, and c that yield the best fit. Finding the best straight line that goes through data would mean finding the values of a and b in the equation $ax + b$.

MATLAB has a function to do this, called **polyfit**. The function **polyfit** finds the coefficients of the polynomial of the specified degree that best fits the data using a least squares algorithm. There are three arguments passed to the function: the vectors that represent the data, and the degree of the desired polynomial. For example, to fit a straight line (degree 1) through the points representing temperatures, the call to the **polyfit** function would be

```
>> polyfit(x,y,1)
ans =
    0.0000   67.6000
```

which says that the best straight line is of the form $0x + 67.6$.

However, from the plot (shown in Figure 15.2), it looks like a quadratic would be a much better fit. The following would create the vectors and then fit a polynomial of degree 2 through the data points, storing the values in a vector called *coefs*.

```
>> x = 2:6;
>> y = [65 67 72 71 63];
>> coefs = polyfit(x,y,2)
coefs =
    -1.8571   14.8571   41.6000
```

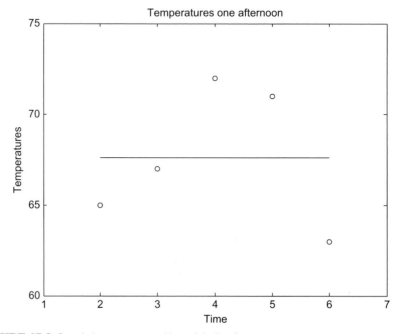

FIGURE 15.2 Sampled temperatures with straight-line fit

This says that MATLAB has determined that the best quadratic that fits these data points is $-1.8571x^2 + 14.8571x + 41.6$. So, the variable *coefs* now stores a coefficient vector that represents this polynomial.

The function **polyval** can then be used to evaluate the polynomial at specified values. For example, we could evaluate at every value in the x vector:

```
>> curve = polyval(coefs,x)
curve =
    63.8857   69.4571   71.3143   69.4571   63.8857
```

This results in *y* values for each point in the x vector, and stores them in a vector called *curve*. Putting all of this together, the following script called *polytemp* creates the x and y vectors, fits a second-order polynomial through these points, and plots both the points and the curve on the same figure. Calling this results in the plot seen in Figure 15.3. The curve doesn't look very smooth on this plot, but that is because there are only five points in the *x* vector.

```
polytemp.m
```
```
% Fits a quadratic curve to temperature data
x = 2:6;
y = [65 67 72 71 63];
coefs = polyfit(x,y,2);
curve = polyval(coefs,x);
plot(x,y,'ro',x,curve)
xlabel('Time')
ylabel('Temperatures')
title('Temperatures one afternoon')
axis([ 1 7 60 75] )
```

PRACTICE 15.1

To make the curve smoother, modify the script *polytemp* to create a new x vector with more points for plotting the curve. Note that the original x vector for the data points must remain as is.

To estimate the temperature at different times, **polyval** can be used for discrete x points; it does not have to be used with the entire x vector. For example, to interpolate between the given data points and estimate what the temperature was at 2:30 pm, 2.5 would be used.

```
>> polyval(coefs,2.5)
ans =
    67.1357
```

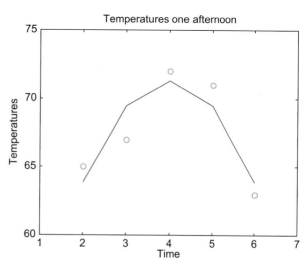

FIGURE 15.3 Sampled temperatures shown with a quadratic curve

Also, **polyval** can be used to extrapolate beyond the given data points. For example, to estimate the temperature at 1 pm:

```
>> polyval(coefs,1)
ans =
   54.6000
```

The better the curve fit, the more exact these interpolated and extrapolated values will be.

Using the **subplot** function, we can loop to show the difference between fitting curves of degrees 1, 2, and 3 to some data. For example, the following script will accomplish this for the temperature data. (*Note:* The variable *morex* stores 100 points so the graph will be smooth.)

polytempsubplot.m

```
% Fits curves of degrees 1-3 to temperature
%   data and plots in a subplot
x = 2:6;
y = [65 67 72 71 63];
morex = linspace(min(x),max(x));
for pd = 1:3
    coefs = polyfit(x,y,pd);
    curve = polyval(coefs,morex);
    subplot(1,3,pd)
    plot(x,y,'ro',morex,curve)
    xlabel('Time')
    ylabel('Temperatures')
    title(sprintf('Degree %d',pd))
    axis([1 7 60 75])
end
```

Executing the script

```
>> polytempsubplot
```

creates the Figure Window shown in Figure 15.4.

15.1.4 Least squares

The **polyfit** function uses the *least squares regression* method. To find the equation of the straight line $y = mx + b$ that best fits using a least squares regression, the equations for m and b are:

$$m = \frac{n\sum x_i y_i - \sum x_i \sum y_i}{n\sum x_i^2 - \left(\sum x_i\right)^2}$$

$$b = \bar{y} - m\bar{x}$$

where n is the number of points in x and y, all summations are from $i = 1$ to n, and $\bar{y}$ and $\bar{x}$ represent the means of the vectors y and x. These equations will not

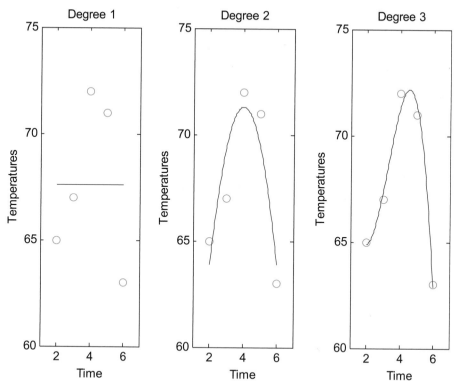

FIGURE 15.4 Subplot to show temperatures with curves of degrees 1, 2, and 3

be derived here; the derivations can be found in the MATLAB help browser by doing a search for "least squares."

Least squares regression is implemented in the following in a function *mylinfit* that receives two vectors x and y, and returns the values of m and b. This is the same algorithm used by the **polyfit** function for a degree 1 polynomial, so it returns the same values.

```
mylinfit.m
```

```
function [m,b] = mylinfit(x,y)
% mylinfit implements a least squares regression for a
%    straight line of the form y = mx+b
% Format: mylinfit(x,y)

n = length(x);   % Assume y has same length

numerator = n * sum(x .* y) - sum(x) * sum(y);
denom = n * sum(x .^ 2) - (sum(x))^2;
m = numerator/denom;

b = mean(y) - m*mean(x);
end
```

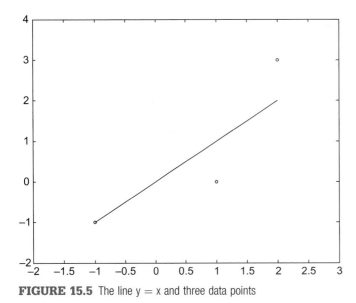

FIGURE 15.5 The line y = x and three data points

```
>> x = [-1 1 2];
>> y = [-1 0 3];
>> [m b] = mylinfit(x,y)
m =
    1.2143
b =
    -0.1429
>> polyfit(x,y,1)
ans =
    1.2143    -0.1429
```

The least squares fit minimizes the sum of the squares of the differences between the actual data and the data predicted by the line. The "best" straight line in this case has been identified as y = 1.2143x − 0.1429.

If we did not know that was the best straight line, we might instead guess that the line that best fits the data is the line y = x. The plot is shown in Figure 15.5.

This straight line goes through one of the points, but "splits" the other two points, in that one is one unit below the line and the other is one above the line. So, it seems as if it fits the data well. However, we will compare this to the line found by **polyfit** and the function *mylinfit*.

Table 15.1 shows the x coordinates, y coordinates of the original points, y coordinates predicted by the line y = x, and the differences (data – predicted).

The sum of the differences squared is 0 + 1 + 1, or 2. According to the least squares algorithm, however, the values using the line y = 1.2143x – 0.1429 are shown in Table 15.2. The sum of the squares of these differences is 1.7857, which is better than (a smaller number than) the sum of the squares of the differences obtained for the straight line above. In fact, **polyfit** minimizes the sum of the squares.

MATLAB has another related function, **interp1**, that does a table look-up to interpolate or extrapolate. There are several ways to call this function (using **help** describes them). The default method that is used is 'linear', which gives a linear interpolation.

Table 15.1 y Coordinates Predicted by Line y = x

x	Data y	Predicted y	Difference
−1	−1	−1	0
1	0	1	−1
2	3	2	1

Table 15.2 y Coordinates Predicted by Least Squares Regression

x	Data y	Predicted y	Difference
−1	−1	−1.3571	0.3571
1	0	1.0714	−1.0714
2	3	2.2857	0.7143

For example, for the previous time and temperature vectors:

```
>> x = 2:6;
>> y = [65 67 72 71 63];
```

The **interp1** function could be used to interpolate between the points; for example,

```
>> interp1(x,y,3.5)
ans =
   69.5000

>> interp1(x,y,2.5)
ans =
   66
```

To extrapolate using the linear interpolation method, which is the default, the strings 'linear' and 'extrap' would also be passed.

```
>> interp1(x,y,1,'linear','extrap')
ans =
   63

>> interp1(x,y,7,'linear','extrap')
ans =
   55
```

15.2 COMPLEX NUMBERS

A *complex number* is generally written in the form

```
z = a + bi
```

where *a* is called the **real part** of the number z, b is the **imaginary part** of z, and i is $\sqrt{-1}$.

We have seen that in MATLAB both **i** and **j** are built-in functions that return $\sqrt{-1}$ (so, they can be thought of as built-in constants). Complex numbers can be created using **i** or **j**, such as "5 + 2i" or "3 – 4j". The multiplication operator is not required between the value of the imaginary part and the constant **i** or **j**.

Note

This is the way mathematicians usually write a complex number; in engineering it is often written as a + bj, where *j* is $\sqrt{-1}$. A complex number is **purely imaginary** if it is of the form $z = bi$ (in other words, if *a* is 0).

QUICK QUESTION!

Is the value of the expression "3i" the same as "3 * i"?

Answer: It depends on whether *i* has been used as a variable name or not. If *i* has been used as a variable (e.g., an iterator variable in a **for** loop), then the expression "3 * i" will use the defined value for the variable, and the result will not be a complex number. Therefore, it is a good idea when working with complex numbers to use 1i or 1j rather than just i or j. The expressions 1i and 1j always result in a complex number, regardless of whether i and j have been used as variables.

```
>> i = 5;
>> i
i =
    5
>> 1i
ans =
        0 + 1.0000i
```

MATLAB also has a function **complex** that will return a complex number. It receives two numbers, the real and imaginary parts in that order, or just one number, which is the real part (so the imaginary part would be 0). The following are examples of creating complex numbers in MATLAB:

```
>> z1  = 4 + 2i
z1 =
   4.0000 + 2.0000i

>> z2 = sqrt(-5)
z2 =
        0 + 2.2361i

>> z3 = complex(3,-3)
z3 =
   3.0000 - 3.0000i

>> z4 = 2 + 3j
z4 =
   2.0000 + 3.0000i

>> z5 = (-4) ^ (1/2)
ans =
   0.0000 + 2.0000i

>> myz = input('Enter a complex number: ')
Enter a complex number: 3 + 4i
myz =
   3.0000 + 4.0000i
```

Note that even when **j** is used in an expression, **i** is used in the result. MATLAB shows the type of the variables created here in the Workspace Window (or using **whos**) as **double (complex)**. MATLAB has functions **real** and **imag** that return the real and imaginary parts of complex numbers.

```
>> real(z1)
ans =
     4

>> imag(z3)
ans =
    -3
```

In order to print a complex number, the **disp** function will display both parts automatically:

```
>> disp(z1)
   4.0000 + 2.0000i
```

The **fprintf** function will only print the real part unless both parts are printed separately:

```
>> fprintf('%f\n', z1)
4.000000

>> fprintf('%f %f\n', real(z1), imag(z1))
4.000000 2.000000

>> fprintf('%f + %fi\n', real(z1), imag(z1))
4.000000 + 2.000000i
```

The function **isreal** returns **logical** 1 for **true** if there is no imaginary part of the argument, or **logical** 0 for **false** if the argument does have an imaginary part (even if it is 0). For example,

```
>> isreal(z1)
ans =
     0

>> z6 = complex(3)
z5 =
     3

>> isreal(z6)
ans =
     0

>> isreal(3.3)
ans =
     1
```

For the preceding variable $z6$, even though it shows the answer as 3, it is really stored as $3 + 0i$, and that is how it is displayed in the Workspace Window. Therefore, **isreal** returns **logical false** since it is stored as a complex number.

15.2.1 Equality for complex numbers

Two complex numbers are equal to each other if both their real parts and imaginary parts are equal. In MATLAB, the equality operator can be used.

```
>> z1 == z2
ans =
     0

>> complex(0,4) == sqrt(-16)
ans =
     1
```

15.2.2 Adding and subtracting complex numbers

For two complex numbers $z1 = a + bi$ and $z2 = c + di$,

$$z1 + z2 = (a + c) + (b + d)i$$

$$z1 - z2 = (a - c) + (b - d)i$$

As an example, we will write a function in MATLAB to add two complex numbers together and return the resulting complex number.

THE PROGRAMMING CONCEPT

In most cases, to add two complex numbers together you would have to separate the real and imaginary parts, and add them to return your result.

addcomp.m

```
function outc = addcomp(z1, z2)
% addcomp adds two complex numbers z1 and z2 &
%    returns the result
% Adds the real and imaginary parts separately
% Format: addcomp(z1,z2)

realpart = real(z1) + real(z2);
imagpart = imag(z1) + imag(z2);
outc = realpart + imagpart * 1i;
end
```

```
>> addcomp(3 + 4i, 2 - 3j)
ans =
   5.0000 + 1.0000i
```

THE EFFICIENT METHOD

MATLAB automatically does this to add two complex numbers together (or subtract).

addcomp2.m

```
function outc = addcomp2(z1,z2)
% addcomp2 adds two complex numbers z1 and z2 &
%    returns the result
% Format: addcomp(z1,z2)
outc = z1 + z2;
end
```

```
>> addcomp2(3 + 4i, 2 - 3j)
ans =
   5.0000 + 1.0000i
```

15.2.3 Multiplying complex numbers

For two complex numbers $z1 = a + bi$ and $z2 = c + di$,

```
z1 * z2 = (a + bi) * (c + di)

        = a * c + a * di + c * bi + bi * di

        = a * c + a * di + c * bi − b * d

        = (a * c − b * d) + (a * d + c * b) i
```

For example, for the complex numbers

```
z1 = 3 + 4i
z2 = 1 − 2i
```

the result of the multiplication would be defined mathematically as

```
z1 * z2 = (3 * 1 − −8) + (3 * -2 + 4 * 1) i = 11 −2i
```

This is, of course, automatic in MATLAB:

```
>> z1 * z2
ans =
   11.0000 − 2.0000i
```

15.2.4 Complex conjugate and absolute value

The *complex conjugate* of a complex number $z = a + bi$ is $\bar{z} = a − bi$. The *magnitude*, or *absolute value*, of a complex number z is $|z| = \sqrt{a^2 + b^2}$. In MATLAB, there is a built-in function **conj** for the complex conjugate, and the **abs** function returns the absolute value.

```
>> z1 = 3 + 4i
z1 =
   3.0000 + 4.0000i

>> conj(z1)
ans =
   3.0000 − 4.0000i

>> abs(z1)
ans =
     5
```

15.2.5 Complex equations represented as polynomials

We have seen that MATLAB represents polynomials as a row vector of coefficients; this can be used when the expressions or equations involve complex numbers, also. For example, the polynomial $z^2 + z − 3 + 2i$ would be represented by the vector $[1 \; 1 \; -3 + 2i]$. The **roots** function in MATLAB can be used to find the roots of an equation represented by a polynomial. For instance, to solve the equation $z^2 + z - 3 + 2i = 0$:

```
>> roots([1 1 - 3 + 2i])

ans =

  -2.3796 + 0.5320i
   1.3796 - 0.5320i
```

The **polyval** function can also be used with this polynomial. For example:

```
>> cp = [1 1 - 3 + 2i]
cp =
   1.0000              1.0000           -3.0000 + 2.0000i

>> polyval(cp,3)
ans =
   9.0000 + 2.0000i
```

15.2.6 Polar form

Any complex number $z = a + bi$ can be thought of as a point (a,b) or vector in a *complex plane* in which the horizontal axis is the real part of z, and the vertical axis is the imaginary part of z. So, *a* and *b* are the Cartesian or rectangular coordinates. Since a vector can be represented by either its rectangular or polar coordinates, a complex number can also be given by its polar coordinates r and θ, where r is the magnitude of the vector and θ is an angle.

To convert from the polar coordinates to the rectangular coordinates:

```
a = r cos θ
b = r sin θ
```

To convert from the rectangular to polar coordinates:

$$r \;\; = |z| \; = \; \sqrt{a^2 + b^2}$$

$$\theta \;\; = \arctan\left(\frac{b}{a}\right)$$

So, a complex number $z = a + bi$ can be written as $r \cos \theta + (r \sin \theta) i$, or

$$z = r (\cos \theta + i \sin \theta)$$

Since $e^{i\theta} = \cos\theta + i\sin\theta$, a complex number can also be written as $z = re^{i\theta}$. In MATLAB, r can be found using the **abs** function, while there is a built-in function to find θ, called **angle**.

```
>> z1 = 3 + 4i;
r = abs(z1)
r =
    5

>> theta = angle(z1)
theta =
    0.9273

>> r * exp(i * theta)
ans =
    3.0000 + 4.0000i
```

15.2.7 Plotting

Several methods are commonly used for plotting complex data:

- Plot the real parts versus the imaginary parts using **plot**.
- Plot only the real parts using **plot**.
- Plot the real and the imaginary parts in one figure with a legend, using **plot**.
- Plot the magnitude and angle using **polar**.

Using the **plot** function with a single complex number or a vector of complex numbers will result in plotting the real parts versus the imaginary parts; for example, **plot(z)** is the same as **plot(real(z), imag(z))**. Another example: For the complex number $z1 = 3 + 4i$, this will plot the point (3,4) (using a large asterisk so we can see it!) as shown in Figure 15.6.

```
>> z1 = 3 + 4i;
>> plot(z1,'*', 'MarkerSize', 12)
>> xlabel('Real part')
>> ylabel('Imaginary part')
>> title('Complex number')
```

PRACTICE 15.2

Create the following complex variables:

```
c1 = complex(0,2);
c2 = 3 + 2i;
c3 = sqrt(-4);
```

Then, carry out the following:

- Get the real and imaginary parts of $c2$.
- Print the value of $c1$ using **disp**.
- Print the value of $c2$ in the form 'a + bi'.
- Determine whether any of the variables are equal to each other.
- Subtract $c2$ from $c1$.
- Multiply $c2$ by $c3$.
- Get the complex conjugate and magnitude of $c2$.
- Put $c1$ in polar form.
- Plot the real part versus the imaginary part for $c2$.

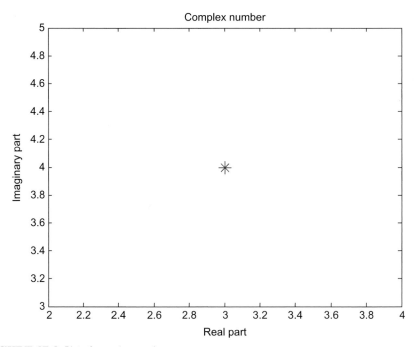

FIGURE 15.6 Plot of complex number

15.3 SYMBOLIC MATHEMATICS

Symbolic mathematics means doing mathematics on symbols (not numbers!). For example, $a + a$ is $2a$. The symbolic math functions are in the Symbolic Math Toolbox in MATLAB. Toolboxes contain related functions and are add-ons to MATLAB. (Therefore, this may or may not be part of your own system.) Symbolic Math Toolbox includes an alternative method for solving equations, and is therefore covered in this chapter.

To find out about the symbolic functions, **help** can be used:

```
help toolbox\symbolic
```

15.3.1 Symbolic variables and expressions

MATLAB has a type called **sym** for symbolic variables and expressions; these work with strings. For example, to create a symbolic variable a and perform the addition just described, first a symbolic variable would be created by passing the string 'a' to the **sym** function:

```
>> a = sym('a');
>> a + a
ans =
2 * a
```

Symbolic variables can also store expressions. For example, the variables b and c store symbolic expressions:

```
>> b = sym('x^2');
>> c = sym('x^4');
```

All basic mathematical operations can be performed on symbolic variables and expressions (e.g., add, subtract, multiply, divide, raise to a power, etc.). The following are examples:

```
>> c/b
ans =
x^2

>> b^3
ans =
x^6

>> c*b
ans =
x^6

>> b + sym('4 * x^2')
ans =
5 * x^2
```

It can be seen from the last example that MATLAB will collect like terms in these expressions, adding the x^2 and $4x^2$ to result in $5x^2$.

The following creates a symbolic expression by passing a string, but the terms are not automatically collected:

```
>> sym('z^3 + 2 * z^3')
ans =
z^3 + 2 * z^3
```

If, on the other hand, z is a symbolic variable to begin with, quotes are not needed around the expression, and the terms are automatically collected:

```
>> z = sym('z');
>> z^3 + 2 * z^3
ans =
3 * z^3
```

If using multiple variables as symbolic variable names is desired, the **syms** function is a shortcut instead of using **sym** repeatedly. For example,

```
>> syms x y z
```

does the same thing as

```
>> x = sym('x');
>> y = sym('y');
>> z = sym('z');
```

15.3.2 Simplification functions

There are several functions that work with expressions, and simplifying the terms. Not all expressions can be simplified, but the **simplify** function does whatever it can to simplify expressions, including gathering like terms. For example,

```
>> x = sym('x');
>> myexpr = cos(x)^2 + sin(x)^2
myexpr =
cos(x)^2 + sin(x)^2
>> simplify(myexpr)
ans =
1
```

The functions **collect**, **expand**, and **factor** work with polynomial expressions. The **collect** function collects coefficients, such as the following:

```
>> x = sym('x');
>> collect(x^2 + 4 * x^3 + 3 * x^2)
ans =
4 * x^2 + 4 * x^3
```

The **expand** function will multiply out terms, and **factor** will do the reverse:

```
>> expand((x + 2) * (x - 1))
ans =
x^2 + x - 2

>> factor(ans)
ans =
(x + 2)* (x - 1)
```

If the argument is not factorable, the original input argument will be returned unmodified.

The **subs** function will substitute a value for a symbolic variable in an expression. For example,

```
>> myexp = x^3 + 3 * x^2 - 2
myexp =
x^3 + 3 * x^2 - 2
>> subs(myexp,3)
ans =
    52
```

If there are multiple variables in the expression, one will be chosen by default for the substitution (in this case, x), or the variable for which the substitution is to be made can be specified:

```
>> syms a b x
>> varexp = a * x^2 + b * x;
>> subs(varexp,3)
ans =
9 * a + 3 * b

>> subs(varexp,'a',3)
ans =
3 * x^2 + b * x
```

With symbolic math, MATLAB works by default with rational numbers, meaning that results are kept in fractional forms. For example, performing the addition $1/3 + 1/2$ would normally result in a **double** value:

```
>> 1/3 + 1/2
ans =
    0.8333
```

However, by making the expression symbolic, the result is symbolic also. Any numeric function (e.g., **double**) could change that:

```
>> sym(1/3 + 1/2)
ans =
5/6
>> double(ans)
ans =
    0.8333
```

The **numden** function will separately return the numerator and denominator of a symbolic expression:

```
>> sym(1/3 + 1/2)
ans =
5/6
```

```
>> [n, d] = numden(ans)
n =
5
d =
6

>> [n, d] = numden((x^3 + x^2)/x)
n =
x^2 * (x + 1)
d =
x
```

15.3.3 Displaying expressions

The **pretty** function will display these expressions using exponents. For example:

```
>> b = sym('x^2')
b =
x^2

>> pretty(b)
                                                    2
                                                   x
```

There are several plot functions in MATLAB with names beginning with "ez" that perform the necessary conversions from symbolic expressions to numbers and plot them. For example, the function **ezplot** will draw a two-dimensional plot in the x-range from -2π to 2π, with the expression as the title. The expression

```
>> ezplot('x^3 + 3 * x^2 - 2')
```

produces the figure that is shown in Figure 15.7. The domain for the **ezplot** function can also be specified; for example, to change the x-axis to the range 0 to π, it is specified as a vector. The result is shown in Figure 15.8.

```
>> ezplot('cos(x)',[0 pi])
```

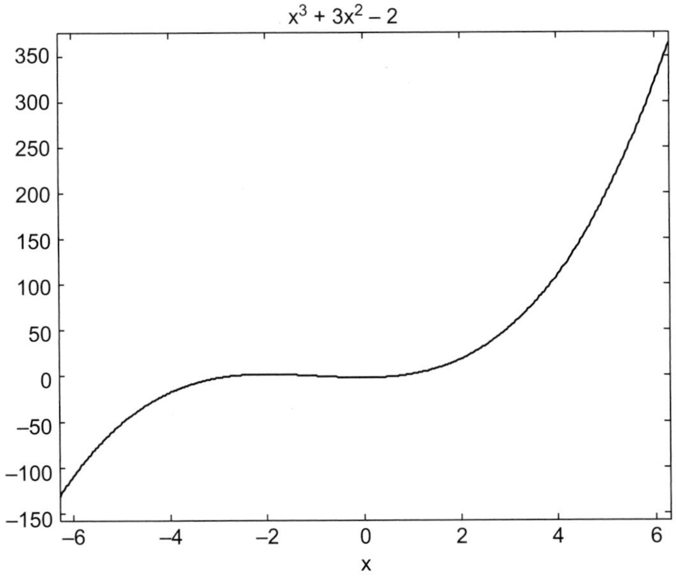

15.3.4 Solving equations

We've seen already several methods for solving simultaneous linear equations, using a matrix representation. MATLAB can also solve sets of equations using symbolic math.

FIGURE 15.7 Plot produced using **ezplot**

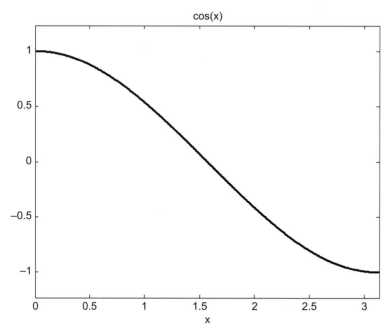

FIGURE 15.8 Result from **ezplot** with custom *x*-axis

The function **solve** solves an equation and returns the solution(s) as symbolic expressions. The solution can be converted to numbers using any numeric function, such as **double**:

```
>> x = sym('x');
>> solve('2 * x^2 + x = 6')
ans =
3/2
 -2

>> double(ans)
ans =
    1.5000
   -2.0000
```

If an expression is passed to the **solve** function rather than an equation, it will set the expression equal to 0 and solve the resulting equation. For example, the following will solve $3x^2 + x = 0$:

```
>> solve('3 * x^2 + x')
ans =
    0
-1/3
```

If there is more than one variable, MATLAB chooses which to solve for. In the following example, the equation $ax^2 + bx = 0$ is solved. There are three variables. As can be seen from the result, which is given in terms of a and b, the equation was solved for x. MATLAB has rules built in that specify how to choose which variable to solve for. For example, x will always be the first choice if it is in the equation or expression.

```
>> solve('a * x^2 + b * x')
ans =
    0
-b/a
```

However, it is possible to specify which variable to solve for:

```
>> solve('a * x^2 + b * x','b')
ans =
-a * x
```

MATLAB can also solve sets of equations. In this example, the solutions for x, y, and z are returned as a structure consisting of fields for x, y, and z. The individual solutions are symbolic expressions stored in fields of the structure.

```
>> solve('4 * x - 2 * y + z = 7','x + y + 5 * z = 10','-2 * x + 3 * y -
z = 2')
ans =
    x: [1 x 1 sym]
    y: [1 x 1 sym]
    z: [1 x 1 sym]
```

To refer to the individual solutions, which are in the structure fields, the dot operator is used.

```
>> x = ans.x
x =
124/41
```

```
>> y = ans.y
y =
121/41
```

```
>> z = ans.z
z =
33/41
```

The **double** function can then be used to convert the symbolic expressions to numbers, and store the results from the three unknowns in a vector.

```
>> double([x y z])
ans =
    3.0244    2.9512    0.8049
```

15.4 CALCULUS: INTEGRATION AND DIFFERENTIATION

MATLAB has functions that perform common calculus operations on a mathematical function f(x), such as integration and differentiation.

15.4.1 Integration and the trapezoidal rule

The integral of a function f(x) between the limits given by x = a and x = b is written as

$$\int_a^b f(x)dx$$

and is defined as the area under the curve f(x) from a to b, as long as the function is above the x-axis. Numerical integration techniques involve approximating this.

One simple method of approximating the area under a curve is to draw a straight line from f(a) to f(b) and calculate the area of the resulting trapezoid as

$$(b-a)\,\frac{f(a)+f(b)}{2}$$

In MATLAB, this could be implemented as a function.

THE PROGRAMMING CONCEPT

Here is a function to which the function handle and limits *a* and *b* are passed:

trapint.m

```
function int = trapint(fnh, a, b)
% trapint approximates area under a curve f(x)
%     from a to b using a trapezoid
% Format: trapint(handle of f, a, b)
int = (b-a) * (fnh(a) + fnh(b))/2;
end
```

To call it, for example, for the function $f(x) = 3x^2 - 1$, an anonymous function is defined and its handle is passed to the *trapint* function.

```
>> f = @ (x) 3 * x .^ 2 - 1;
approxint = trapint(f, 2, 4)
approxint =
    58
```

THE EFFICIENT METHOD

MATLAB has a built-in function **trapz** that will implement the trapezoidal rule. Vectors with the values of x and y = f(x) are passed to it. For example, using the anonymous function defined previously:

```
>> x = [2 4];
>> y = f(x);
>> trapz(x,y)
ans =
    58
```

An improvement on this is to divide the range from a to b into n intervals, apply the trapezoidal rule to each interval, and sum them. For example, for the preceding if there are two intervals, you would draw a straight line from f(a) to f((a+b)/2), and then from f((a+b)/2) to f(b).

THE PROGRAMMING CONCEPT

The following is a modification of the previous function to which the function handle, limits, and the number of intervals are passed:

trapintn.m

```
function intsum = trapintn(fnh, lowrange,highrange, n)
% trapintn approximates area under a curve f(x) from
%    a to b using a trapezoid with n intervals
% Format: trapintn(handle of f, a, b, n)
intsum = 0;
increm = (highrange - lowrange)/n;
for a = lowrange: increm : highrange - increm
    b = a + increm;
    intsum = intsum + (b-a) * (fnh(a) + fnh(b))/2;
end
end
```

For example, this approximates the integral of the previous function that is given with two intervals:

```
>> trapintn(f,2,4,2)
ans =
    55
```

THE EFFICIENT METHOD

To use the built-in function **trapz** to accomplish the same thing, the x vector is created with the values 2, 3, and 4:

```
>> x = 2:4;
>> y = f(x)
>> trapz(x,y)
ans =
    55
```

In these examples, straight lines, which are first-order polynomials, were used. Other methods involve higher-order polynomials. The built-in function **quad** uses Simpson's method. Three arguments are normally passed to it: the handle of the function, and the limits a and b. For example, for the previous function:

```
>> quad(f,2,4)
ans =
    54
```

15.4.2 Differentiation

The derivative of a function $y = f(x)$ is written as $\frac{dy}{dx} f(x)$ or $f'(x)$, and is defined as the rate of change of the dependent variable y with respect to x. The derivative is the slope of the line tangent to the function at a given point.

MATLAB has a function **polyder**, which will find the derivative of a polynomial. For example, for the polynomial $x^3 + 2x^2 - 4x + 3$, which would be represented by the vector [1 2 −4 3], the derivative is found by:

```
>> origp = [1 2 −4 3];
>> diffp = polyder(origp)
diffp =
    3    4    -4
```

which shows that the derivative is the polynomial $3x^2 + 4x - 4$. The function **polyval** can then be used to find the derivative for certain values of x, such as for $x = 1$, 2, and 3:

```
>> polyval(diffp, 1:3)
ans =
    3    16    35
```

The derivative can be written as the limit

$$f'(x) = \lim_{h \to 0} \frac{f(x+h) - f(x)}{h}$$

and can be approximated by a difference equation.

Recall that MATLAB has a built-in function, **diff**, which returns the differences between consecutive elements in a vector. For a function $y = f(x)$ where x is a vector, the values of $f'(x)$ can be approximated as **diff(y)** divided by **diff(x)**. For example, the equation $x^3 + 2x^2 - 4x + 3$ can be written as an anonymous function. It can be seen that the approximate derivative is close to the values found using **polyder** and **polyval**.

```
>> f = @ (x) x .^ 3 + 2 * x .^ 2 - 4 * x + 3;
>> x = 0.5 : 3.5
x =
    0.5000    1.5000    2.5000    3.5000
>> y = f(x)
y =
    1.6250    4.8750    21.1250    56.3750
>> diff(y)
ans =
    3.2500    16.2500    35.2500
>> diff(x)
ans =
    1    1    1
>> diff(y) ./ diff(x)
ans =
    3.2500    16.2500    35.2500
```

15.4.3 Calculus in Symbolic Math Toolbox

There are several functions in Symbolic Math Toolbox to perform calculus operations symbolically (e.g., **diff** to differentiate and **int** to integrate). To learn about the **int** function, for example, from the Command Window:

```
>> help sym/int
```

For instance, to find the indefinite integral of the function $f(x) = 3x^2 - 1$:

```
>> syms x
>> int(3 * x^2 - 1)
ans =
x^3 - x
```

To instead find the definite integral of this function from $x = 2$ to $x = 4$:

```
>> int(3 * x^2 - 1, 2, 4)
ans =
54
```

Limits can be found using the **limit** function. For example, for the difference equation described previously:

```
>> syms x h
>> f
f =
    @ (x) x .^3 + 2 .*x.^2 - 4 .* x + 3

>> limit((f(x+h) - f(x))/h,h,0)
ans =
3 * x^2-4 + 4 * x
```

To differentiate, instead of the anonymous function we write it symbolically:

```
>> syms x f
>> f = x^3 + 2 * x^2 - 4 * x + 3
f =
x^3 + 2 * x^2 - 4 * x + 3

>> diff(f)
ans =
3 * x^2 - 4 + 4 * x
```

PRACTICE 15.3

For the function $2x^2 - 5x + 3$:

- Find the indefinite integral of the function.
- Find the definite integral of the function from $x = 2$ to $x = 5$.
- Approximate the area under the curve from $x = 2$ to $x = 5$.
- Find its derivative.
- Approximate the derivative.

SUMMARY

Common Pitfalls
- Forgetting that the **fprintf** function by default only prints the real part of a complex number
- Extrapolating too far away from the data set

Programming Style Guidelines
- The better the curve fit, the more exact interpolated and extrapolated values will be.
- When working with symbolic expressions, it is generally easier to make all variables symbolic variables to begin with.

MATLAB Functions and Commands		
sym2poly	conj	numden
poly2sym	angle	pretty
roots	polar	ezplot
polyval	sym	solve
polyfit	syms	trapz
interp1	simplify	quad
complex	collect	polyder
real	expand	int
imag	factor	limit
isreal	subs	

Exercises

1. Express the following polynomials as row vectors of coefficients:

$$2x^3 - 3x^2 + x + 5$$
$$3x^4 + x^2 + 2x - 4$$

2. Find the roots of the equation $f(x) = 0$ for the following function. Also, create x and y vectors and plot this function in the range -3 to 3 to visualize the solution.

$$f(x) = 3x^2 - 2x - 5$$

3. Evaluate the polynomial expression $3x^3 + 4x^2 + 2x - 2$ at $x = 4$, $x = 6$, and $x = 8$.

4. Sometimes the roots of polynomial equations are complex numbers. For example, create the polynomial row vector variable *pol*:

```
>> pol = [ 3  6  5];
```

Use the **roots** function to find the roots. Also, use **ezplot(poly2sym(pol))** to view a plot. Then, change the last number in *pol* from 5 to -7 and again find the roots and view the plot.

5. Create a vector x that contains the integers 1 through 20. Create a vector y that stores 20 random integers, each in the range -2 to $+2$. Fit a straight line through these points. Plot the data points and the straight line on the same graph.

6. The compliance or elasticity of the lung is defined as:

$$Compliance = \frac{\Delta \text{Volume}}{\Delta \text{Pressure}}$$

In a biomedical engineering physiology lab, a spirometer was used to measure the volume of the lungs at a given pressure, which was measured by a pressure transducer. The following data were collected:

Pressure	Volume
0 cm H2O	1.750 L
5 cm H2O	2.500 L
10 cm H2O	3.750 L
15 cm H2O	4.000 L
20 cm H2O	4.750 L

Write a script that creates vectors to represent these data. Next, the script will find the straight line that best fits the data, and plots the data as circles and also the straight line on the same graph. The slope of this line is the actual compliance of the lung and chest wall. Label your axes and put a title on it.

7. The voltage in a circuit is determined at various times, as follows:

```
time:         1    2    3    4    5    6    7
voltage:    1.1  1.9  3.3  3.4  3.1  3.3  7.1
```

Fit a straight line through the data points, and then plot this line along with the sample voltages. According to your straight line, determine at what time the voltage would be 5.

8. Write a script that will generate a vector of 10 random integers, each in the range 0 to 100. If the integers are evenly distributed in this range, then when arranged in order from lowest to highest, they should fall on a straight line. To test this, fit a straight line through the points and plot both the points and the line with a legend. For example, when tested, the random integers might be

   95   23   61   49   89   76   46   2   82   44

and the plot might look like the one in Figure 15.9.

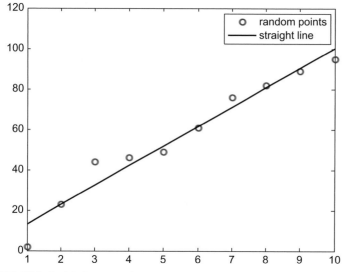

FIGURE 15.9 Straight line curve fit to random integers

9. Write a function that will receive data points in the form of x and y vectors. If the lengths of the vectors are not the same, then they can't represent data points, so an error message should be printed. Otherwise, the function will fit a polynomial of a random degree through the points, and will plot the points and the resulting curve with a title specifying the degree of the polynomial. The degree of the polynomial must be less than the number of data points, n, so the function must generate a random integer in the range from 1 to n − 1 for the polynomial degree.

10. Temperature readings were performed every hour (starting at 1 pm, but the end time could vary) and stored in a vector called *readings*. Write a function called *halffit* that receives this vector as an argument and uses a quadratic interpolation (second order) to determine what the temperature was every half hour between the actual recorded temperatures. The function then plots, on one graph, the original temperature

readings (using a circle for the points), the interpolated temperatures at the half hours (using a + for these points), and the quadratic curve that was used for the interpolation. Place a legend on the graph to distinguish the curves. The number of hours that was used for the original vector may not be assumed. For example, the function might be called as follows:

```
>> readings = [33, 40, 42, 41, 39, 32];
>> halffit(readings)
```

The Figure Window would look like Figure 15.10.

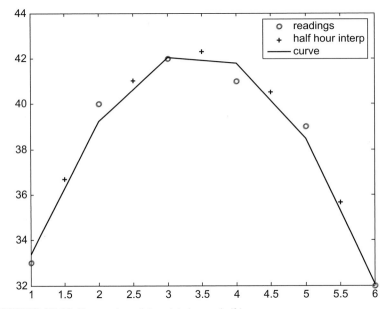

FIGURE 15.10 Temperatures interpolated every half hour

11. Some data points have been created in which the *y* values rise to a peak and then fall again. However, instead of fitting a quadratic curve through these points, what is desired is to fit two straight lines through these points: one that goes through all points from the beginning through the point with the largest *y* value, and another that starts with the point with the largest *y* value through the last point. Write a function **fscurve** that will receive as input arguments the *x* and *y* vectors, and will plot the original points as red stars (*) and the two lines (with default colors, line widths, etc.). Figure 15.11 shows the Figure Window resulting from an example of calling the function.

```
>> y = [2  4.3  6.5  11.11  8.8  4.4  3.1];
>> x = 1:length(y);
>> fscurve(x, y)
```

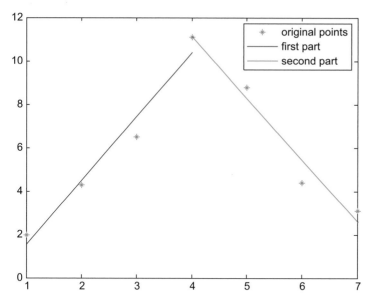

FIGURE 15.11 Two straight lines

You may not assume that you know anything about the data except that you may assume that they do rise to a peak and then fall again.

12. Vectors x and y have been created to represent x and y points. The vectors have the same length (let's call this n). Write a function called *fitsubs* that receives these vectors and graphically displays the difference between fitting polynomials of degree 1, 2, 3, ... $n - 1$ to the data. For example, *if* the data are as shown in the following, the Figure Window would look like Figure 15.12.

```
>> x = 1:4;
>> y = [23   35   46   39];
>> fitsubs(x,y)
```

13. Create vectors for four points. Fit a straight line through the points, and also a quadratic. Plot both of these, and the points, on one figure with legends.

14. Create a data file that stores data points (the x values in one column and then the y values in a second column). Write a script that will execute the following:
   ▣ Read the data points.
   ▣ Fit a straight line to the points.
   ▣ Create a vector *diffv* that stores for every x value the difference between the actual y value and the y value predicted by the straight line.
   ▣ Find and print the standard deviation of the vector *diffv*.
   ▣ Plot the original data points and the line.
   ▣ Print how many of the actual y values were greater than the predicted.
   ▣ Print how many of the actual data *y* values were within 1 (+ or −) of the predicted y value.

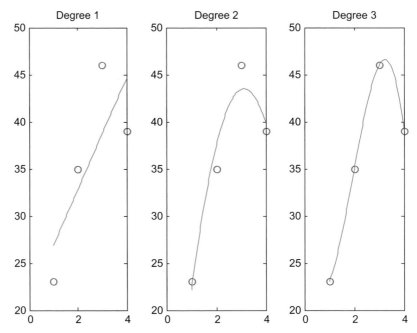

FIGURE 15.12 Subplot to demonstrate curves of increasing degrees

15. The temperature (in degrees Fahrenheit) was recorded every three hours for a day at a particular location. Using a 24-hour clock where midnight is 0, for example, the data might be:

```
Time:    0    3    6    9    12    15    18    21
Temp:    55.5 52.4 52.6 55.7 75.6  77.7  70.3  66.6
```

- Create vectors for the data.
- Plot the data.
- Find a curve that fits the data.
- At what time(s) was it 60 degrees? 65 degrees?

Data on the flow of water in rivers and streams are of great interest to civil engineers, who design bridges, and to environmental engineers, who are concerned with the environmental impact of catastrophic events such as flooding.

16. The Mystical River's water flow rate on a particular day is shown in Figure 15.13 and the table that follows it. The time is measured in hours and the water flow rate is measured in cubic feet per second. Write a script that will fit polynomials of degree 3 and 4 to the data and create a subplot for the two polynomials. Plot also the original data as black circles in both plots. The titles for the subplots should include the degree of the fitted polynomial. In addition, include appropriate x and y labels for the plots.

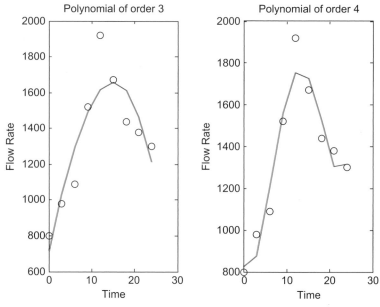

FIGURE 15.13 Mystical River flow rates

Time	0	3	6	9	12	15	18	21	24
Flow Rate	800	980	1090	1520	1920	1670	1440	1380	1300

17. Write a function that will receive x and y vectors representing data points. You may assume that the vectors are the same length, and that the values in the x vector are all positive, although not necessarily integers. The function will fit polynomials of degrees 2 and 3 to these points. It will plot the original data points with black stars (*), and also the curves (with 100 points each in the range given by the x vector so that the curves look very smooth). It will also generate one random integer x value and use the curves to interpolate at that x value. The range of the random integer must be within the range of the original x vector so that it is interpolating, not extrapolating (e.g., in the following example the x values range from 0.5 to 5.5 so the random integer generated is in the range 1 to 5). The interpolated values should be plotted with red stars (*), and the mean of the two should be plotted with a red circle (the axes and the colors for the curves are defaults, however). For example, the plot in Figure 15.14 was generated by calling the function and passing x and y vectors (and the random integer was 4).

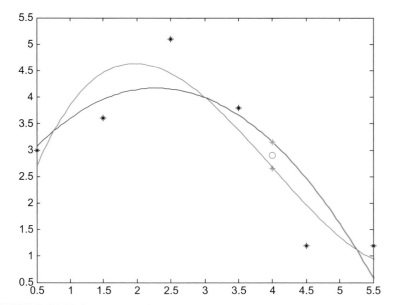

FIGURE 15.14 Random interpolated value

18. The distance (in miles) and speed of a car (in miles per hour) are measured at several points along a highway and are to be stored in a file and then read into a variable called *tripdata*. For example, *tripdata* MIGHT contain:

```
  1      44
 10      45
 50      65
100      60
150      55
```

It may be assumed that there are two columns; the first is the distance, and the second is the speed. It may not be assumed that the number of rows is known. The algorithm is:

- Create the data file and load it into a matrix variable, then separate the data into two vectors.
- Fit a straight line to the data.
- Plot the data points and the line on the same graph, with appropriate labels on the axes (not just 'x' and 'y'!).

19. Write a function that will receive *x* and *y* vectors representing data points. The function will create, in one Figure Window, a plot showing these data points as circles and also in the top part a second-order polynomial that best fits these points and on the bottom a third-order polynomial. The top plot will have a line width of 3 and will be a grey color. The bottom plot will be blue, and have a line width of 2. For example, the Figure Window might look like Figure 15.15.

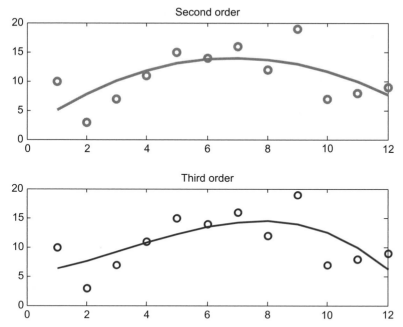

FIGURE 15.15 Subplot of second- and third-order polynomials with various line properties

The axes are the defaults. Note that changing the line width also changes the size of the circles for the data points. You do not need to use a loop.

20. The depth of snow in inches has been measured in a very cold location every week since the snow began accumulating. At this point, the season has changed and it is getting warmer, so the pile of snow is beginning to recede but it hasn't all gone away yet. The depths that have been recorded every week are stored in a file called *snowd.dat*. For example, it might contain the following:

8   20   31   42   55   65   77   88   95   97   89   72   68   53   44

Write a script that will predict in which week the snow will be totally gone by fitting a quadratic curve through the data points. This will be called the "snow gone week number" and will be rounded up. For example, if the data are as previously shown, the snow would be gone by week number 18. The script will produce a plot in the format shown in Figure 15.16, showing the original data points from the file and also the curve (from week 1 through the snow-gone week). The snow-gone week number will also be printed in the title. The x-axis should range from 0 to the snow-gone week number, and the y-axis from 0 to the maximum snow accumulation.

21. A data file called *acme.dat* stores the Acme Products Corporation's costs and sales for every quarter last year. There are four lines in the file, each consisting of the costs

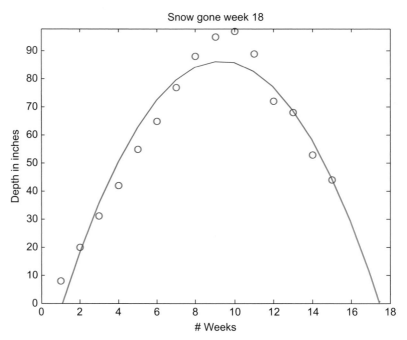

FIGURE 15.16 Prediction of snow melt

for the quarter followed by the sales. For example, the file might contain the following:

```
  2.2    4.4
  4      3.8
  6.5    6.5
 11.1   10.5
```

Write a script that will load this into a matrix and then separate the costs and sales into vectors. Create a Figure Window that shows bar charts for the costs, sales, and profits for the four quarters. Next, extrapolate to determine what the costs are likely to be in the next quarter (assuming a linear progression).

22. Store the following complex numbers in variables, and print them in the form a + bi.

$$\frac{3 - 2i}{\sqrt{-3}}$$

23. Create the following complex variables:

```
c1 = 2 − 4i;
c2 = 5 + 3i;
```

Perform the following operations on the variables:

- Add them.
- Multiply them.
- Get the complex conjugate and magnitude of each.
- Put them in polar form.

24. Represent the expression $z^3 - 2z^2 + 3 - 5i$ as a row vector of coefficients, and store this in a variable *compoly*. Use the **roots** function to solve $z^3 - 2z^2 + 3 - 5i = 0$. Also, find the value of *compoly* when $z = 2$ using **polyval**.

25. Determine how to use the **polar** function to plot the magnitude and angle of a complex number in polar form.

26. The real parts and imaginary parts of complex numbers are stored in separate variables. For example:

```
>> rp =   [1.1 3 6];
>> ip = [2 0.3 4.9];
```

Determine how to use the **complex** function to combine these separate parts into complex numbers, such as:

```
1.1000 + 2.0000i   3.0000 + 0.3000i   6.0000 + 4.9000i
```

27. Solve the simultaneous equations $x - y = 2$ and $x^2 + y = 0$ using **solve**. Plot the corresponding functions, $y = x - 2$ and $y = -x^2$, on the same graph with an x range from -5 to 5.

28. For the following set of equations,

$$2x_1 + 2x_2 + x_3 = 2$$
$$x_2 + 2x_3 = 1$$
$$x_1 + x_2 + 3x_3 = 3$$

write it in symbolic form and solve using the **solve** function. From the symbolic solution, create a vector of the numerical (**double**) equivalents.

29. For the following system of equations,

$$4x_1 - x_2 + 3x_4 = 10$$
$$-2x_1 + 3x_2 + x_3 - 5x_4 = -3$$
$$x_1 + x_2 - x_3 + 2x_4 = 2$$
$$3x_1 + 2x_2 - 4x_3 = 4$$

use the **solve** function to solve it. Verify the answer using any other method (in MATLAB!!).

30. Biomedical engineers are developing an insulin pump for diabetics. To do this, it is important to understand how insulin is cleared from the body after a meal. The concentration of insulin at any time t is described by the equation

$$C = C_0\, e^{-30t/m}$$

where C_0 is the initial concentration of insulin, t is the time in minutes, and m is the mass of the person in kilograms. Use **solve** to determine for a person whose mass is 65 kg how long it will take an initial concentration of 90 to reduce to 10. Use **double** to get your result in minutes.

31. To analyze electric circuits, it is often necessary to solve simultaneous equations. To find the voltages Va, Vb, and Vc at nodes a, b, and c, the equations are:

$$2(Va - Vb) + 5(Va - Vc) - e^{-t} = 0$$
$$2(Vb - Va) + 2Vb + 3(Vb - Vc) = 0$$
$$Vc = 2\sin(t)$$

Find out how to use the **solve** function to solve for Va, Vb, and Vc so that the solution will be returned in terms of t.

32. The reproduction of cells in a bacterial colony is important for many environmental engineering applications such as wastewater treatments. The formula

$$\log(N) = \log(N_0) + t/T \log(2)$$

can be used to simulate this, where N_0 is the original population, N is the population at time t, and T is the time it takes for the population to double. Use the **solve** function to determine the following: If $N_0 = 10^2$, $N = 10^8$, and $t = 8$ hours, what will be the doubling time T? Use **double** to get your result in hours.

33. Using the symbolic function **int**, find the indefinite integral of the function $4x^2 + 3$, and the definite integral of this function from $x = -1$ to $x = 3$. Also, approximate this using the **trapz** function.

34. Use the **quad** function to approximate the area under the curve $4x^2 + 3$ from -1 to 3. First, create an anonymous function and pass its handle to the **quad** function.

35. Use the **polyder** function to find the derivative of $2x^3 - x^2 + 4x - 5$.

Index

Note: Page numbers followed by *b* indicate boxes, *f* indicate figures, and *t* indicate tables.